Benchmark Papers in Geology Series

Editor: Rhodes W. Fairbridge—Columbia University

A selection from the published volumes in this series

METAMORPHISM AND PLATE TECTONIC REGIMES / *W. G. Ernst*
GEOCHEMISTRY OF IRON / *Henry Lepp*
SUBDUCTION ZONE METAMORPHISM / *W. G. Ernst*
METALLOGENY AND GLOBAL TECTONICS / *Wilfred Walker*
CRYSTAL FORM AND STRUCTURE / *Cecil J. Schneer*
STATISTICAL ANALYSIS IN GEOLOGY / *John M. Cubitt and Stephen Henley*
MINERAL DEPOSITS, CONTINENTAL DRIFT AND PLATE TECTONICS / *J. B. Wright*
MEGACYCLES: Long-Term Episodicity in Earth and Planetary History / *G. E. Williams*
OROGENY / *John G. Dennis*
ECONOMIC EVALUATION OF MINERAL PROPERTY / *Sam L. VanLandingham*
MINING GEOLOGY / *Willard C. Lacy*
MINERAL EXPLORATION / *Willard C. Lacy*

STABILITY OF HEAVY MINERALS IN SEDIMENTS

Edited by

GRETCHEN LUEPKE
U.S. Geological Survey
Menlo Park, California

A Hutchinson Ross Benchmark® Book

VNR SAE **Van Nostrand Reinhold Company**
Scientific and Academic Editions

New York Cincinnati Stroudsburg
Toronto London Melbourne

Benchmark Papers in Geology, Volume 81
Library of Congress Catalog Card Number: 83-27421
ISBN: 0-442-25925-5

Manufactured in the United States of America.

Published by Van Nostrand Reinhold Company Inc.
135 West 50th Street
New York, New York 10020

Van Nostrand Reinhold Company Limited
Molly Millars Lane
Wokingham, Berkshire RG11 2PY, England

Van Nostrand Reinhold
480 Latrobe Street
Melbourne, Victoria 3000, Australia

Macmillan of Canada
Division of Gage Publishing Limited
164 Commander Boulevard
Agincourt, Ontario MIS 3C7, Canada

15 14 13 12 11 10 9 8 7 6 5 4 3 2 1

Library of Congress Cataloging in Publication Data
Main entry under title:
Stability of heavy minerals in sediments.
(Benchmark papers in geology; 81)
English and German.
"A Hutchinson Ross benchmark book."
Includes indexes.
1. Heavy minerals—Addresses, essays, lectures. 2. Sediments (Geology)—Addresses, essays, lectures.
I. Luepke, Gretchen. II. Series.
QE364.2.H4S68 1984 549 83-27421
ISBN 0-442-25925-5

CONTENTS

SERIES EDITOR'S FOREWORD

The philosophy behind the Benchmark Papers in Geology is one of collection, sifting, and rediffusion. Scientific literature today is so vast, so dispersed, and, in the case of old papers, so inaccessible for readers not in the immediate neighborhood of major libraries that much valuable information has been ignored by default. It has become just so difficult, or so time consuming, to search out the key papers in any basic area of research that one can hardly blame a busy person for skimping on some of his or her "homework."

This series of volumes has been devised, therefore, as a practical solution to this critical problem. The geologist, perhaps even more than any other scientist, often suffers from twin difficulties—isolation from central library resources and immensely diffused sources of material. New colleges and industrial libraries simply cannot afford to purchase complete runs of all the world's earth science literature. Specialists simply cannot locate reprints or copies of all their principal reference materials. So it is that we are now making a concerted effort to gather into single volumes the critical materials needed to reconstruct the background of any and every major topic of our discipline.

We are interpreting "geology" in its broadest sense: the fundamental science of the planet Earth, its materials, its history, and its dynamics. Because of training in "earthy" materials, we also take in astrogeology, the corresponding aspect of the planetary sciences. Besides the classical core disciplines such as mineralogy, petrology, structure, geomorphology, paleontology, and stratigraphy, we embrace the newer fields of geophysics and geochemistry, applied also to oceanography, geochronology, and paleoecology. We recognize the work of the mining geologists, the petroleum geologists, the hydrologists, and the engineering and environmental geologists. Each specialist needs a working library. We are endeavoring to make the task of compiling such a library a little easier.

Each volume in the series contains an introduction prepared by a specialist (the volume editor)—a "state of the art" opening or a summary of the object and content of the volume. The articles, usually some twenty to fifty reproduced either in their entirety or in significant extracts, are selected in an attempt to cover the field, from the key papers of the last century to fairly recent work. Where the original works are in foreign languages, we have endeavored to locate or commission translations. Geologists, because of their global subject, are often acutely aware of the oneness of our world. The selections cannot therefore be restricted to any one country, and whenever possible an attempt is made to scan the world literature.

To each article, or group of kindred articles, some sort of "highlight commentary" is usually supplied by the volume editor. This commentary should serve to bring that article into historical perspective and to emphasize its particular role in the growth of the field. References, or citations, wherever possible, will be reproduced in their entirety—for by this means the observant reader can assess the background material available to that particular author, or, if desired, he or she too can double-check the earlier sources.

A "benchmark," in surveyor's terminology, is an established point on the ground that is recorded on our maps. It is usually anything that is a vantage point, from a modest hill to a mountain peak. From the historical viewpoint, these benchmarks are the bricks of our scientific edifice.

RHODES W. FAIRBRIDGE

PREFACE

The literature on heavy minerals in sediments is far too extensive to do justice to the many facets of research in a single book. *Stability of Heavy Minerals in Sediments* covers the early work in the field plus specific research on transportation, deposition, and stability of heavy minerals. A second book will focus on quantitative analysis and economic aspects, including correlation and provenance studies. This division, as in all scientific studies, is somewhat arbitrary, and aspects dealt with primarily in one book will certainly be touched upon in the other.

The study of detrital heavy minerals has sometimes been considered synonymous with sedimentary petrology; in truth, it is but one facet, although the science of sedimentary petrology did not come into its own until the study of heavy minerals became fashionable. Therefore, in selecting papers to be included in this volume, any papers that represent other important facets of sedimentary petrology, such as grain size or shape, were not considered for publication if they did not focus primarily on heavy minerals. Papers dealing strictly with methods were also excluded. Authigenesis, the process by which new minerals form in place within a sediment or sedimentary rock, is a sufficiently complex subject in its own right; papers covering this aspect have been excluded.

In selecting papers for publication, every effort was made to examine papers from all parts of the world. The following colleagues helped to insure this was done: Dr. Robert Folk, University of Texas at Austin; Dr. Joseph Schreiber, University of Arizona; Dr. Harvey Blatt, University of Oklahoma; Dr. Orrin Pilkey, Duke University; Dr. Ralph Hunter, U.S. Geological Survey, Menlo Park, California; Dr. German Müller, Universität Heidelberg, Germany; and Dr. André Cailleux, Ecole Practique des Hautes Etudes, St. Maur, France. Special thanks are due to the staff of the U.S. Geological Survey library—in particular, to William Sanders and Nancy Blair in Menlo Park, who helped to obtain many hard-to-find publications.

The following people reviewed one or more sections of the manuscript and helped to improve it: Dr. R. Dana Russell, Santa Rosa, California; Dr. Francis Pettijohn, Johns Hopkins University; Dr. Tjeerd van Andel, Stanford University; and Dr. John Hubert, University of Massachusetts. Dr. Waldemere Bejnar, Socorro, New Mexico, translated the Polish paper that appears in this book. Dr. Robert Arnal, U.S. Geological Survey, Menlo Park, translated the necessary French correspondence and other material.

I thank all the authors and publishers for granting me the permission to reprint their works. Material for the biographical notes was gathered from memorials and standard references; I regret that I was unable to get complete information on all the authors whose papers appear in this volume. Finally, I thank the series editor, Dr. Rhodes Fairbridge, for encouragement and guidance throughout this project.

GRETCHEN LUEPKE

CONTENTS BY AUTHOR

STABILITY OF HEAVY MINERALS IN SEDIMENTS

INTRODUCTION

*Minéraux lourds, Schwermineralien, Minerale pesante, Minerál'ny tyazhyoli, Mineraly cięzkie, Ju kobutsu, Zhòng kuàngwù**—whatever the language, heavy minerals comprise "the portion of density greater than 2.8 . . . which are of the greatest interest, beauty, and value from a stratigraphic point of view" (Boswell, 1916, 106). The term *accessory mineral* is sometimes used synonymously with heavy mineral, but the former term implies a minor rock constituent whose presence or absence does not affect the rock's petrologic analysis; it is not commonly used when referring to detrital minerals in sediments.

The earliest use of the term *minéraux lourds* (heavy minerals) appears to be in René Breón's paper on mineral separation (1880). At this time, heavy liquids, possessing a specific gravity of 2.8 or greater, came into widespread use. The term *heavy minerals,* applied to all minerals that did not float in these liquids, was hardly sophisticated. In some early papers the term was bracketed with quotation marks, a practice usually reserved for informal or slang terms. Ralph Daniel Reed (1924, 731) speaks of the "impropriety" of the name "heavy mineral" but does not state why he considers the name to be improper. Since its introduction in 1880, the term *heavy mineral* has gained worldwide usage and will surely not disappear.

EARLY WORK

The earliest paper (1718) on the petrography of sands was a short paper by René Antoine Ferchault de Réaumur (Boswell, 1933). This paper is extensively quoted in a footnote in J. W. Retgers's classic paper (Paper 2). De Réaumur, a famous entomologist and inventor of the Réaumur temperature scale (a forerunner of the Celsius scale), was also a pioneer in the investigation of the crystallinity of metals (Burke, 1966, 27). He correctly proved that the black sands of the Rhine and Rhone rivers contain iron minerals attracted by a magnet. But he called the red minerals hyacinth, topaz, sapphire, ruby, and emerald; in fact, they are zircon, rutile, garnet, and epidote.

*Languages represented are French, German, Italian, Russian, Polish, Japanese, and Chinese.

These misidentifications are not the serious errors they appear to be. Before the founding of the science of crystallography in 1801 (with the publication of *Traité de Minéralogie* by Abbé René-Just Haüy), most minerals, particularly gemstones, were identified exclusively by color (Adams, 1954, 145). Indeed, given the state of knowledge of minerals at the time, de Réaumur's work is remarkably accurate.

The sciences of chemistry and crystallography were well established by the early 1800s, but the science of pertrology was not. Minerals at this time could not be correctly identified unless they showed definite crystallographic features or could be concentrated for chemical analysis. Moreover, any early work on mineralogy of sands centered almost exclusively on fluvial sands, which are normally coarse grained and easy to examine. Before the invention of thins sections (1849) and heavy liquids (c. 1880), heavy minerals in sizes less than 0.25 mm were not extensively studied.

All research on heavy minerals prior to 1870 was purely descriptive; published papers did not begin to appear regularly until the middle 1800s. In Holland, Pieter Harting (1852) made the first attempt to determine the mineralogy of soils. Augustin Alexis Damour (1856) described the diamond-bearing sands of Limöero, Bahia, Brazil, using both chemical analyses and mineralogic descriptions. It is interesting that Henry Clifton Sorby, the inventor of thin sections and founder of the science of sedimentary petrology, never wrote a paper that dealt primarily with heavy minerals.

During the 1870s, some workers were already going beyond mere descriptions of minerals to discuss their sources but only if the sources were obvious. The first detailed account of mineralogic characters of sands is contained in a study of the Tyrrhenian coast by Gustavo Uzielli (1876). The famous mineralogist Auguste Michel-Lévy (1878) published a short descriptive paper in the first issue of the *Bulletin Societé Minéralogique de France,* in which he added some discussion of sources.

With the invention of thin sections, the petrographic microscope, and heavy liquids, more detailed descriptions were forthcoming. The number of species identified in sediments increased significantly; many rare species were recorded for the first time. Julien Thoulet developed techniques using heavy liquids for separating heavy minerals from bulk sands. His classic paper on the Sahara (1881) is the first description of red desert sands and was long consulted as the definitive paper on methods of separation. Arturo Issel and Carlo de Amezaga (1883) first examined marine samples collected during the *Washington* expedition of 1881 in the Mediterranean Sea.

Hans Thürach (1884) began the study of heavy minerals in Germany

(Sindowski, 1949) with descriptions of common heavy minerals in sandstones, including zircon, rutile, anatase, brookite, staurolite, garnet, glaucophane, and apatite. He emphasized the importance of varietal forms of minerals and their significance as to source. His was also the first attempt to use heavy minerals to study ancient deposits. A. B. Dick (Paper 1) wrote a short note that is frequently cited as the first significant heavy-mineral paper (Milner, 1962, 23).

The last decade of the nineteenth century saw the publication of many works that stand as classics in thoroughness of description. These include the papers of Ettore Artini (1898), a description of river sands in the Veneto region of northeastern Italy, the most thorough description of river sands ever undertaken and the first study to treat heavy minerals quantitatively; Lucien Cayeux (1897), a complete description of the Jurassic, Cretaceous, and Eocene sediments of northern France; William Fraser Hume (1897), a stratigraphic account of the Chalk and Upper Greensand formations of County Antrim, Northern Ireland, with detailed heavy-mineral descriptions; and J. W. Retgers (Paper 2), investigations of dune sands at Scheveningen, the Netherlands.

During the first fifteen years of the twentieth century, studies of heavy minerals in sediments flourished. While descriptive papers continued to constitute the bulk of these papers (e.g., Thoulet, 1908; Lacroix, 1909), many geologists went further and speculated on direction of transport and source for the sediments they described. Rocks of all ages were now under scrutiny, and many workers began to note varying proportions of the minerals, initiating the beginnings of quantitative analysis. The best representative paper of this period is by H. H. Thomas (Paper 3). Other noteworthy papers include those of Francesco Salmojraghi (1905), Aladár Vendl (1910), Italo Chelussi (1913), and Franjo Kučan (1913–1914).

From 1915 to 1930, descriptions of heavy minerals began to place a stronger emphasis on varietal characteristics. Interdependence among other geologic disciplines became more common. Alfred Brammall's paper on heavy minerals from the Dartmoor granite (1928) tied the igneous source of the heavy minerals to sedimentary rocks in many parts of the United Kingdom. This period also marked the beginning of the use of heavy minerals for correlation (Boswell, 1915; Reed, 1924). Other important papers from this period include those of Marcus Isaac Goldman (1916), Louis Déverin (1920), Henry Brewer Milner (1922), and Hans Wieseneder (1930).

While descriptions of heavy minerals were a necessary first phase, true understanding of the hows and why of their existence in sediments could not begin until geologists had an understanding of the principles of transportation, deposition, and stability of these minerals within the containing sediments. The next sections trace the developments in these important research areas.

TRANSPORTATION AND DEPOSITION

Sorting of sand was undoubtedly noticed by many people long before anyone tried to explain the mechanism. One of the earliest descriptions of heavy-mineral sorting in beach sand, written by Colonel Brydges R. Branfill (1885, 732), refers to a deposit at Cape Comorin, the southernmost tip of India:

> The sand around the rocks at the present point is of three kinds, and lies curiously levigated, or arranged in separate beds and layers. First there are the coarse water-worn grains of white quartz and feldspar, like grains of very large rice; next we see large beds of pure garnet sand; and thirdly, streaks of fine black sand just like diamond gunpowder. These three kinds lie touching one another, yet quite distinct and unmixed.

Not until thirty-seven years later (Kennedy, 1922) was there any attempt to examine the processes responsible for this layering. The principal work upon which all subsequent studies of heavy-mineral distribution are based is the classic paper by William Mackie (Paper 4). However, W. W. Rubey (Paper 5) first presented a comprehensive basis for the study of transportation and deposition of heavy minerals —namely, hydraulic equivalence.

A hydraulic equivalent is defined as the number of Udden grain sizes between the size of a given mineral grain and the size of a quartz grain with which it is deposited or to which it is hydraulically equivalent (the size of a larger or smaller grain that settles with the given mineral under the same conditions). Through this principle, light minerals of a given size are transported and deposited with heavy minerals of a different size. The term *hydraulic equivalence* was first used by Gordon Rittenhouse (Paper 6), though the term from which it is derived, *hydraulic value (hydraulischer Werth)*—the diameter of a quartz sphere having the same settling velocity as a given particle in water—was first introduced by Emil Schöne (1868). Others who studied hydraulic equivalence were Boris Brajnikov (1944), Lev Borisovich Rukhin (1937), and Wolf von Engelhardt (1940). The latter two papers sought to distinguish wind- and water-laid sediments based on grain-size distributions of heavy minerals.

Since the publication of Rittenhouse's paper, numerous studies of hydraulic equivalence have been conducted. Investigators quickly discovered that hydraulic equivalence did not seem to be valid in all circumstances. For some time this was attributed primarily to the lack of a given grain size of a mineral in the sediment population. But

Tjeerd H. van Andel (1950) and Louis I. Briggs (1965) disputed this idea. Van Andel suggested the depositional environment to be a major factor. Briggs attributed variations in observed hydraulic equivalences to be proportional to both the range of available minerals and the range of hydraulic selectivity.

Shape of heavy-mineral grains, certainly one factor in observed discrepancies in hydraulic equivalence, has received less attention than it deserves. In most studies of settling equivalence, all grains have been assumed to be spherical. R. Dana Russell (1939, 37–38) made the following observation on effects of shape on mineral composition:

> Progressive sorting according to size . . . tends toward an increase in the relative amounts of the initially smaller minerals. . . . Zircon, apatite, magnetite, ilmenite, and other accessory minerals of crystalline rocks appear to belong in this group. . . . Progressive sorting according to shape leads to a concentration of low-sphericity minerals . . . when transportation is by suspension. When movement is by rolling, on the other hand, the more equidimensional minerals tend to be concentrated in the direction of transportation. Progressive sorting according to specific gravity may result in a decrease in the percentage of the heavier minerals, as these tend to lag behind the lighter ones.

Briggs et al. (1962) made an experimental study of the dynamic shape factor (DSF)—the squared ratio of the particle's fall velocity to the fall velocity of its nominal sphere—and concluded that shape can influence a mineral's hydraulic properties in similar magnitude as variation in specific gravity among heavy mineral species. The most recent experimental studies on the influence of transport on shape and roundness of heavy minerals is by Volker Dietz (1973).

Others have credited discrepancies in settling equivalences to differences in entrainment potential between grains of the same settling velocity but of different size and density (von Englehardt, 1940; McIntyre, 1959; Hand, 1967; Lowright et al., 1972). Neil S. Grigg and Ronald E. Rathbun (1969) examined hydraulic equivalence and the reentrainment process; Lawrence L. Brady and Harvey E. Jobson (1973) studied heavy-mineral segregation under conditions of alluvial flow. The use of differential entrainment of grains in heavy-mineral-concentrating processes on beaches has been studied by Eihard Cordes (1966), James P. May (1973), and F. W. Stapor (Paper 7). Stapor, on the one hand, postulated two concentrating processes, one at work in the higher energy environment of the open coast and the other operating in the lower energy regime of a protected bay. May, on the other hand, attributed the cocentrating processes in bays and sounds to differentials in the entrainment capabilities between landward and seaward pulses generated by shoaling waves.

From these apparently contradicting studies, it is evident that hydraulic (settling) and entrainment equivalences are related but not the same. R. L. Slingerland (Paper 8) developed a quantitative theory and experimental support for these deviations from settling equivalence as a result of differential mineral entrainment. A. H. Sallenger (Paper 9) introduces a third equivalence associated with sedimentary deposits—namely, dispersive equivalence, which seems the prevalent process sorting the heavy minerals in inversely graded grain-flow deposits.

From now on, investigators should redefine hydraulic equivalence to mean much more than the simple settling equivalence it originally encompassed. Entrainment and dispersive equivalences should be subdivisions of hydraulic equivalence along with settling equivalence. It is possible that more kinds of equivalences are operating and to be discovered.

STABILITY

Stability of heavy minerals in sediments has long been a major focus of research. Unlike igneous and metamorphic settings, sedimentary environments impose external forces not directly concerned with the initial formation of the mineral (except for authigenesis, with which I do not deal in this book).

Orville Adelbert Derby (1891) appears to have been the first to recognize the importance of heavy-mineral stability. Retgers (Paper 2) discussed the relative stability of each mineral he found in the Holland dune sands. Thoulet (1913, 8) was the first to compile an orc'er of resistance to destruction:

> Les minéraux sains sont diversement sensibles aux agents de destruction. L'ordre croissant de leur résistance, pour les principaus d'entre ceux d'espèce déterminée, est: *péridot, pyroxènes, amphiboles, apatite, plagioclases, orthose, biotite, muscovite,* puix, en dernier, *le quartz, le rutile, le zircon* et *le corindon,* qui sont presque indestructibles, qu'on trouve à peu près partout et qui témoignent ainsi, sous une nouvelle forme, de la lutte pour l'existence et de la survie du plus résistant, vraie pour tout ce qui fait partie de la nature. (Emphasis added.)

Thoulet's order of destruction is markedly similar to the well-known reaction series devised by Norman Levi Bowen (1922). In his classic paper on rock weathering, Samuel S. Goldich (1938) recognized this order of stability, thereby suggesting a principle governing stability in sedimentary environments. The Russian geochemist Aleksandr Evgenevich Fersman (quoted in Pettijohn, 1975, 500) formalized this relationship of mineral stability to certain thermodynamic principles:

> The greater amount of energy evolved by an ion during its passage into the crystalline state, the more stable is the crystal obtained,

> the more difficult it is to reduce it to a dispersed state, to dissolve and to melt, or to divide again the atom of the lattice into free ions; the more stable such a system of minerals, the higher its ability to accumulate during natural processes and the less is it subject to destruction, melting and dissolution.

After Thoulet's order of resistance was published, others attempted to refine the list. Stability tables for heavy-mineral species have been offered by Goldich (1938); Boswell (1923, 1942); and Pettijohn (Paper 18). These tables, however, continue to be relative, particularly to climate.

Research on weathering of heavy minerals may focus on a single species at a time—for example, garnet (Bramlette, 1929); amphiboles and pyroxenes (Edelman, 1931; Edelman and Doeglas, Paper 10); staurolite (Edelman and Doeglas, 1934; Bond, Paper 11); uraninite (Davidson and Cosgrove, Paper 13); zircon (Carroll, Paper 12); and ilmenite (Bailey et al., 1956; Carroll, 1960). Comparative rates of weathering of heavy minerals were studied by Lincoln and Clarissa Dryden (Paper 14) and by Richard Weyl (1951). Weyl focused on both mineral abundance and degree of grain etching to determine stability differences.

Directly related to stability is resistance to wear during transport, especially by water. The first study to examine this problem quantitatively was made by F. W. Freise (Paper 16). Later studies of sands of the Mississippi (Russell, 1937) and the Rhine (van Andel, 1950) concluded that abrasion is not significant in long-distance river transport, a fact earlier suspected by Thoulet (1907).

Another major focus in stability research has been the effects of intrastratal solution, the phenomenon in which minerals disappear from the sediment after they have been deposited. The fact of its occurrence has never been in question, but the relative importance has been constantly debated.

The British sedimentary petrographer Percy George Hamnel Boswell (1923) was one of the first to observe that the complexity of heavy-mineral suites increases with decreasing age of the deposit; many less resistant heavy minerals are conspicuously absent in sediments older than Cenozoic ages. F. J. Pettijohn (Paper 18) quantifies Boswell's generalization with a study based on 315 published papers (Pettijohn, personal communication, 1982) and attributes the dearth of unstable minerals in pre-Cenozoic sediments primarily to intrastratal solution.

In contrast, P. D. Krynine (Paper 19) believes that provenance changes are far more important in judging changes in heavy-mineral composition, with intrastratal solution playing a local role only. Maria Turnau-Morawska (Paper 20) thinks that with enough data, the absolute number of heavy-mineral species identified in sediment of various ages will not be significantly different, although there may well be more of the less resistant minerals in Holocene sediments. Van Andel

(1959) suggests that there was a bias in heavy-mineral studies prior to 1941 (when Pettijohn's study was published) in that Cenozoic studies had concentrated on geosynclinal sediments with their diverse heavy-mineral assemblages and that pre-Cenozoic studies had focused on cratonic sediments with their impoverished heavy-mineral assemblages. The survival or loss of heavy minerals seems to be indirectly related to the age of strata but directly related to the depth of burial and length of time below sea level (Gazzi, 1965; Morton, 1979).

Researchers have recognized that the presence or absence of a particular heavy-mineral species is more than a matter of one mineral being more resistant than another. M. N. Bramlette (Paper 17) describes the presence of hornblende in sandstone concretions and its corresponding absence in the surrounding sediment. Apatite tends to be preserved in a carbonate-rich environment, while garnet is attacked (Lemcke et al., 1953). In the older Molasse north of the Alps, only cemented samples from deep wells contain staurolite (Füchtbauer, 1964). In an experimental study of mineral dissolution, Einhart Nickel (1973) found that the minerals dissolved incongruently and formed residual layers of varying composition and thickness, depending upon the pH and dissolution time. Oil-bearing sands commonly contain higher percentages of heavy minerals than the corresponding non-oil-bearing sands (Yurkova, 1970; Morton, 1979).

In a study of sandstone-shale pairs, Harvey Blatt and Berry Sutherland (Paper 21) found that the heavy-mineral suites from the fine-grained facies are richer in number of species than the corresponding coarse facies. They thus make a strong case for examining heavy minerals in fine-grained sediments, a practice which in general has been underutilized and would be most helpful in ascertaining the importance of intrastratal solution. Mapping of paleohydrologic regimes offers a way to examine the extent of intrastratal solution and diagenesis on a regional scale (Pettijohn et al., 1972, 405).

Mineral dissolution plus diurnal changes in temperature and moisture further complicate the study of heavy minerals in soils. Heavy minerals are particularly important in soils because they may contain traces of rare elements important in plant, animal, and human nutrition or in geochemical prospecting (Carroll, 1970). Harold J. Lutz (1941, 345), in discussing a study of virgin and cultivated soils in Michigan and New Hampshire, states:

> It has been found that cultivated soils contain a consistently higher percentage of heavy minerals in the A horizon than do the comparable virgin soils. The results indicate that the vertical distribution of heavy minerals in the upper horizons of soil profiles may serve as an index of past agricultural use of land. This criterion may prove useful in studies concerned with past history of land utilization.

Extreme weathering processes, like laterization, can lead not only to mineral destruction but also to authigenesis (Dill and Zech, 1980). In a study of the index minerals quartz, zircon, and garnet, J. B. Raeside (Paper 15) cautions that climate, crystalline inclusions, and crystal strain must be considered whenever these stable minerals are used to evaluate changes in soil profiles.

Studies of stability of heavy minerals in both sediments and soils promise to be even more fruitful in the future. With the advent of the scanning electron microscope, studies of individual grain textures have become an almost indispensible part of any heavy-mineral study (e.g., Rahmani, 1973; Hemingway and Tamar-Agha, 1975; Simpson, 1976; Morton, 1982).

REFERENCES

Adams, F. D., 1954, *The Birth and Development of the Geological Sciences,* Dover Publications, New York, 506p.

Artini, E., 1898, Intorno alla composizione mineralogica della sabbie di alcuni fiumi de Veneto, con applicazione terreni di trasporto, *Riv. Mineral. e. Cristall. Ital., Padova* **19:**33-94.

Bailey, S. W., E. N. Cameron, H. R. Spedden, and R. J. Weege, 1956, The Alteration of Ilmenite in Beach Sands, *Econ. Geology* **51:**263-279.

Boswell, P. G. H., 1915, The Stratigraphy and Petrology of the Lower Eocene Deposits of the Northwestern Part of the London Basin, *Geol. Soc. London Quart. Jour.* **71:**536-591.

Boswell, P. G. H., 1916, The Application of Petrological and Quantitative Methods to Stratigraphy, *Geol. Mag.* **53:**105-111, 163-169.

Boswell, P. G. H., 1923, Some Aspects of the Petrology of Sedimentary Rocks, *Liverpool Geol. Soc. Proc.* **13:**231-303.

Boswell, P. G. H., 1933, On the Mineralogy of Sedimentary Rocks, Thomas Murby & Co., London, 393p.

Boswell, P. G. H., 1942, The Stability of Minerals in Sedimentary Rocks, *Geol. Soc. London Quart. Jour.* **97:**lvi-lxxv.

Bowen, N. L., 1922, The Reaction Principle in Petrogenesis, *Jour. Geology* **30:**177-198.

Brady, L. L., and H. E. Jobson, 1973, An Experimental Study of Heavy-Mineral Segregation under Alluvial-Flow Conditions, *U.S. Geol. Survey Prof. Paper 562-K,* 38p.

Brajnikov, B., 1944, Sur l'importance de la granulométrie pour l'étude quantitative des minéraux lourds dans un sédiment, *Soc. Géol. France Bull.* ser. 5, **14:**381-390.

Bramlette, M. N., 1929, Natural Etching of Detrital Garnet, *Am. Mineralogist* **14:**336-337.

Brammall, A., 1928, The Dartmoor Detritals—A Study in Provenance, *Geologists Assoc. London Proc.* **39:**27-48.

Branfill, B. R., 1885, Notes on the Physiography of Southern India, *Royal Geog. Soc. Proc.* **7:**719-735.

Breón, R., 1880, Séparation des minéraux microscopiques lourds, *Soc. Minéral. France Bull.* **3:**46-56.

Briggs, L. I., 1965, Heavy Mineral Correlations and Provenances, *Jour. Sed. Petrology* **35:**939-955.

Briggs, L. I., D. S. McCulloch, and F. Moser, 1962, The Hydraulic Shape of Sand Particles, *Jour. Sed. Petrology* **32:**645-656.

Burke, J. G., 1966, *Origin of the Science of Crystals,* University of California Press, Berkeley and Los Angeles, 198p.

Carroll, D., 1960, Ilmenite Alteration under Reducing Conditions in Unconsolidated Sediments, *Econ. Geology* **55:**618-619.

Carroll, D., 1970, *Rock Weathering,* Plenum Press, New York-London, 203p.

Cayeux, L., 1897, Contribution à l'étude micrographique des terrains sédimentaires, *Soc. Géol. du Nord Mém.* **5**(2):1-589.

Chelussi, I., 1913, Nuove ricerche petrografiche sopra alcuni fondi di mare del Mediterraneo, *Soc. Geol. Italiana Bol.* **32:**143-159.

Cordes, E., 1966, Aufbau und Bildungsbedingungen der Schwermineralseifen bei Skagen (Danemark), *Meyniana* **16:**1-35.

Damour, A. A., 1856, Nouvelles recherches sur les sables diamantifères, *Soc. Géol. France Bull.* ser. 2, **13:**542-554.

de Réaumur, R. A. F., 1718, Essais de l'historie des rivières et des ruisseaux du Royaume qui roulent des Paillettes d'Or, *Acad. Royale Sci. Paris Mém.,* pp. 68-88.

Derby, O. A., 1891, On the Separation and Study of the Heavy Accessories of Rocks, *Rochester Acad. Sci. Proc.* **1:**198-206.

Déverin, L., 1920, Analyse minéralogique de quelques sédiment arénacés, *Soc. Vaudoise Sci. Nat. Bull.* **53:**1-21.

Dietz, V., 1973, Experiments on the Influence of Transport on Shape and Roundness of Heavy Minerals, *Contr. Sedimentology* **1:**103-125.

Dill, H., and W. Zech, 1980, Schwermineralverteilung in einigen bayerisches Deckschicht- und Bodenprofilen, *Geol. Jahrb.,* ser. D, **41:**3-22.

Edelman, C. H., 1931, Diagenetische Umwandlungserscheinungen an detritischen Pyroxenen und Amphibolen, *Fortschr. Mineralogie Krystall. Petrog.* **16:**323-324.

Edelman, C. H., and D. J. Doeglas, 1934, Über Umwandlungserscheinungen an detritischen Staurolith und anderen Mineralien, *Tschermaks Mineralog. u. Petrog. Mitt.,* new series, **45B:**225-234.

Füchtbauer, H., 1964, Sedimentpetrographische Untersuchungen in der älteren Molasse nördlich der Alpen, *Eclogae Geol. Helvetiae* **57:**157-298.

Gazzi, P., 1965, On the Heavy Mineral Zones in the Geosyncline Series—Recent Studies in the Northern Apennines, *Jour. Sed. Petrology* **35:**109-115.

Goldich, S. S., 1938, A Study in Rock Weathering, *Jour. Geology* **46:**17-58.

Goldman, M. I., 1916, The Petrography and Genesis of the Sediments of the Upper Cretaceous of Maryland, *Maryland Geol. Survey, Upper Cretaceous* **1:**111-182.

Grigg, N. S., and R. E. Rathbun, 1969, Hydraulic Equivalence of Minerals with a Consideration of the Reentrainment Process, *U.S. Geol. Survey Prof. Paper 650-B,* pp. 77-80.

Hand, B. M., 1967, Differentiation of Beach and Dune Sands, Using Settling Velocities of Light and Heavy Minerals, *Jour. Sed. Petrology* **37:**514-520.

Harting, P., 1852, Over de Bodem onder Amsterdam onderzocht en beschreven, *K. Inst. Amsterdam Verhand.,* ser. 3, **5:**73-232.

Hemingway, J. E., and M. Y. Tamar-Agha, 1975, The Effects of Diagenesis on

Some Heavy Minerals from Sandstones of the Middle Limestone Group in Northumberland, *Geol. Soc. Yorkshire Proc.* **40:**537-546.

Hume, W. F., 1897, The Cretaceous Strata of County Antrim, *Geol. Soc. London Quart. Jour.* **53:**540-606.

Issel, A., and C. de Amezaga, 1883, *Esame sommario dei saggi di fondo raccolti dalla spedizioni idrografica del R. Piroscafo* Washington *(nella campagna del 1881)*, Pubb. Uff. Idrograf. R. Mar Italia, Genova, pp. 1-19.

Kennedy, N. W., 1922, Natural Panning of Minerals in Littoral Deposits, *Geol. Soc. Liverpool Proc.* **13:**161-165.

Kučan, F., 1913-1914, Pijesak u hrvatskoj (Über den Sand in Kroatien): Glasnik Hrvatskoga Prirodoslovnogo Drustva, *Glasnik Soc. Sci. Nat. Croatica* **25**:63-70, 107-114, 171-177, 229-239; **26:**1-7.

Lacroix, A., 1909, Sur l'existence de sables monazités à Madagascar, *Soc.* Minéral. France Bull. **32:**313-317.

Lemcke, K., W. von Englehardt, and H. Füchtbauer, 1953, Geologische und sedimentpetrographische Untersuchungen im Westteil der ungefalteten Molasse des süddeutschen Alpenvorlandes, *Geol. Jahrb. Beihefte* **11:**1-110.

Lowright, R., E. G. Williams, and F. Dachille, 1972, An Analysis of Factors Controlling Deviation in Hydraulic Equivalence in Some Modern Sands, *Jour. Sed. Petrology* **42:**635-645.

Lutz, H. J., 1941, The Vertical Distribution of Heavy Minerals in Virgin and Cultivated Soils, *Science* **94:**344-345.

McIntyre, D. D., 1959, The Hydraulic Equivalence and Size Distributions of Some Minerals Grains from a Beach, *Jour. Geology* **67:**278-301.

May, J. P., 1973, Selective Transport of Heavy Minerals by Shoaling Waves, *Sedimentology* **20:**203-211.

Michel-Lévy, A., 1878, Note sur quelques minéraux contenus dans les sables du Mesvrin, près Autun, *Soc. Minéral. France Bull.* **1:**39-41.

Milner, H. B., 1922, The Nature and Origin of the Pliocene Deposits of the County of Cornwall and their Bearings on the Pliocene Geography of the South-West of England, *Geol. Soc. London Quart. Jour.* **78:**348-377.

Milner, H. B., 1962, *Sedimentary Petrography,* 4th ed., vols. 1 and 2, Allen & Unwin, London, vol. 1, 643p; vol. 2, 715p.

Morton, A. C., 1979, Depth Control of Intrastratal Solution of Heavy Minerals from the Paleocene of the North Sea, *Jour. Sed. Petrology* **49:**281-286.

Morton, A. C., 1982, The Provenance and Diagenesis of Paleogene Sandstones of Southern England as Indicated by Heavy Mineral Analysis, *Geol. Assoc. London Proc.* **93:**263-274.

Nickel, E., 1973, Experimental Dissolution of Light and Heavy Minerals in Comparison with Weathering and Intrastratal Solution, *Contr. Sedimentology* **1:**1-68.

Pettijohn, F. J., 1975, *Sedimentary Rocks,* 3rd. ed., Harper & Bros., New York, 628p.

Pettijohn, F. J., P. E. Potter, and R. Siever, 1972, *Sand and Sandstone*, Springer-Verlag, New York, 618p.

Rahmani, R. A., 1973, Grain Surface Etching Features of Some Heavy Minerals, *Jour. Sed. Petrology* **43:**882-888.

Reed, R. D., 1924, Role of Heavy Minerals in the Coalinga Tertiary Formations, *Econ. Geology* **19:**730-749.

Rukhin, L. B., 1937, A New Method for the Determination of the Conditions of Deposition of Ancient Sands, *Problems of Soviet Geology* **7:**953-959. (In Russian.)

Russell, R. D., 1937, Mineral Composition of Mississippi River Sands, *Geol. Soc. Amer. Bull.* **48:**1307-1348.

Russell, R. D., 1939, Effects of Transportation on Sedimentary Particles, in *Recent Marine Sediments,* P. D. Trask, ed., American Association of Petroleum Geologists, Tulsa, Okla. pp. 32-47.

Salmoraghi, F., 1905, Sulla continuita sotteranea del fiume Timavo—contributo mineralogico, *Soc. Ital. Sci. Nat. Atti* **44:**115-152.

Schöne, E., 1868, Über einen neuen Apparat für die Schlämmanalyse, *Fresenius Zeitschr. Analytische Chemie* **7:**29-47.

Simpson, G. S., 1976, Evidence of Overgrowths on, and Solution of, Detrital Garnets, *Jour. Sed. Petrology* **46:**689-693.

Sindowski, K. H., 1949, Results and Problems of Heavy Mineral Analysis in Germany—A Review of Sedimentary Petrological Papers, 1936-1948, *Jour. Sed. Petrology* **19:**3-25.

Thoulet, J., 1881, Étude minéralogique d'un sable du Sahara, *Soc. Minéral. France Bull.* **4:**262-268.

Thoulet, J., 1907, Sur la marche des sables de long de rivages, *Acad. Sci. C. R.,* **144:**938-940.

Thoulet, J., 1908, Origin éolienne des minéraux fins contenus dans le fonds marins, *Acad. Sci. C. R.,* **146:**1346-1348.

Thoulet, J., 1913, Notes sur lithologie sous-marins, *Inst. Océanog. Monaco Ann.* **5:**1-14.

Thürach, H., 1884, Über des Vorkommen mikroskopischer Zirkone und Titan-Mineralien in den Gesteinen, *Verhandlungen Phys.-Medic. Gesell. Würzburg* **18:**203-284.

Uzielli, G., 1876, Sopra lo zircone della Costa Tirrena, *Accad. Lincei Atti Reale* **3:**862-877.

van Andel, Tj. H., 1950, Provenance, Transport and Deposition of Rhine Sediments—A Heavy Mineral Study, Ph.D. dissertation, Rijksuniversiteit Groningen, 129p.

van Andel, Tj. H., 1959, Reflections on the Interpretation of Heavy Mineral Analyses, *Jour. Sed. Petrology* **29:**153-163.

Vendl, A., 1910, *Adatok a Duna homokjának ásrámytani ismertetéséhez,* Franklin, Budapest, 30p.

von Englehardt, W., 1940, Die Unterscheidung wasser- und windsortierter Sande auf Grund der Korngrössenverteilung ihrer leichten und schweren Gemengteile, *Chemie Erde* **12:**445-465.

Weyl, R., 1951, Schwermineralverwitterung in schleswig-holsteinischen Böden, *Naturw. Ver. Schleswig-Holstein Schr.* **25:**157-165.

Wieseneder, H., 1930, Studien an Sanden des niederösterreichischen Marchfeldes, *Tschermaks Mineralog. u. Petrog. Mitt.* **40:**303-324.

Yurkova, R. M., 1970, Comparison of Postsedimentary Alteration of Oil-, Gas-, and Water-bearing Rocks, *Sedimentology* **15:**53-68.

Part I

EARLY WORK

Editor's Comments on Papers 1 Through 4

1 **DICK**
On Zircons and Other Minerals Contained in Sand

2 **RETGERS**
Excerpts from *Ueber die mineralogische und chemische Zusammensetzung der Dünensande Hollands und über die Wichtigkeit von Fluss- und Meeressanduntersuchungen im Allgemeinen*

3 **THOMAS**
The Mineralogical Constitution of the Finer Material of the Bunter Pebble-Bed in the West of England

4 **MACKIE**
The Principles that Regulate the Distribution of Particles of Heavy Minerals in Sedimentary Rocks, as Illustrated by the Sandstones of the North-East of Scotland

Allan Brugh Dick (1833–1926) was a Scottish scientist who made noteworthy contributions in chemistry, microscopy, metallurgy, and mineralogy. He not only laid the foundation for the systematic investigation of heavy minerals in sedimentary rocks but also, with his friend Henry Clifton Sorby, aided significantly in the development of the petrographic microscope. Dick worked at the Royal School of Mines in London from 1851 to 1856 and with private industry thereafter. In 1924 he was elected honorary fellow of the Edinburgh Geological Society.

Although Boswell (1933) points out that it was not the first paper about heavy minerals, Dick's short note in *Nature* (Paper 1) is frequently cited as such (Milner, 1962).* Prior to the publication of this paper, all heavy-mineral studies were purely descriptive. Dick goes much further by noting the smaller size and enrichment of zircons and rutiles relative to other minerals in the sand. He also speculates that this enrichment could occur under natural conditions and thereby produce economic deposits of these minerals.

*Milner lists the title of this paper as "Bagshot sand of Hampstead Heath." This is puzzling because the title is clearly evident in the original paper.

Jan Willem Retgers (1856-1896), a Dutch scientist born in the Dutch East Indies, made his scientific contributions primarily in the field of physical chemistry, specifically isomorphism. He successfully combined his training in chemistry with geology to make significant contributions in research on heavy liquids. Later, he used all of this knowledge to put together his comprehensive paper on the dune sands of Holland that stands as a classic in thoroughness nearly a century after its publication (excerpts presented as Paper 2). This paper turned out to be his crowning achievment; Retgers died the year after its publication, two months before his fortieth birthday. In addition to the detailed chemical analyses and mineralogical descriptions, this study also discusses sources for the dune sands, notes the absence of mica in dune sands, and measures the average grain size at 0.25 mm, a grain-size average for dune sands repeatedly confirmed in later studies (Ahlbrandt, 1979). Due to space limitations, the sections describing research methods, mineralogical composition, and chemical analyses are not reproduced here.

Retgers's introduction offers an excellent summary of heavy-mineral investigations up to 1895, including a summary of de Réaumur's paper in an extended footnote. In the final discussion, Retgers remarks on the lack of quantitative investigations in studies of sands. One critical observation is that one expects to see stable, but not unstable, minerals in dune sands, and therefore many unstable mineral species present in dune sands have been overlooked.

Among other points recognized are the value of medium- and small-size rivers draining limited source areas; the value of studies of marine sands, which could differentiate between local and hinterland sources; and the possibility of currents moving sand to and from other coastal areas. Up to this time, studies of river sands were numerous but not quantitative, either physically or microchemically.

After the publication of Retgers's paper, heavy-mineral investigators began in earnest to go beyond the mere description of the minerals and the sediments containing them and discuss source areas and transport directions. Herbert Henry Thomas (1876-1935) was a British geologist with H.M. Geological Survey whose early work focused on stratigraphy and heavy minerals in sediments. His 1902 paper (Paper 3) is most commonly cited as the best early paper representing this group of investigators. Mineralogic descriptions were still important, as illustrated by the concise descriptions in this paper. Although descriptions constitute the bulk of the paper, the really important points concern transport direction and ultimate source areas. Based on the distribution of the various mineral species, Thomas postulates a main current from the Triassic Armorican massif to the south and a secondary current probably from the highlands of Devon and Cornwall in the west.

Later, as a petrographer, Thomas's interests centered on igneous petrology and structural geology. He won the Sedgewick Prize for an essay on the mineralogy of the Bunter Pebble Beds, an essay similar to his 1902 paper on the same subject, and in 1915 he was awarded the Murchison Medal. In 1927 he was elected a fellow of the Royal Society. He served as secretary of the Geological Society of London from 1912 to 1922 and as vice president from 1922 to 1924 and from 1931 to 1933.

William Mackie (1856-1932), the famous Scottish physician/geologist, maintained a general medical practice in Elgin, Scotland, for over thirty-five years and pursued researches in geology and mineralogy in his leisure time. His work laid the foundation for detailed studies on the mechanisms regulating the distribution of heavy minerals in sediments—that is, problems of transportation, deposition, and mineral stability. These problems had been hinted at earlier, but Mackie's paper (Paper 4) focused the attention that brought the study of heavy minerals in sediments to its maturity. This paper, based on a lifetime of work in northeastern Scotland, covers a wide range of subjects, from a simple experiment on the effect of specific gravity on minerals to discussions of selective action of surface tension, running water, and wind on heavy-mineral concentrations on beaches. Causes of disappearances of minerals—specifically, intrastratal solution—are discussed in some detail. Mackie then gives detailed illustrations of each topic, using the rocks of northeastern Scotland.

The quality of Mackie's geologic work was such that he was elected a fellow of the Edinburgh Geological Society in 1897 and served as the society's president from 1925 to 1927. In addition, the Geological Society of London presented him with an award from the Murchison Geological Fund in 1917, and the University of Aberdeen conferred on him the Ll.D. degree in 1923. In his lifetime he published some forty papers on geological problems in northern Scotland.

REFERENCES

Ahlbrandt, T. S., 1979, Textural Parameters of Eolian Deposits, *U.S. Geol. Survey Prof. Paper 1052-B,* pp. 21-51.

Boswell, P. G. H., 1933, *On the Mineralogy of Sedimentary Rocks,* Thomas Murby & Co., London, 393p.

Milner, H. B., 1962, *Sedimentary Petrography,* 4th ed., vols. 1 and 2, Allen & Unwin, London, vol. 1, 643p; vol. 2, 715p.

1

Reprinted by permission from *Nature* **36:**91-92 (1887)

ON ZIRCONS AND OTHER MINERALS CONTAINED IN SAND

A. B. Dick

OWING to recent attempts to employ the brilliant light of incandescent zirconia and allied earths instead of that of gas or electricity, attention has been drawn to the sources of the rarer earths.

It occurred to me that an inexhaustible supply might be obtained from the small crystals of zircon so widely distributed in rocks, sands, and soils, if an easy method of concentration could be found. I wished particularly to examine some of the larger deposits of sand, especially the fine-grained deposits, but was stopped by want of the requisite appliances. The matter having been mentioned by a friend to the Director-General of the Geological Survey, I at once received permission to use, for this investigation, the petrological laboratory attached to the Survey. The following results were obtained therein.

The denser minerals were separated from the sand by means of Sonstadt's solution of density 2·86, and borotungstate of cadmium of the density of 3·2.

			Per cent.
Dense minerals in	Lower Bagshot sand,	Hampstead	3·85
,,	,,	another sample	3·60
,,	,,	another sample from apparently a different deposit near the Fire-engine Office	1·04
,,	,,	High Beeches, Essex	1·32
,,	Glacial drift sand, Bagillt, North Wales		0·30
,,	Casting sand, Bullwell, near Nottingham		0·16
,,	Lower Pagshot sand, near Otterburne, Hants ...		0·12

These heavy minerals were found to be composed principally of matter attracted by a strong magnet, and of zircons, rutiles, tourmalines, and other grains, the relative proportions varying considerably. It will be observed that the richest and the poorest samples were from deposits of the Lower Bagshot sands. The drift sand from North Wales contained, in addition, many garnets. [Since this was written I have received a sample from High Beech Reservoir, 359 feet above Ordnance datum, containing 4 per cent. of dense minerals.]

The sand to which I devoted most attention, and of which I propose to give an account, is represented in the foregoing table by the first two samples. The samples were taken by me from a cutting recently made for the construction of a drain to connect "The Spaniards" and North-End, Hampstead. It was 2 or 3 feet wide, and 10 or 12 feet deep through great part of its length. It passed through a fine-grained yellowish sand, which was thrown out of the cutting or tunneled through in order to lay the drain-pipes. I made an average sample with as much care as though the sand had been a valuable ore which I was sampling for sale or purchase. It represented at least 1000 tons of sand in sight or cut through, and probably many acres of such sand on either side of the cutting. It consisted of grains so small as to pass almost entirely through a piece of cambric, 120 holes per square inch, stretched on a frame and used as a sieve. The average size of the grains was about the two-hundreth of an inch.

By means of dense solutions, vanning, analysis, and the constant use of the microscope, it was found that the sand had the following composition. It is given as an approximation to the truth, and does not, I think, vary more from it than must be expected in cases where it is not possible to get the minerals pure enough for weighing. Only under the microscope could the mixtures and impurities be estimated.

		Per cent.
Quartz, with 1 or 2 per cent. of flint fragments ...	about	75
Feldspar	,,	20
Grains attracted by a strong magnet	,,	2
Clay	,,	1
Zircons in recognisable crystals	,,	$\frac{1}{2}$
Grains, more or less opaque, probably zircons	,,	$\frac{1}{4}$
Rutiles	,,	$\frac{1}{6}$
Tourmalines	,,	$\frac{1}{10}$

In addition to the above, there were about 1 per cent. of grains over the density of 3·2, in regard to the composition of which I cannot at present say anything. A few appeared to be cleavage flakes of cyanite, but the majority were opaque earthy-looking bodies of various colours.

Feldspars.—These were nearly all of density equal to or lower than that of quartz. The majority were more or less cloudy, but some were quite transparent, showing the structure of microcline. A few showed the banded structure of plagioclase. Owing to the small size of the particles, and to their cavities and inclosures, I found it impracticable to get a satisfactory separation by dense solutions; and as it was impossible to distinguish the grains in all cases under the microscope, recourse was had to analysis of grains (consisting principally of quartz and feldspar) floated off from the denser minerals in a solution in which quartz floated and anorthite sank. It was found that they contained 94 per cent. of silica and 4·6 of alumina, with an unweighable trace of lime. The alkalies were not determined. This would correspond to about 20 per cent. of feldspar, but the estimate is perhaps rather high, as amongst the matter floated by the solution were found some particles which looked like glauconite covered by a transparent covering of varying thickness; also a little mica and some opaque grains of doubtful origin. It was interesting to note how the small fragments of feldspar have remained unchanged since the parent rocks were formed, and as they survived the disintegration of those rocks so they have continued unchanged in the sands. I saw several containing each a small zircon, and some contained what I think were microliths of apatite. The feldspars were mostly in angular fragments like the quartz.

The zircons are generally transparent and colourless prisms with double terminations of various kinds; many are more or less rounded and some wholly rounded as by attrition.

The rutiles are oblong and rounded grains, but many are sharply edged prisms, and a few have double terminations. Twins are not common, but both knee-shaped and kite-shaped twins occur. These rutiles resemble those of the metamorphic rocks, in which rounded grains and sharply defined crystals are met with side by side.

The tourmalines are generally in flat plates, more or less rounded, but some are perfect crystals with double terminations. They vary much in colour and power of absorption. Various methods of concentrating the zircons were tried. The simplest found was to sift the sand in air or water through a sieve with 120 holes to the inch, that being the smallest mesh I could meet with. In coarse-grained sands, such as the drift sand of North Wales, a considerable enrichment is thereby at once effected, but only a partial enrichment takes place in working with sands so small-grained as to pass entirely through the sieve.

Thus when 8 ounces of the Hampstead sand were sifted till 4 ounces had passed, it was found that the zircons and rutiles, being smaller, smoother, and heavier than the sand, passed through the sieve faster. By again sifting the 4 ounces which had passed till 2 ounces passed, a further enrichment was found to have taken place. The 2 ounces were again sifted on the same sieve till 1 ounce had passed, which was again sifted till ¼ of an ounce passed. This was examined quantitatively, and found to consist of 54 per cent. of quartz and feldspar, and 46 per cent. of dense minerals. By vanning and submitting the residue to the action of a strong magnet, almost nothing but zircons and rutiles remained. It is needless to say that much remained with the sand, especially in the latter parts of the operation.

A current separater was tried, but it seemed more difficult to work, though perhaps it might answer better on a larger scale, where it might be set to work automatically. There was no difficulty in getting a considerable enrichment, but it was evident that a great deal of care would be required in "sizing" the particles before a good separation could be effected. The most hopeful method seemed to be that of washing away the medium-sized grains of sand and afterwards sifting the sediment.

Perhaps in some of the streams running through the Bagshot or other sands natural eddies may be found or formed artificially from which enriched sand may be dredged, or it may be got on the sea-shore under sand-cliffs.

The object of the present communication is to draw attention to the matter in hopes that some deposit richer in zircons than the Hampstead sand may be found. Much care must be taken in sampling, because the sands, having been deposited from currents, must vary in composition. A trial is easily made by anyone accustomed to use a microscope and who knows the minerals by sight under such circumstances. A thimbleful of such sand as exists at Hampstead is enough for a trial by vanning, but if one of the dense liquids be used, from 10 to 20 grains by weight of the sand will give a good microscopic slide of the dense minerals. It may of course prove that the Hampstead sand is a residue of denudation in which the denser minerals have accumulated. In that case it is not improbable that other similar deposits may be found, some, perhaps, much more zirconiferous.

On the whole it appears that the matter is worthy of further attention. In some future communication I hope to be able to give an account of the composition of the matter attracted by a strong magnet, and also of the grains of earthy-looking minerals over the density of 3·2, and of any richer deposit of zirconiferous sand of which I can obtain reliable samples.

2

Reprinted from pages 16-24 and 65-74 of *Neues Jahrb. Mineralogie, Geologie u. Paläontologie* **1**:16-74 (1895)

Ueber die mineralogische und chemische Zusammensetzung der Dünensande Hollands und über die Wichtigkeit von Fluss- und Meeressanduntersuchungen im Allgemeinen.

Von

J. W. Retgers.

1. Einleitung.

Vor einiger Zeit ist von mir eine Untersuchung der Mineralien des Dünensandes der Westküste Hollands publicirt worden, auf welche später eine chemische Erforschung dieses Sandes gefolgt ist[1].

Da beide Untersuchungen vielleicht den Leser dieses Jahrbuches interessiren werden, so erlaube ich mir, hier in etwas abgekürzter Form ihre Resultate wiederzugeben.

Allgemein bekannt ist, dass der Dünensand Hollands zum grössten Theil aus Quarzkörnern besteht. Dass er jedoch ausschliesslich aus solchen bestehen würde, ist schon a priori unwahrscheinlich, weil die festen Gesteine, welche den Quarz geliefert haben, jedenfalls auch andere Mineralien enthalten haben, welche ebenso hart sind wie Quarz und ebenso widerstandsfähig gegen mechanische und chemische Einflüsse.

---

[1] J. W. Retgers, De Samenstelling van het Duinzand van Nederland. Verhandelingen der koninklijke Akademie van Wetenschappen te Amsterdam. 1891. — J. W. Retgers, Essai d'une Analyse chimique du Sable des Dunes. Annales de l'Ecole Polytechnique de Delft **7**. 161—186. 1892. — Beide Aufsätze vereinigt im Recueil des Travaux chimiques des Pays-Bas. **11**. 169—257. 1893.

Ausserdem braucht man nur den Dünensand auf einen Bogen weissen Papiers auszustreuen, um schon mit unbewaffnetem Auge zwischen den zahlreichen farblosen oder weissen Quarzkörnern mehrere braune oder schwarze Körner zu entdecken, welche offenbar anderen Mineralien angehören.

Mineralogische Sanduntersuchungen haben bekanntlich schon mehrmals stattgefunden. Ich verweise auf die der Mineralien des Sandes des Mesvrinflusses bei Autun in Frankreich von Michel-Lévy [1], vom vulkanischen Sande Santorins von Fouqué [2], vom Sande der Sahara-Wüste von Thoulet [3], ausserdem von zahlreichen, durch bestimmte Mineralien interessante Sande, wie Gold- und Platinsande, Chromeisensteinsande, schwarze magnetitreiche Sande, die diamantführenden Monazitsande Brasiliens, mehrerer Gletschersande [4], die Leucit- und Zirkon-führenden Küstensande West-Italiens [5] u. s. w. [6]

---

[1] A. Michel-Lévy, Notes sur quelques Minéraux contenus dans les Sables du Mesvrin, près Autun. Bull. soc. min. **1**. 39. 1879.

[2] F. Fouqué, Santorin et ses Eruptions. 1879.

[3] J. Thoulet, Etude minéralogique d'un Sable du Sahara. Bull. soc. min. **4**. 262. 1881.

[4] H. Wichmann, Mineralogische Zusammensetzung eines Gletschersandes. Tschermak's Mineral. Mitth. **7**. 452. 1886; dies. Jahrb. 1887. **2**. -85-.

[5] G. Uzielli, Sopra lo zircone della Costa tirrena. Atti R. Accad. Lincei. **3**. 1876; dies. Jahrb. 1877. -303-. Besonders soll auf diese Arbeit als erstes Beispiel einer mineralogischen Meeressanduntersuchung die Aufmerksamkeit gelenkt werden. Uzielli fand in dem Meeressande der tyrrhenischen Küste von Neapel bis Civita-vecchia Quarz, Plagioklas, Orthoklas, Augit, Olivin, Leucit, Magnetit, Apatit, Calcit und Zirkon.

[6] Eine der ersten mikroskopischen Untersuchungen von Sanden ist die bekannte Arbeit von H. C. Sorby, On the microscopical Characters of Sands and Clays. (The monthly microscopical Journal. March 1877; dies. Jahrb. 1880. **1**. -216-.) Er wies neben dem Überfluss der Quarzkörner noch Feldspath-, Hornblende- und Turmalinkörner nach in den Sanden. — Mikroskopische Untersuchungen von Sandsteinen (wobei jedoch auch noch mehrere für eigentliche „Sande" zutreffende Beobachtungen beschrieben sind) fanden statt von F. A. Anger (Mikr. Studien über klastische Gesteine. Tschermak's Miner. Mitth. 1875. 153; dies. Jahrb. 1876. -213-) und bekanntlich auf sehr gründliche Weise von G. Klemm (Mikr. Untersuchungen über psammitische Gesteine. Zeitschr. der deutschen geolog. Ges. **34**. 771. 1882; dies. Jahrb. 1883. **2**. -71-. Weiter schrieben hierüber A. Daubrée, Etudes synthétiques de Géologie expérimentale. 248—289. 1878; dies. Jahrb. 1880. **2**. -168-; J. A. Philips, On the Constitution and History of Grits and Sandstones (Quart. Journ. Geol. Soc. **37**. 6.

Bei all diesen Sanden hat man sich gewöhnlich begnügt mittelst Schlämmung einige interessante Mineralien zu isoliren und zu beobachten. Von einer systematisch durchgeführten sowohl qualitativen als quantitativen Untersuchung war hierbei keine Rede. — Beim Dünensande[1] hat man um so weniger an eine derartige Untersuchung gedacht, als die selteneren Mineralien desselben nur einige Procente betragen und sozusagen verloren sind in einer übergrossen Menge Quarzkörner und deshalb das directe Aussuchen der ersten fast unausführbar ist und eine Anreicherung mittelst Schlämmen stets mit grossem Verluste verbunden ist.

1881; dies. Jahrb. 1882. **2**. -371-); E. KALKOWSKY, Über die Erforschung der archäischen Formationen; dies. Jahrb. 1880. **1**. 1—28. H. O. LANG, Über Sedimentär-Gesteine aus der Umgegend von Göttingen (Zeitschr. d. deutsch. geolog. Gesellsch. **33**. 217. 1881; dies. Jahrb. 1882. **2**. -68-). Die jüngste petrographische Untersuchung von Sandsteinen ist wohl die von J. CHELUSSI, welche im eben erschienenen 4. Heft des 4. Bandes des Giornale di Mineralogia 1894 veröffentlicht ist. Er untersuchte einige Sandsteine der Provinz Aquila (Abruzzen) und wies mikroskopisch Quarz, Orthoklas, Plagioklas, Biotit, Muscovit, Chlorit, Granat, Epidot, Turmalin, Zirkon, Calcit und Magnetit hierin nach. Im Jahre 1892 erschien eine Arbeit von C. RIVA über einige Sandsteine der Appeninen im Giornale di Mineralogia **3**. 250. 1892 (dies. Jahrb. 1893. **2**. -494-). Er traf ausser den ebengenannten Mineralien noch Pyrit und Apatit an. Ebenfalls von einem italienischen Forscher, L. RICCIARDI (Ricerche sulle sabbie delle coste adriatiche. Atti Soc. Ital. d. sc. nat. **33**. 1—22. 1890; dies. Jahrb. 1893. **2**. -49-) wurden die Sande der adriatischen Küste von der Po-Mündung bis nach Bari untersucht, und ausser Quarz, Feuerstein, Kalkstein noch Augit nachgewiesen. Es darf wohl bemerkt werden, dass nicht bloss, wie im Referate nachdrücklich betont worden ist, die Untersuchungen die weitgehenden Schlüsse des Autors keineswegs rechtfertigen, sondern dass dieselbe auch in mineralogischer Hinsicht äusserst mangelhaft sind, indem ohne Zweifel viel mehr Mineralien in dem adriatischen Meeressand vorkommen werden. — Eine sehr gründliche Untersuchung des goldhaltigen Flusssandes des Tessin fand statt von E. ARTINI (Giorn. di Miner. **2**. 1891; dies. Jahrb. 1892. **1**. 515.)

[1] Ich mache bei dieser Gelegenheit aufmerksam auf eine schon im Jahre 1884 in russischer Sprache erschienene Arbeit über Dünen von Herrn Landesgeologen SOKOLOW in Petersburg, welche nicht bloss die Stranddünen, sondern auch die Fluss- und Binnenland- (Wüsten-) Dünen behandelt. Eine deutsche Übersetzung dieser wichtigen Arbeit, welche Herr Prof. ARZRUNI bearbeitet, wird bald erscheinen.

Die einzige Art und Weise, bei welcher die Untersuchung Erfolg verspricht, ist die vollständige Trennung in mehrere Gruppen von den leichtesten bis zu den schwersten Mineralien mit schweren Flüssigkeiten und die optische und chemische Untersuchung der hierdurch isolirten Mineralien.

Ich habe auf diese Weise den Dünensand der Westküste Hollands bei Scheveningen untersucht. Die Proben wurden sämmtlich auf kurze Entfernung südlich des bekannten Badeortes, unweit des Leuchtthurmes, am Westabhange der Dünenreihe gesammelt.

Indem ich für die Beschreibung der Methoden und der nachgewiesenen Mineralien auf die folgenden Seiten verweise, erlaube ich mir vorher noch einige Bemerkungen.

Man wird unter den dem Quarz beigemischten Mineralien zwar andere ebenfalls harte und feste gesteinbildende Mineralien, wie Granat, Spinell, Turmalin, Magnetit, erwarten, welche auch der Verwitterung gut Widerstand leisten können, jedoch keine leicht spaltbare, wie Feldspath und Hornblende, oder leicht zersetzbare und leicht lösliche, wie Olivin und Kalkspath.

Die Resultate der Untersuchung zeigten, dass diese Vermuthung nicht richtig war. Zwar bilden die harten Mineralien, wie Granat, Augit, Staurolith, Spinell, Zirkon, Rutil, Turmalin, Epidot die Hauptmasse; neben ihnen kommen jedoch nicht bloss leicht spaltbare, wie Feldspath und Hornblende, leicht zersetzbare, wie Olivin und Apatit, sondern auch sogar sehr weiche und leichtlösliche, wie der Kalkspath, vor, und zwar bilden diese ebenso grosse Körner wie die Quarzkörner selbst. Es ist dies wohl eine Folge davon, dass die Sandkörner keine gewaltsame Zertrümmerung erlitten haben, sondern mehr eine gegenseitige Abreibung, welche verhindert, dass die Körner unter ein gewisses Minimum der Grösse herabsinken.

Die Körner sind einer Abnutzung sowohl im Wasser als auch trocken (durch den Wind) ausgesetzt. Die Wirkung des letzteren ist bedeutend stärker als die des ersteren. Die abschleifende Kraft der Körner wird durch das Wasser selbst stark gehemmt[1]. Dagegen ist die abnutzende Kraft einer

[1] Ich verweise hier auf die bekannten Versuche von Daubrée (Etudes synth. de géol. expér. 1879; Formation des galets, du sable et du limon,

durch Wind bewegten Sandmasse eine sehr starke. Am besten lässt sich dies am Dünenfuss selbst sehen, nicht bloss daran, dass Körper, wie z. B. Glasscherben, nach einigen Tagen vollkommen matt geschliffen werden, wenn sie auf trockenem Dünensande liegen, sondern auch daran, dass die härtesten Basalt- und Granitpflastersteine bald stark von vom Winde bewegtem Sande angegriffen werden. Auf dieser stark abnutzenden Kraft der vom Winde fortgeführten Sandkörner (welche sogar im Gesicht Schmerz verursachen, wenn man bei starkem Winde längs des Strandes geht) beruht ja die technische Methode des Glasätzens mittelst des Sandgebläses. Herr J. BOSSCHA Jzn.[1] hat sogar experimental bewiesen, dass diese Kraft so stark ist, dass Quarzkörner bei sehr lebhafter Fortbewegung mittelst eines Luftstroms durch ihre gegenseitigen Stösse in Stücke zerspringen, was unmöglich im Wasser stattfinden kann. Dies erklärt vielfach die oft nachgewiesenen scharfkantigen Quarzfragmente im Flugsand (z. B. in dem des Rheinalluviums, dies. Jahrb. 1892. **1**. 224) und die eckigen Quarzsplitter im Löss, was ebenfalls für aeolische Bildung spricht. — Vergleicht man den Meeressand mit dem Dünensand, so ist ersterer stets bedeutend grobkörniger und die Körner sind eckiger. — Die bedeutendste Zerkleinerung erleiden die Körner somit im

---

248—258), welche deutlich die äusserst geringe Abnutzung im Wasser beweisen. Quarzsand bleibt eckig, sobald er nur aufgewirbelt werden kann, und verliert dann nicht mehr an Gewicht. (Auch die versuchsweise während langer Zeit in einem mit Wasser gefüllten gedrehten Cylinder herumgewirbelten Quarzstücke blieben eckig.) Deshalb die zahlreichen Sandsteine mit eckigen Sandkörnern. Solche traf ich zwar im Fluss- und Meeressande, jedoch nicht im Dünensande an. Windtransport rundet also bedeutend. Auch THOULET (Bull. soc. min. **4**. 262. 1881) macht auf den auffallenden Unterschied des rundkörnigen Sahara-Sandes gegenüber dem eckigkörnigen Meeressand aufmerksam und schreibt auch hier dem Windtransport eine bedeutend stärkere abnutzende Wirkung als dem Wassertransport zu. Über die eigentlichen Sanddünen der Saharawüste handelt eine Arbeit von C. ROLLAND, Sur les grandes dunes de sable du Sahara (Bull. soc. géol. **10**. 30. 1882).

[1] J. BOSSCHA Jzn., Het Zanddiluvium in Nederland. 1879. Leiden. S. 35—43. Ausser der bekannten Einwirkung des Wassers, des Windes und des Frostes auf die Sandkörner, will BOSSCHA auch die Sonnenhitze am Strande (Zersprengen der Quarzkörner durch Erhitzung der Einschlüsse vom liquiden CO_2) annehmen als Ursache der Zertrümmerung.

Windtransport. Recht merkwürdig ist die Thatsache, dass diese Zerkleinerung durchaus nicht zur vollkommnen Zerreibung führt, sondern dass, wie gesagt, der Durchmesser der Dünensandkörner nicht unter eine bestimmte Grösse sinkt. Es scheint, als ob dieser Durchmesser eine bestimmte Function der mittleren Windstärke ist. Ich habe Dünensande von sehr abweichenden Orten der holländischen Küste (z. B. Scheveningen, Zandvoort, Helder, Vlieland, Schiermonnikoog) untersucht und immer dieselbe Grösse angetroffen. Ja, sogar an allen möglichen Stellen der Dünen (Innen- und Aussenseite, Fuss und Scheitel) hatte der Dünensand ungefähr gleiche Grösse (welche man im Mittel rund zu $\frac{1}{4}$ mm Durchmesser annehmen kann)[1]. Die Abreibung scheint also im Laufe der Zeit nicht weiter zu gehen. Obwohl dies a priori nicht leicht einzusehen ist und man eher das Gegentheil annehmen sollte, wird es vielleicht erklärlich durch den Vergleich mit der Zerkleinerung eines Stabes, den man leicht in Stücke zerbrechen kann, wenn er lang ist, was jedoch immer schwieriger wird, wenn die Stücke kleiner werden, sodass sich schliesslich zwischen Körperkraft und Länge der Fragmente eine deutliche Abhängigkeit herausstellt. Ein derartiges Verhältniss besteht vielleicht auch zwischen mittlerer Windstärke und Sandkörner-Durchmesser[2].

---

[1] Bosscha (l. c. 48) hat Messungen an dem Scheveninger Dünensand angestellt: die Körnergrösse wechselte von $\frac{1}{3}$ bis $\frac{1}{10}$ mm. Bei weitem die meisten Körner hatten $\frac{1}{4}$ bis $\frac{1}{8}$ mm Durchmesser. Die Scheveninger Meeressande waren deutlich grobkörniger. Die Körner des Saharasandes hatten nach Thoulet (l. c. 263) 0,2 bis 0,7 mm Diam., waren also grösser als die des Dünensands.

[2] Schon Daubrée (l. c. 256—257) constatirte ein derartiges Dimensions-Minimum („Dimension limite") für in Wasser bewegte Sandkörner, welches Minimum eine Function ist von der Geschwindigkeit des Wassers und welche die auffallende Gleichheit der Grösse der Sandkörner in Sandsteinen erklärt. Ob dies auch für Dünensand und Wüstensand gilt, ist fraglich. Während Flusssand kommt im Meere wieder zur Ruhe und deshalb die eckigen Formen behält, ist Dünen- und Wüstensand in fortwährender Bewegung. Sowohl Land- als Meeresdünen verschieben sich und werden gründlich umgewühlt. Hier ist also die Existenz eines Dimensions-Minimums viel unerklärlicher. — Richtige Feinreibung, also Zertrümmerung von Gesteinskörnern bis zu der äussersten Grenze, findet eigentlich nur in dem stark zerriebenen Gletschersande statt. Jedoch selbst hier muss man immer ein Minimum

Die von mir im Dünensande nachgewiesenen Mineralien lassen sich, was die Häufigkeit ihres Vorkommens betrifft, eintheilen in:

a) Hauptminerale: Quarz, Granat, Augit, Hornblende, Turmalin, Epidot, Staurolith, Rutil, Zirkon, Magnetit, Ilmenit, Orthoklas, Kalkspath, Apatit.

b) Untergeordnete Minerale: Plagioklas, Mikroklin, Cordierit, Titanit, Sillimanit, Olivin, Disthen, Korund, Spinell.

Sehr leicht kann man am von Meereswasser bei Ebbe und Fluth abwechselnd bespültem Strande (z. B. in Scheveningen) die drei (nach dem Quarz) wichtigsten Minerale an ihrer Farbe unterscheiden: die rothen (bei auffallendem Lichte sogar kupferrothen) Granatkörner, die grünen Augitkörner und die schwarzen Eisenerzkörner häufen sich durch Schlämmung in Pfützen an und ziehen leicht die Aufmerksamkeit auf sich.

Die meisten Körner dieser Mineralien sind stark abgerundet, obwohl vollkommene Kugeln nicht vorkommen[1].

---

der Körnergrösse haben, welches von der angewandten Kraft abhängt. Kann man doch sogar Quarz in einem Achatmörser nicht unter einer gewissen Dimension zerkleinern, wie die Untersuchung des Pulvers u. d. M. lehrt.

[1] G. Klemm giebt in seiner bekannten Arbeit über psammitische Gesteine (Zeitschr. d. deutsch. geol. Ges. **34**. 771. 1882) an, dass einer der untrüglichsten Charaktere für klastische Mineralkörner in den „zerhackten Umrissen liege, welche die Durchschnitte klastischer Körner in Folge der während des Transports erlittenen Oberflächenverletzungen zeigen," auch soll die Oberfläche selbst dieses „verletzte" Äussere besitzen. Ich habe dies Äussere an den Quarzkörnern des Dünensandes nicht deutlich beobachten können. Auch kommt mir diese „Einkerbung" der Oberfläche durch Verletzung etwas fraglich vor. Die Körner erleiden während des Wassertransportes in die Flüsse und durch die Brandung am Strande eine ziemlich sanfte gegenseitige Abnutzung. Von einem starken Einschneiden oder Verletzen kann kaum die Rede sein. Anstatt ein „zerhacktes Äussere" zu bekommen, werden die Körner durch Wassertransporte immer runder und glatter. Ja, sogar der Windtransport, der, wie vorhin erwähnt, bedeutend stärker die Oberfläche der Körner angreift, giebt durchaus nicht zu derartigen Einkerbungen Veranlassung. Ich glaube, dass, wo die Körner das „zerhackte Äussere" zeigen, dies vielleicht eine Folge ist der früheren eckigen und unregelmässig bröckeligen Oberfläche, welche der Gesteinsgrus immer zeigen wird. Diese wird durch die Abnutzung eher verschwinden als zunehmen. Nur wenn man die Körneroberfläche mit absolut glatten und spiegelnden Krystallflächen vergleicht, z. B. bei Zirkon, sind die abgerundeten Theile immer etwas rauher. Von „zerhackt" ist hierbei jedoch kaum die Rede.

Auch dies ist wiederum ein Beweis, dass die Abreibung durchaus nicht ihr Maximum erreicht hat. Manchmal begegnet man im Gegentheil noch schönen Krystallen, an welchen die Facetten sehr gut erhalten sind. Am auffallendsten ist dies bei den Zirkonen der Fall, welche oft höchstens eine Kantenabrundung erlitten haben. Ausserdem zeigen auch Spinell (Oktaëder), Magneteisen (Oktaëder) und manchmal Granat (Rhombendodekaëder) ihre früheren Krystallformen.

Die specifischen Gewichte der verschiedenen Mineralien sind sehr abweichend: von den leichten Feldspath- und Quarzkörnern (2,5—2,7) bis zu den schwereren Augit-, Hornblende- und Granatkörnern (3,0—4,2) und den schwersten Rutil-, Zirkon- und Eisenerzkörnern (4,2—5,2) sind alle Abstufungen der Dichte vertreten [1]. Dennoch trifft man im Dünensande durchaus keine Separation nach der Dichte an. Wenn eine solche stattfinden könnte, würde man sie leicht in den Dünen selbst beobachten können, da die schwersten Mineralien durch ihren Eisenreichthum die am dunkelsten gefärbten sind und man also z. B. schwarze magnetitreichere Schichten alternirend mit weissen magnetitarmen Schichten antreffen müsste. Vom Winde wird jedoch alles aufgewirbelt und ohne Ordnung abgesetzt. Auch dies ist eine der Thatsachen, die man a priori nicht erwartet hätte, indem man sicher Luftseparation annehmen würde. Auch am Strande im Meereswasser selbst findet keine eigentlich gründliche Separation nach der Dichte statt (zwar stellenweise kleine Separationen, wie z. B. die vorhin erwähnte in Pfützen, jedoch niemals in dicken Schichten). Die für eine Separation nöthige langdauernde Bewegung des ganzen Sandes (wie sie z. B. in Setzkästen in Aufbereitungsanstalten stattfindet) fehlt eben am Strande.

Erwähnen möchte ich noch, dass im Dünensande keine Glimmerblättchen vorkommen, welche man jedoch wohl im

---

[1] Nach dem zunehmenden specifischen Gewichte geordnet sind die Mineralien des Dünensandes: Orthoklas (2,56), Mikroklin (2,56), Plagioklas (2,6—2,7), Cordierit (2,6—2,7), Quarz (2,65), Kalkspath (2,71), Apatit (3,2), Amphibol (3,1—3,3), Turmalin (3,0—3,3), Pyroxen (3,3—3,6), Epidot (3,3—3,5), Titanit (3,4—3,6), Sillimanit (3,2—3,3), Olivin (3,3—3,5), Granat (3,6—4,2), Staurolith (3,4—3,8), Disthen (3,5—3,7), Korund (3,9—4,0), Spinell (3,6—4,1), Rutil (4,2—4,4), Zirkon (4,4—4,8), Ilmenit (4,8—5,0), Magnetit (5,0—5,2).

Meeressande antrifft. Ich kann dies nicht anders erklären, als dadurch, dass bei der kräftigen Abnutzung beim Windestransport die zarten Glimmerblättchen zerbrochen und ihre Fragmente vom Winde weit zerstreut werden.

[*Editor's Note:* Material has been omitted at this point.]

5. Die Wichtigkeit von Fluss- und Meeressanduntersuchungen.

Ich möchte die oben beschriebene Untersuchung des holländischen Dünensandes zugleich als Beispiel einer systematischen Sandanalyse im Allgemeinen verwenden.

Der Sand ist überhaupt gegenüber den festen Gesteinen von den Petrographen in auffallender Weise vernachlässigt worden. Man möchte sich die Frage erlauben, ob jetzt, wo die mikroskopischen Untersuchungen der Dünnschliffe ihren Höhepunkt erreicht oder vielleicht sogar überschritten haben, und wo oft eine ermüdende Monotonie in den sich immer mehr ähnlich sehenden Beschreibungen von Gesteinssuiten herrscht, es nicht an der Zeit wäre, dass auch die Petrographie oder wenigstens ein Theil derselben andere Bahnen der Forschung aufsuchte.

Eine solche neue Bahn dürfte vielleicht in der mineralogischen Untersuchung der Sande zu finden sein.

Eine systematische Untersuchung des Fluss- oder Meeressandes ist nicht bloss mineralogisch hochinteressant[2] und

[2] Als interessant in historischer Hinsicht, obwohl nur geringen wissenschaftlichen Werth besitzend, mögen hier zwei sehr alte Abhandlungen über den Gehalt an Gold und Mineralien der Flusssande erwähnt werden. Die jüngste dieser Abhandlungen stammt von dem italienischen Gelehrten L. Bossi, welcher dieselbe im Jahre 1804 in der französischen Akademie der Wissenschaften vorlas (Observations sur l'or natif en paillettes que l'on trouve dans les sables, par M. Louis Bossi, de Milan; lues le 18 germinal, an 12). Hierin wurde besonders der Goldreichthum des Tessin behandelt, welcher schon seit alten Zeiten als goldführend bekannt war (neuerdings wurde, wie vorhin erwähnt, der Tessinosand von E. Artini mineralogisch untersucht und auch wiederum das Gold darin nachgewiesen.

Interessanter für uns wegen der Untersuchung der Mineralien in den Flusssanden ist die ältere Abhandlung, welche schon aus dem Jahre bietet wegen der völlig abweichenden Methode nicht nur eine angenehme Abwechselung von den Dünnschliffbeobachtungen dar, sondern verspricht auch in geologischer Beziehung recht fruchtbar zu werden.

Erstens hat der Sand eines Flusses auf jedem bestimmten Punkte des Stromlaufes seinen eigenen mineralogischen Charakter, welcher immer abhängig ist von der petrographischen Beschaffenheit des vom Fluss durchlaufenen Landes.

Recht auffallend zeigt sich dies bei uns in Holland, dem typischen Lande der Flussmündungen. In dem Boden der

---

1718 stammt und von dem damaligen berühmten französischen Gelehrten Réaumur in den „Mémoires de l'Académie royale des sciences de Paris" publicirt wurde unter dem Titel: „Essai historique des rivières et des ruisseaux du royaume, qui roulent des paillettes d'or, avec des observations sur le sable avec lequel elles sont mêlées." — Réaumur unterscheidet in dem Rhone- und Rheinsande schon sehr richtig die drei Hauptgruppen: den weissen, den rothen und den schwarzen Sand (also den Quarz-, den Granat- und den Eisenerzsand). Er begnügte sich jedoch durchaus nicht mit dieser rohen Unterscheidung, sondern untersuchte jeden dieser Sande mikroskopisch. Wegen des Interesses, welche diese älteste mikroskopische Sanduntersuchung für die Geschichte der Mineralogie besitzt, mögen hier einige Stellen wörtlich citirt werden:

„Le sable avec lequel ces paillettes (d'or) sont mêlées, est lui-même une espèce de richesse, mais qui ne peut toucher que les seuls observateurs de la nature. Les laveurs le jettent comme inutile. Nous en distinguerons de trois sortes en les distinguant par rapport a leurs couleurs; savoir: un sable blanc, un sable rougeâtre et un sable noir. Le blanc est celui qui est emporté par les premières lotions: observé au microscope, il paroît composé de cristaux pareils à ceux des sables les plus communs; aussi n'est-il qu'un sable ordinaire.

„Mais le sable rougeâtre, vu au microscope, et même à la loupe, offre le plus joli spectacle du monde: c'est un amas de toutes les pierres transparentes et colorées, connues dans la jouaillerie. Il n'est que rubis, saphirs, émeraudes, jacinthes, etc. Les pierres qui y sont les plus communes sont celles dont les nuances tiennent depuis la couleur du rubis balais, jusqu'à celles de toutes sortes de jacinthes: de là vient que la couleur de ce sable est rougeâtre pour la vue simple. Les saphirs, topazes, émeraudes, y sont plus rares, quoi qu'on y en découvre de très-belles couleurs.

„Pour le sable noir, il est presque tout de fer, et aussi attirable par le couteau aimanté, que la limaille même de fer. Il y a beaucoup plus de ce sable noir parmi celui du Rhône, que parmi celui du Rhin. J'ai tiré du premier, avec le couteau aimanté, près du tiers en fer.

„Il reste pourtant du sable noir, sur lequel le couteau aimanté n'a point de prise, et qui apparemment n'est point du fer, car après avoir été

Niederlande kann man, insoweit er aus Sand besteht, vier grosse Factoren unterscheiden, nämlich den Boden selbst und die Sande der drei Hauptflüsse. Der eigentliche Boden besteht als westliches Ende der norddeutschen Tiefebene wohl hauptsächlich aus skandinavischem Material. Die grosse Verbreitung skandinavischer Gesteine (Findlinge) lässt schon für die Herkunft der Hauptmasse des holländischen Diluvialsandes kaum eine andere Erklärung zu, als die Abstammung aus nordischen Gegenden. Ausserdem spricht das auffallend reichliche Vorkommen von echten Gneiss- und Granitmineralien in den von mir untersuchten Dünensanden im höchsten Grade hierfür.

---

exposé au feu pendant quelque tems, il n'en devient pas plus attirable. La pesanteur de ce sable semble cependant prouver qu'il est métallique; mais comme il y en a peu de celui-ci mêlé avec beaucoup de sable rouge, il n'est pas aisé d'en faire l'essai.

„Le sable rouge, ou ce sable qui n'est qu'un amas de petits grains de différentes et vives couleurs, est aussi d'une pesanteur approchante de celles des sables métalliques, puisque par les lotions il ne peut être séparé des grains de fer, quoique la grosseur de ses grains ne surpasse guère celles de leurs; il tient aussi apparemment des parties métalliques.

„Les grains du sable du Rhin ont des couleurs plus foncées que ceux du sable du Rhône: ceux-ci n'ont souvent qu'une légère teinte de couleur de chair, comme les rubis balais les moins colorés: on y voit pourtant aussi des topazes, des émeraudes, etc.“

Es ist wohl gewiss, dass die mineralogischen Bestimmungen Réaumur's nicht richtig sind, wie dies zur damaligen Zeit, wo die Wissenschaft der Mineralogie noch geschaffen werden musste, wohl kaum anders möglich war. Seine rothen Sande, dessen Körner so schwer waren, dass sie kaum von den Eisenerzkörnern zu trennen waren, sind zweifelsohne hauptsächlich aus Granat zusammengesetzt, ein Mineral, das er auffallenderweise nicht nennt. Auch die fleischrothe Farbe des Rhônesandes weist auf Granat. Er nennt sie „Hyacinthe“ oder „Rubine“. Die Bezeichnung „rubis balais“ hat mit Rubin nichts zu thun, sondern wurde zur damaligen Zeit von den Edelsteinhändlern entweder an rothe Spinelle oder an rothe Topase gegeben (s. Haüy, Traité de Minér. Tome 2. 156. 1822). Auch die übrigen Bezeichnungen Réaumur's: Rubine, Sapphire, Topase, Smaragde etc. beruhen wohl hauptsächlich auf Phantasie. Die rothen „Rubine“ werden Granaten, die grünen „Smaragde“ grüne Augite oder Hornblenden, die „gelben Topase“ gelbe Epidote gewesen sein. Die blauen Körner mögen Sapphire gewesen sein, welches Mineral, wie vorhin erwähnt, auch von Michel-Lévy im Sande des Mesvrinflusses nachgewiesen wurde. Merkwürdig ist, dass Réaumur schon den magnetischen Magnetit von dem nichtmagnetischen Titaneisen trennte.

Wenn zufälligerweise die drei grossen Flüsse Rhein, Maas und Schelde nicht in Holland ausmündeten (z. B. wenn, wie es ganz gut der Fall hätte sein können, der Rhein in Deutschland bei Emden und die beiden anderen in Belgien oder Frankreich in die See strömten), so würden wir in Holland fast ausschliesslich auf skandinavischem Sande wohnen. Jetzt aber, da die drei Flüsse zusammen in einem kleinen Lande ausmünden und jeder von ihnen seinen eigenen, von dem skandinavischen ganz abweichenden Sand führt, wird der Sandboden mineralogisch bedeutend bunter. Der Rheinsand wird am wenigsten von dem skandinavischen Sand abweichen: aus der Schweiz und dem südlichen Deutschland bringt der Rhein viel Granit- und Gneissmaterial, dagegen durchströmt er keine grosse Kalksteingebiete; sein Sand wird also arm an Kalkspath sein. Die Maas und die Schelde dagegen durchlaufen gar keine Granit- und Gneissterrains, jedoch ausgedehnte Kalksteingebiete verschiedenen geologischen Alters[1]. Schon hierdurch haben wir einen scharf ausgesprochenen, leicht zu bestimmenden und wahrscheinlich zu interessanten Schlussfolgerungen brauchbaren Unterschied. Bei näherer Erforschung werden sich wahrscheinlich noch andere brauchbare Unterschiede auffinden lassen. Die mineralogische Entwirrung der Sande Hollands bildet den Gegenstand einer grösseren Arbeit, mit welcher ich seit einiger Zeit beschäftigt bin.

Es sei dies nur als Beispiel angeführt, um zu zeigen, wie interessant derartige mineralogische Sanduntersuchungen sein können und wie fruchtbar bisweilen ihre Resultate sind. Gefahr, dass dergleichen Arbeiten in ein gedankenloses Detailstudium ausarten werden, ist also durchaus nicht vorhanden. Es lässt sich diese Untersuchung der Flusssande sogar von einem allgemeinen Standpunkte aus betrachten.

Die Hauptfrage hierbei ist immer diese: Inwieweit können wir in einem beliebigen Punkte des Stromlaufs eines Flusses in seinem Sande qualitativ und quantitativ die zwei Hauptbestandtheile: 1) die localen Mineralien der beiden

[1] Die Maas durchströmt die devonischen und carbonischen Kalksteine Belgiens, die Schelde, die cretaceischen und tertiären Kalksteine und Kreideablagerungen Nordfrankreichs und Belgiens.

Flussufer und 2) die aus dem Hinterlande zugeführten Mineralkörner unterscheiden.

Der einfachste Fall ist, dass beide identisch sind. Strömt ein Fluss von Anfang bis zu Ende nur durch Granit oder Gneiss (wie es z. B. bei mehreren Flüssen Skandinaviens der Fall sein mag), so wird in irgend einem Punkt des Stromlaufs die mineralogische Zusammensetzung des Flusssandes nicht oder kaum nennenswerth[1] abweichen von jeder anderen und deshalb die Untersuchung wenig interessant sein.

Umgekehrt wird das Problem, wenn es zu complicirt ist, kaum entwirrbar. Bei einem grösseren Strom, wie der Rhein oder die Donau, wird es äusserst schwierig sein, von jedem Mineral im Flusssande die Herkunft zu bestimmen. Das durchlaufene Gebiet (oder, besser gesagt, das ganze Stromgebiet, denn die Sache wird durch die Nebenflüsse noch complicirter) ist geologisch zu bunt, um eine vollkommene Entwirrung noch durchführbar zu machen, obwohl eine theilweise Bestimmung der Herkunft immer noch möglich erscheint. So werden z. B. die Olivinkörner im Rheinsande wohl den zahlreichen Basaltvorkommnissen im Stromgebiete des Rheines zuzuschreiben sein.

Am günstigsten gestaltet sich der Fall einer mittleren Complication. Am einfachsten ist es z. B. bei zwei scharf contrastirenden Gesteinen: Ein Fluss durchschneidet im Oberlauf Basalt, im Unterlauf Kalkstein. Der ursprünglich schwarze Basaltsand des Flusses wird, nachdem dieser die Gesteinsgrenze überschritten hat, allmählich weisser werden und der Kalksteingehalt stets reichlicher, wie es ausser an der Farbe leicht durch Extraction mittelst Salzsäure nachzuweisen wäre.

Überhaupt hat man die für die Untersuchung und die Schlussfolgerungen günstigsten Fälle bei Flüssen mittlerer Grösse, die zwei oder drei mineralogisch gut trennbare Gebiete durchfliessen. In kleinen Flüssen resp. Bächen wird

---

[1] Ich brauche wohl nicht zu betonen, dass man bei solchen Untersuchungen keine mathematische Genauigkeit zu erwarten hat; Spielraum muss immer gelassen werden. Deshalb sind sie nur bei starken mineralogischen Contrasten zu empfehlen.

gewöhnlich der Sand zu einförmig, in grossen Flüssen, wie dem Rhein, der Elbe etc., der Sand zu bunt gemischt sein.

Ein äusserst günstiges Land für solche Flusssanduntersuchungen scheint mir z. B. Italien zu sein. Die schmale Halbinsel hat fast nur kleine Flüsse (Arno, Tiber) und einen geologisch resp. petrographisch einförmigen Charakter: tertiäre und cretaceische Sandsteine und Kalksteine, theilweise durch eruptive Massen unterbrochen (Gabbro und Serpentin im Arnogebiet, vulcanische Gesteine im Tibergebiet). Alle Minerale dieser Gesteine sind im Flusssande wahrscheinlich verhältnissmässig leicht auseinander zu halten und machen die Untersuchung zu einer sehr lohnenden.

Ein ähnlich geeignetes Land würde vielleicht auch Griechenland sein.

Überhaupt werden sich Halbinsel und Insel mit nicht zu einförmiger und nicht zu complicirter petrographischer Zusammensetzung zu diesen Untersuchungen eignen.

Dagegen sind massive Continentalländer mit bunter geologischer Karte, z. B. von Deutschland oder Österreich, viel weniger geeignet.

Noch interessanter werden die Sanduntersuchungen, wenn man die Meeressande mit einbegreift. In Italien, um bei dem ersten Beispiel zu bleiben, welches Land meistens felsige Küsten besitzt, wird man auch bei den Meeressanden unterscheiden können: a) die localen, von den Küsten selbst stammenden Sandkörner und b) die aus dem Hinterlande von den Flüssen zugeführten Mineralkörner. Vielleicht, jedoch nicht immer sicher, gesellt sich noch ein dritter Factor hinzu: die von Meeresströmungen längs der Küste fortbewegten oder die von anderen Theilen der Küste ins Meer gerathenen und von der Brandung zurückgespülten Körner[1].

[1] Als Beispiel, wie interessant gerade die Meeresküste Italiens für derartige Sanduntersuchungen ist, mag die schon am Anfang meiner Arbeit citirte Untersuchung Uzielli's über die Leucit-, Augit-, Zirkon- etc. führenden Sande der Westküste Italiens erwähnt werden. Würde diese schon mit beschränkten Hilfsmitteln ziemlich ergebnissreiche Untersuchung vervollkommnet zu einer systematischen Sanduntersuchung und vereinigt mit einer solchen der Sande der an dieser Küste ausmündenden Flüsse, so würde man wahrscheinlich zu wichtigen Resultaten kommen können. Es würden gerade derartige Untersuchungen der italienischen Küstensande

Diesen letzten Factor würde man z. B. am besten studiren an einer Kalksteininsel, die im Meere nicht weit von einer Granit- oder Trachytküste entfernt ist.

Ausserdem möchte ich noch auf die Wichtigkeit der Sanduntersuchungen bei Entdeckungsreisen hinweisen. Manchmal, z. B. bei Nordpolfahrten oder überhaupt bei Seereisen, kann man mit dem Schiffe nur die Küste berühren, während das Hinterland unerreichbar ist. Eine Untersuchung des Sandes der Küste oder einer Flussmündung wird (besonders durch Vergleiche mit dem mineralogischen Charakter der anstehenden felsigen Meeresküste) oft Aufschlüsse geben können. Besteht z. B. die Küste aus Gneiss und trifft man in dem Sand der Flussmündung viel Kalkspath- oder Serpentinkörner an, so darf man sich sichere Schlüsse über das Vorkommen derselben im Innern des Landes erlauben. Sind es doch die Flüsse, welche uns durch ihre Sande Kunde bringen von den Gesteinen, welche sie tief im Innenlande durchströmt haben.

Aber auch, abgesehen von diesen recht interessanten Schlussfolgerungen, ist die einfache Kenntniss der Zusammensetzung des Flusssandes eine, nach meiner Ansicht nothwendige Ergänzung unserer Kenntnisse der Eigenschaften der Flüsse selbst.

Fast von jedem Fluss in Deutschland oder überhaupt in Westeuropa ist der Lauf, das Gefälle, die Stromgeschwindigkeit genau bekannt, ja sogar die Flussflora und -Fauna sind studirt; der mineralogischen Zusammensetzung der Flusssande ist jedoch noch niemals die gehörige Beachtung geschenkt worden. Dennoch ist sie — wie gesagt — auf jedem einzelnen Punkt des Stromlaufs eine ziemlich constante oder bewegt sich höchstens innerhalb ziemlich enger Grenzen. Nicht bloss qualitativ, sondern auch quantitativ lässt sich dieselbe ausdrücken, so dass — wenn der Ausdruck erlaubt ist — der Fluss in jedem Punkte eine feste mineralogische Sandformel besitzt.

Schon qualitativ ist oft die Untersuchung der Flusssande

---

empfehlenswerth sein, als es eben italienische Petrographen sind, welche sich durch Untersuchungen von Sanden und Sandsteinen ihres Vaterlandes verdienstlich gemacht haben, wie die schon Anfangs citirten Forscher C. Riva, J. Chelussi, L. Ricciardi und besonders E. Artini durch ihre Arbeiten bewiesen haben.

recht interessant, besonders wenn sie auffallende mineralogische Charakterzüge besitzen. Ich erinnere z. B. an die granat- und chromitreichen Sande der Serpentingebiete, wie das schlesische, strömenden Flüsse, an die Platinsande Russlands, an die an schönen Mineralien (Hauyn, Nosean, Sanidin etc.) reichen Sande des Laacher-Seegebietes etc. Hier wird man meistens eine unerwartet reiche mineralogische Ausbeute antreffen: ein solcher Sand ist oft eine mikromineralogische Sammlung, wie sie reicher kaum gedacht werden kann, und würde schon als Gegenstand einer zwar oft schweren, doch reizvollen Übung im Bestimmen dieser Mineralien auf physikalischem und mikrochemischem Wege höchst anziehend sein.

Mit welcher Absicht man jedoch den Sand auch untersucht, immer soll man zunächst durch eine systematische Sonderung nach dem specifischen Gewichte mittelst allmählich schwerer werdenden Flüssigkeiten Licht zu schaffen suchen, und nicht — wie es jetzt oft geschieht — durch directe mikroskopische Beobachtung des Sandes, so wie er aus dem Flusse geschöpft wird, oder nach nur oberflächlichem Ausschlämmen der schwersten Mineralien mittelst Wasser. Dieses Verfahren dürfte kaum die Bezeichnung einer wissenschaftlichen Methode verdienen. Durch die directe Beobachtung übersieht man leicht die meisten Mineralien; erst durch Beseitigung der zahlreichen Quarzkörner und Concentration der übrigen Mineralien in besondere Gruppen entgeht kein einziges der wichtigen Mineralien der Wahrnehmung. Beim Schlämmen und Waschen dagegen geht erfahrungsgemäss viel verloren.

Schliesslich sei es mir noch erlaubt, auf die Wichtigkeit derartiger Sanduntersuchungen in geologischer Beziehung hinzuweisen. Sind wir durch fortgesetztes Studium des Sandes z. B. der europäischen Meeresküsten und durch Vergleichung desselben mit dem Sand der verschiedenen Flüsse zu genügenden Kenntnissen gekommen, so werden wir übergehen können zu dem Studium geologisch älterer Sande, z. B. diluvialer und tertiärer, ja sogar fester mesozoischer Sandsteine, sofern sie sich (nöthigenfalls nach vorhergehender Auslaugung des Bindemittels durch Säuren) zu dem ursprünglichen Sande leicht zerreiben und in den Bereich der Untersuchung ziehen lassen.

Lassen z. B. die Untersuchungen der recenten Meeressande an der Mündung eines Flusses oft recht deutlich eine local beschränkte mineralogische Abweichung des Sandes erkennen, so wird vielleicht auch umgekehrt der Nachweis solcher Flussmündungen in tertiären oder älteren Sandsteinen auf mineralogischem Wege gelingen können.

Allgemein verbreitet ist die Ansicht, dass die mineralogische Untersuchung von Gesteinen fast ausschliesslich auf Eruptivgesteine und krystallinische Schiefer beschränkt ist, während bei den versteinerungsführenden Sedimentärgesteinen selten eine solche sich lohnt und nur ausschliesslich die palaeontologische Untersuchung hier Nutzen bringt. Ich glaube, dass dies nicht der Fall zu sein braucht, wenn man die Sande auf angegebene Weise untersucht und sich nicht beschränkt auf einen nichtssagenden Sandsteindünnschliff[1].

Ob sich die Methode noch ausdehnen liesse auf andere Sedimente, ist vorläufig noch fraglich. Auf reine Thone glaube ich kaum, dass sie anwendbar ist. Der eigenthümlichen Eigenschaft des Lehmes, lange in Wasser suspendirt bleiben zu können, steht die Sonderung der Bestandtheile mittelst schwerer Flüssigkeiten sehr im Wege. Ein Körper, wie z. B. der Löss, wird sich wahrscheinlich kaum auf diesem Wege richtig in seine mineralogischen Bestandtheile trennen lassen. Ich will jedoch nicht behaupten, dass die Trennung absolut unmöglich ist; vielleicht wird man sie durch recht langes ruhiges Stehenlassen des Gemenges dennoch erreichen können. Man braucht hier auch nicht das Äusserste zu erreichen und kann vielleicht durch eine unvollkommene Trennung schon brauchbare Resultate erhalten[2].

---

[1] Eine derartige Untersuchung psammitischer Gesteine mittelst Dünnschliffe ist bekanntlich von G. Klemm (Zeitschr. d. d. geol. Ges. **34**. 771. 1882) vorgeschlagen. Nach meiner Ansicht hat diese Methode wenig Zukunft. Der Dünnschliff ist n i c h t geeignet für psammitische Gesteine, weil eben das Charakteristische derselben dadurch nicht zur Anschauung kommt. Erst nach der Entfernung des meist 90 und mehr Procent betragenden Hauptminerals (des Quarzes) kommen die weit wichtigeren Nebenminerale zur Geltung.

[2] Man vergleiche über diesen Gegenstand (die Suspension feiner Mineraltheilchen in Wasser) die Arbeit des Herrn Bodländer in dies. Jahrb. 1893. **2**. 147.

Was natürlich wohl zu erreichen ist, ist die Untersuchung des nach vorheriger Wegschlämmung des feinen thonigen Materials zurückgebliebenen sandigen Restes, den man ganz wie gewöhnlichen Sand in schweren Flüssigkeiten trennen kann.

Bei vielen Kalksteinen würde man das $CaCO_3$ durch Säuren lösen können und den zurückbleibenden Sand trennen, und also auch hier die Methode anwenden. Ein Beispiel, wo dies zufälligerweise fruchtbar sein kann, liefert uns die Insel Java. Man unterscheidet hier im Tertiär petrographisch zwei grosse Gruppen: das vorvulcanische oder (nach dem Hauptgesteine genannte) prae-andesitische Tertiär (Eocän oder vielleicht theilweise Jung-Cretaceisch) und das nachvulcanische oder post-andesitische Tertiär (wahrscheinlich Miocän oder Pliocän). Beide haben ihre eigenen Kalksteine. Durch Lösen in HCl und Untersuchung des sandigen Rückstandes würde man sofort entscheiden können, ob er nur aus quarzigem Material (dem altkrystallinischen oder vielleicht granitischen Urboden Javas, der fast ohne Ausnahme von jungen Schichten bedeckt ist, entstammend) besteht oder reichlich vulcanisches (Augit, Plagioklas, Magnetit und Glaskörner führendes) Material enthält. Auch hier wäre die systematische Trennung in Gruppen mit steigender Dichte vielleicht in beiden Fällen erfolgreich anzuwenden.

Ich glaube deshalb jedenfalls die beschriebene Trennungsmethode als eine in zahlreichen Fällen fruchtbare empfehlen zu können.

[*Editor's Note:* In footnote 1 on page 18 of the article the complete reference is: Sokolow, N. A., 1894, *Die Dünen—Bildung, Entwicklung, und innere Bau,* Julius Springer, Berlin, 298p.]

3

Reprinted from *Geol. Soc. London Quart. Jour.* **58:**620–631 (1902)

THE MINERALOGICAL CONSTITUTION OF THE FINER MATERIAL OF THE BUNTER PEBBLE-BED IN THE WEST OF ENGLAND

H. H. Thomas

I. Introduction.

Much work has already been done on the pebbles and larger fragments of rock of the Budleigh-Salterton Pebble-Bed, but little or none on the finer material forming the matrix of this most interesting deposit. It occurred to me that a microscopic investigation of this finer material might bring out a few points of geological as well as mineralogical interest. Therefore it has been my endeavour to make as complete a mineralogical analysis of the sands as possible.

Specimens were collected at intervals along the strike of the Pebble-Bed, from Budleigh Salterton on the southern coast of Devon to Fitzhead, north of Milverton, in Somerset; other sands, for comparison, were taken from the Red Rocks above and below.

After treatment with acids, so as to remove the coating of iron-oxides from the grains, the sands were separated by means of heavy liquids into three parts:—

(*a*) Heavy residue, with a specific gravity greater than 2·80.
(*b*) The bulk of the quartz; and
(*c*) The lightest part, with most of the alkali-felspar.

II. Petrography of the Sands: Description and Distribution of the Mineral Species.

At the southern end of the outcrop of the Pebble-Bed, the matrix consists of incoherent material, each sand-grain being covered with a coating of red iron-oxide. On tracing the deposit northward, however, to the borders of Somerset, the bed loses much of its red colour, and in this area, owing to the presence of many masses of Culm-Measure limestone in the vicinity, the grains are cemented in a calcareous matrix, the deposit ultimately taking on the form of the Dolomitic Conglomerate.

The sands contain, on the whole, a very small percentage of mineral-grains the specific gravity of which is greater than 2·80; while the ratio of material over 2·58 to that under 2·58 is about 70 or 80 per cent. to 30 or 20 per cent. In some cases the heavier material rises to as much as 88 per cent. (at Milverton).

The following is the percentage of material of specific gravity greater than 2·58, at various localities along the outcrop of the Pebble-Bed, taken from south to north :—

	Per cent.
Budleigh Salterton	75
Fair Mile, near Tallaton..............	73
Uffculme	71
Burlescombe..............................	78
Milverton	88
Fitzhead	83

It is interesting to note that the percentage of heavy material and the percentage of light material are almost constant throughout the thickness of the bed at Budleigh Salterton, being 75 and 25 respectively. This, unfortunately, was the only locality where there was a section from which I could be sure of collecting a series of samples at regular intervals, from the top to the bottom of the deposit. Rounding of the grains has taken place, but the degree of rounding is by no means constant, even in the same sample. However, taking into consideration any one mineral, we notice that the rounding becomes more complete as we travel from south to north.

The chief mineral species that occur in the sands forming the matrix of the Pebble-Bed, arranged according to their crystal-systems, are :—

Cubic.
Fluorspar.
Garnet.
Magnetite.

Rhombohedral.
Apatite.
Ilmenite.
Tourmaline.
Quartz.

Monoclinic.
Biotite.
Muscovite.
Orthoclase.
Sphene (titanite).

Tetragonal.
Anatase.
Cassiterite.
Rutile.
Zircon.

Orthorhombic.
Brookite.
Sillimanite.
Staurolite.

Triclinic.
Cyanite (disthene).
Microcline.

Under the head of amorphous minerals and compound grains may be mentioned :—

Felsite, quartzite, and chert.
Shimmer-aggregates, pinite, and mica-schist.
Leucoxene, and other decomposition-products.

The following comprises a description of the species and grains mentioned above, giving in some instances the chief characters by which the minerals were identified.

Fluorspar.—This mineral occurs in colourless isotropic grains, usually angular, with the octahedral cleavage {111} well shown. They reach a size of 0·40 millimetre in greatest length. The index of refraction is low, and this character is useful for purposes of identification.

Fluorspar has been found in the sands from Berry Head [1] to the west of Torbay, from Budleigh Salterton, and Woodbury Castle; but it has not been noticed in the Pebble-Bed north of the last-named locality, nor does it seem to occur in the so-called 'Permian Breccias,' or in the beds above the Pebble-Bed.

Garnet.—The garnets of these sands occur in irregular but somewhat rounded isotropic grains, usually not more than 0·25 millimetre across, but sometimes reaching as much as 0·30 millimetre.

They are of a pale pink colour, and if in large crystals would undoubtedly have been of a very deep red. Crystal-outline is generally wanting.

In the Pebble-Bed they are confined to the North of Devon and Somerset, occurring only along the line from Uffculm northward to Fitzhead. They have not been found to the south, but are common, in fact abundant, in all the red beds above the Pebble-Bed.

Anatase [2] (Pl. XXXII, figs. 1–5).—This mineral occurs in minute, but usually perfect, colourless, transparent tetragonal crystals (rarely exceeding 0·10 millimetre in greatest width). The habit is in most cases tabular, the base being large and the pyramids narrow.

An optically negative axis emerges normally to the basal plane (001). Many of the larger grains are turbid and of a yellow-brown colour: this turbidity seems to be due to a superficial decomposition, possibly resulting in the formation of rutile.

One example only of an octahedral crystal [3] has been noted, and that from the sand of Fitzhead. The basal plane on this crystal was very small.

This mineral is not very plentiful in the Pebble-Bed, but is fairly constant in its occurrence. It is more plentiful in the sands of the Upper Bunter, where it occurs with the same tabular habit.

Rutile [4] (Pl. XXXI, figs. 10–13).—This mineral is very prevalent throughout the whole of the Pebble-Bed, and is no less common in the beds above and below. Simple crystals are seen to consist of the simple tetragonal prism {110} capped by a pyramid {111}. The commonest occurrence, however, is in the form of rounded prisms with indistinct terminations, and grains.

The colour is from yellow to amber-red or brown.

The mineral is often twinned, these twins being mostly poly-

[1] This sand exists as an outlying patch in a fissure in Devonian rocks; from analogy in composition to the finer material of the Pebble-Bed, I should suspect it to be made up of resorted material derived from that deposit.

[2] W. F. Hume, 'Chem. & Micro-mineralog. Researches on the Upper Cretaceous of S. England' 1893, p. 97; J. J. H. Teall, 'Brit. Petrogr.' 1888, pl. xliv, fig. 6.

[3] H. L. Bowman, Mineral. Mag. vol. xii (1900) p. 362.

[4] J. J. H. Teall, 'Brit. Petrogr.' 1888, pl. xliv, fig. 4.

synthetic (figs. 11 & 12), or more rarely geniculate (fig. 13). The long axes of the individuals make an angle of about 65° one with the other. Inclusions are not common.

Although this mineral, as has been stated, is common all through the Red Rocks, it is more plentiful to the north of Uffculm.

Cassiterite is difficult to distinguish from rutile, but it occurs in small quant ty in the heavy residues from some of these sands. There seem to be two modes of occurrence: (i) in pale-yellow, well-shaped crystals, and (ii) in more or less rounded brownish grains.

The crystals are generally short tetragonal prisms, with an ordinary pyramid. A basal plane is usually present. The grains and crystals differ considerably from the rutile in the same sample. The prism seems to be less elongate, the colour is paler, twinning is less frequent, and the lustre is submetallic.

The crystals have been compared with stream-tin from Devon, in order to establish their identity.

This mineral so far has only been obtained from the sands north of Uffculm, and does not appear to be present in the sands south of that locality.

Zircon[1] (Pl. XXXI, figs. 1-9).—Zircon is one of the most abundant minerals in the heavy residues of these sands, and occurs generally in colourless, highly refracting, prisms, in all essentials identical with those so commonly met with. The crystals seldom attain a length of more than 0·25 millimetre.

An interesting form has been discovered by my friend, Mr. J. B. Scrivenor, in the heavy residue from Fitzhead. It consists of a 'capped' crystal (Pl. XXXI, fig. 3)[2], the two individuals being connected by their basal planes and in perfect optic continuity. This is the only example of the kind that has been met with.

The basal plane (001) is a rare face on microscopic zircons, but it has been noticed to be fairly common on the crystals from Burlescombe, Milverton, and Fitzhead. Although this mineral is usually colourless, a pink variety occurring in stumpy prisms, capped seemingly by a simple pyramid, has been noted from Newton Poppleford in Devon (fig. 9). Colourless, and pale-brown stumpy crystals also occur. These last-mentioned varieties have all the optical properties of zircon, and evidently belong to that group: if, however, they are not true zircons, they must belong to an allied species, possibly xenotime.[3]

Many crystals of zircon show a beautiful zonal structure, which is very characteristic.

Inclusions in this mineral are common, and are seen to consist of colourless needles, probably apatite, and of a dark-brown, highly pleochroic mineral, crystallizing in stumpy prisms, with

[1] J. J. H. Teall, 'Brit. Petrogr.' 1888, pl. xliv, fig. 1.
[2] A. B. Dick, 'Nature' vol. xxxvi (1887) p. 91.
[3] O. A. Derby, Mineral. Mag. vol. xi (1897) p. 308.

straight extinction. These seem to be rhombohedral, but as yet I have come to no definite conclusion regarding the mineral species to which they belong. Negative crystals are common, as also are gas- and fluid-cavities. The inclusions and cavities usually occur along definite lines, which are either parallel to the prism, to the pyramid, or to both.

With regard to the distribution of this mineral, it is abundant everywhere; but the zircons with the basal plane well-developed are more common in the sands north of Uffculm, than from localities farther south.

Ilmenite.—This mineral was probably abundant, for leucoxene is a prevalent constituent, and may replace original black grains, of which only a few are left.

It is opaque, or only slightly transparent, but of a whitish-yellow or greenish-brown colour by incident light. It has a granular structure, and the grains are usually round.

Micaceous ilmenite[1] occurs in minute flat plates of a bright reddish-brown colour, enclosing rutile-needles arranged symmetrically at angles of 60° one with the other, taking on the form of sagenite.

Here it might be well to mention another ore of iron which also occurs in some cases, and that is a deep red mineral in thin flakes, made up of minute spheres with a radiate structure: Each sphere gives a perfect black cross between crossed nicols. It is probably a form of hæmatite.

Tourmaline[2] (Pl. XXXI, figs. 14–17).—This mineral is abundant in all the residues, and occurs in various forms. The most plentiful are stumpy crystals, consisting of a prism terminated by the simple rhombohedra μ {100} and μ {$\bar{1}$00}. In these the pleochroism is intense but variable, even in one and the same crystal, the dominant colour being brown. The true form of these crystals is somewhat obscured by the rounding which the majority have undergone; indeed, some are almost spherical.

At the southern end of the Pebble-Bed outcrop they reach 0·25 millimetre in diameter, but are smaller in the north. Another variety occurs, of a rich blue colour, in radial aggregates of acicular crystals (Pl. XXXII, fig. 8). Inclusions usually consist of minute opaque grains, without crystal-outline.

It is worth noting that the blue tourmaline, which is so common in the so-called 'Permian Breccias,' and also occurs in moderate amount in the sands both above and below the Pebble-Bed, is extremely rare in the Pebble-Bed itself. Only a grain or two has been met with in samples from the southern portion of the outcrop. It is, however, more plentiful in the Pebble-Bed sands north of Uffculm.

[1] H. Rosenbusch [transl. Iddings], 'Microscop. Physiogr. of Rock-making Minerals' 4th ed. (1898) p. 166.

[2] J. J. H. Teall, 'Brit. Petrogr.' 1888, pl. xliv, fig. 3.

Quartz.—The grains of this mineral vary considerably, both in size and in their degree of rounding. The more angular, however, seldom exceed 0·6 millimetre, and the perfectly rounded ones are seldom more than 0·5 millimetre in diameter.

Secondary growths of silica on the grains have not been met with, but Dr. Irving[1] records it as of common occurrence in the sands above the Pebble-Bed.

Fluid-cavities are common, occurring both in lines and irregularly distributed through the grain. Inclusions are numerous, and consist of tourmaline, usually blue, and often arranged in radiating masses simulating luxullianite; rutile and apatite, in fine needles; and what I take to be sillimanite,[2] in colourless fibrous crystals.

Much of the quartz gives undulose extinction.

Brookite (Pl. XXXII, fig. 6).[3]—This mineral is rare in the Pebble-Bed of this district, and has been found only as far north as Newton Poppleford. It is associated with anatase. In the Upper Bunter[4] Sands of Ladram Bay, west of Sidmouth, it is seemingly more common.

It occurs in small, tabular, rhombic crystals (0·15 millimetre), flattened parallel to the face (100). This face is characteristically striated, parallel to the vertical axis. A positive bisectrix emerges at right angles to the tabular face. The index of refraction is high, and the double refraction strong. The crystals have a pale yellow colour with feeble pleochroism. The lustre by incident light may be said to be submetallic.

Staurolite[5] (Pl. XXXII, fig. 7).—This mineral is of a yellow to amber-red colour when the grains attain much thickness. It occurs in angular and rough grains up to 0·5 millimetre in greatest diameter, bounded presumably by the cleavage (010) and that parallel to the prisms (110). A conchoidal fracture also occurs. The index of refraction is high, and the pleochroism distinct, ranging from pale to deep yellow or brown. The double refraction is strong, and the interference-tints brilliant. Some grains show a biaxial figure with a positive bisectrix, and an angle between the optic axes varying between 80° and 90°. The plane of the optic axes lies in the plane bisecting the acute angle formed by the intersection of the prismatic cleavages.

The grains are mostly very fresh, there being hardly a trace of decomposition. Occasionally, however, grains may be seen passing into a micaceous decomposition-product.

[1] Rev. A. Irving, Quart. Journ. Geol. Soc. vol. xlviii (1892) p. 71.

[2] H. Rosenbusch [transl. Iddings], 'Microscop. Physiogr. of Rock-making Minerals' 4th ed. (1898) pl. xvi, fig. 3.

[3] W. F. Hume, 'Chem. & Micro-mineral. Researches on the Upper Cretaceous of S. England' 1893, p. 97.

[4] E. Hull, Quart. Journ. Geol. Soc. vol. xlviii (1892) p. 65.

[5] H. Rosenbusch, *op. supra cit.* p. 201.

In some cases inclusions are numerous, consisting mostly of quartz, and of opaque black particles of doubtful nature; other fragments of staurolite occur which are quite free from inclusions of any kind.

This is one of the most abundant of the heavy minerals in the Pebble-Bed, forming fully 20 per cent. of the heavy residue in the Budleigh-Salterton area; but it gradually decreases in quantity northward, till at Fitzhead in Somerset the percentage is relatively small. The greatest percentage was noted from the sand of Berry Head (footnote, p. 622). This mineral is mainly confined to the Pebble-Bed, although a few grains do occur both above and below that horizon.[1]

The Micas.

White mica is fairly abundant, and occurs in colourless flakes, usually about 0·45 to 0·50 millimetre across. A negative bisectrix emerges normally, but the angle between the optic axes is variable. Much of the mica gives undulose extinction. Inclusions are common, and consist of zircon in well-shaped crystals, needles and knee-shaped twins of rutile, and fine, colourless, curved and branched needles of obscure nature (Newton Poppleford).

Blue and brown pleochroic needles of tourmaline are also common, which lie in intersecting rows and radiating groups. Pleochroic halos may be noticed rarely round the more minute inclusions.

Brown mica is not very abundant, but its presence has been proved in many of the sands by the flakes of gelatinous silica, left after treatment with acids. Where the mineral has escaped decomposition, it is of a pale-brown colour with intense pleochroic halos round inclusions.

The micas, so far as one can judge, seem less abundant in the Pebble-Bed than in the beds above and below it.

Felspars.—The majority of the felspar is of the ordinary potash variety orthoclase; some microcline, however, occurs, easily recognized by its characteristic cross-hatching. Decomposition has in many cases gone on to such an extent that the grains have lost most of their original characters. Even the freshest grain is usually much clouded by decomposition-products. The orthoclase sometimes shows the characteristic Carlsbad twinning, but no undoubted plagioclase has been met with.

Felspar is fairly abundant in the Pebble-Bed sands, reaching, as far as one can estimate it, about 6 per cent. of the total mass.

Sphene occurs in rounded, irregular, feebly transparent, brown grains, reaching 0·20 millimetre in diameter, which are traversed by rough cracks. The pleochroism is poor. This mineral is not very common, but is fairly constant in its distribution; it occurs also in the Upper Bunter and Keuper Marls.

Grains have been noticed, enclosed in a highly refracting crypto-crystalline mass which exhibits a rudely radiate structure. This enclosing material is probably a decomposition-product.

[1] It may be recollected that staurolite has been found in the dune-sands of Holland.

Cyanite.[1]—This mineral occurs as the unaltered colourless cores of some of the shimmer-aggregates (see below) which are so common. The cleavage is distinct, the refractive index high, and some grains show in convergent light a negative bisectrix perpendicular to the good cleavage (100).

Compound Grains.

As might be expected in a deposit of this nature, compound grains are very prevalent. They consist of the following:—

Quartzite, which occurs in more or less rounded grains; it is of a fine texture, and may be referred to the quartzite that forms the mass of pebbles of which the bed is chiefly composed.

Chert in rounded grains.

Felsites in cryptocrystalline masses, sometimes showing micropegmatitic structure.

Mica-schist in rounded or oval grains, consisting largely of white mica with a parallel arrangement. In some cases a well-marked knotted structure is noticeable. Included minerals are common, and consist of rutile and tourmaline in fine needles, grains of staurolite, as well as many other minerals which, owing to their minute size, are difficult to identify.

Shimmer-Aggregates.[2]—These aggregates, which consist of a felted cryptocrystalline mass of a white micaceous mineral, with no parallel structure to speak of, are one of the commonest constituents of these sands. They are crowded with inclusions of great variety. These aggregates are evidently decomposition-products from aluminous silicates, probably from cyanite, andalusite, and possibly others.

Cyanite has been recognized as an unaltered core in some cases, while altered andalusite from the Skiddaw Slates presents an exactly similar structure.

Among the inclusions the following may be observed:—zircon in its usual form, tourmaline and rutile in fine needles, crystals and grains of magnetite, and grains of staurolite and sphene. Thin hexagonal and rounded grains of a reddish-brown mineral also occur, and in some instances are very plentiful: it is probably a garnet, possibly melanite.

These shimmer-aggregates are slightly more abundant in the south.

Pinite.[3]—Cryptocrystalline aggregates of a green micaceous mineral, usually not more than 0·30 millimetre across, have been noted; but they are rare.

They include grains of an opaque mineral, probably some ore of iron, and from their general characters would seem to be a decomposition-product of some mineral, probably cordierite. No unaltered cordierite has been met with.

[1] J. J. H. Teall, 'Brit. Petrogr.' 1888, pl. xliv, fig. 2.
[2] G. Barrow, Quart. Journ. Geol. Soc. vol. xlix (1893) p. 349 & pl. xvi, fig. 5.
[3] H. A. Miers, Mineral. Mag. vol. xi (1897) p. 283, and A. Lacroix, 'Minéralogie de la France' vol. i (1893–95) pp. 517–18.

III. Possible Directions of the Sediment-bearing Currents, and Sources of the Mineral-Grains.

There are many peculiarities in the distribution of the foregoing minerals, which point more or less distinctly to the direction in which the material was carried. From these we may, I think, deduce the existence of a southerly main current, over the whole line from the southern coast of Devon to Fitzhead in Somerset, a distance of about 30 miles; and at the same time it will be seen that this main current was joined to the north of Uffculm by a minor current, probably from the west.

The evidence for the foregoing statement is as follows:—Taking into consideration the percentage of material with a specific gravity greater than 2·58 (see p. 621), we see that from 75 per cent. at Budleigh-Salterton it falls gradually northward until at Uffculm (20 miles to the north) the percentage is 71. It now begins to rise, and continues to do so until it reaches as much as 88 per cent. in the neighbourhood of Milverton in Somerset, north of which the percentage again falls. The gradual diminution of the percentage of heavier material as we pass from south to north, would in itself suggest a transport of material in that direction; while the rise in the neighbourhood of Burlescombe would point to a fresh influx of material. This view is strengthened by the distribution of the individual minerals. For instance, staurolite, which is so plentiful in the extreme south, diminishes considerably in quantity in the more northerly part of the outcrop of the Pebble-Bed, and a gradual diminution in the size of the rounded tourmaline-grains has been noted in the same direction. Fluorspar does not extend northward farther than about 4 or 5 miles from the coast. Associated with the rise in percentage of the heavy material at Burlescombe and the other localities near, an assemblage of minerals occurs which differs markedly from that of the normal, more southerly type.

Several minerals, such as garnet and cassiterite, make their appearance for the first time, and the blue variety of tourmaline becomes more common. It thus seems probable that the main current was from the south, which is in accordance with the conclusion arrived at by Mr. Ussher[1] from a study of the distribution of the pebbles.

The confluent westerly current is simply one of many minor currents such as are met with in any drainage-system.

When we consider what are the minerals which have been recorded, we can hardly refrain from concluding that an area differing widely from any in the South or West of England must have been undergoing denudation to supply them. The occurrence of minerals such as staurolite, cyanite, sillimanite, microcline, and probably cordierite, and the prevalence of 'shimmer-aggregates,' suggest some great metamorphic area as the source of much of the material. So far as is known at present, staurolite does not occur in the metamorphic rocks round the Devon-Cornwall[2] granite-masses.

[1] Mem. Geol. Surv. 'Exeter' 1902 (in the press). See also W. A. E. Ussher, Quart. Journ. Geol. Soc. vol. xxxiv (1878) pp. 246–47.

[2] K. Busz, Neues Jahrb. Beilage-Band xiii (1899) p. 90.

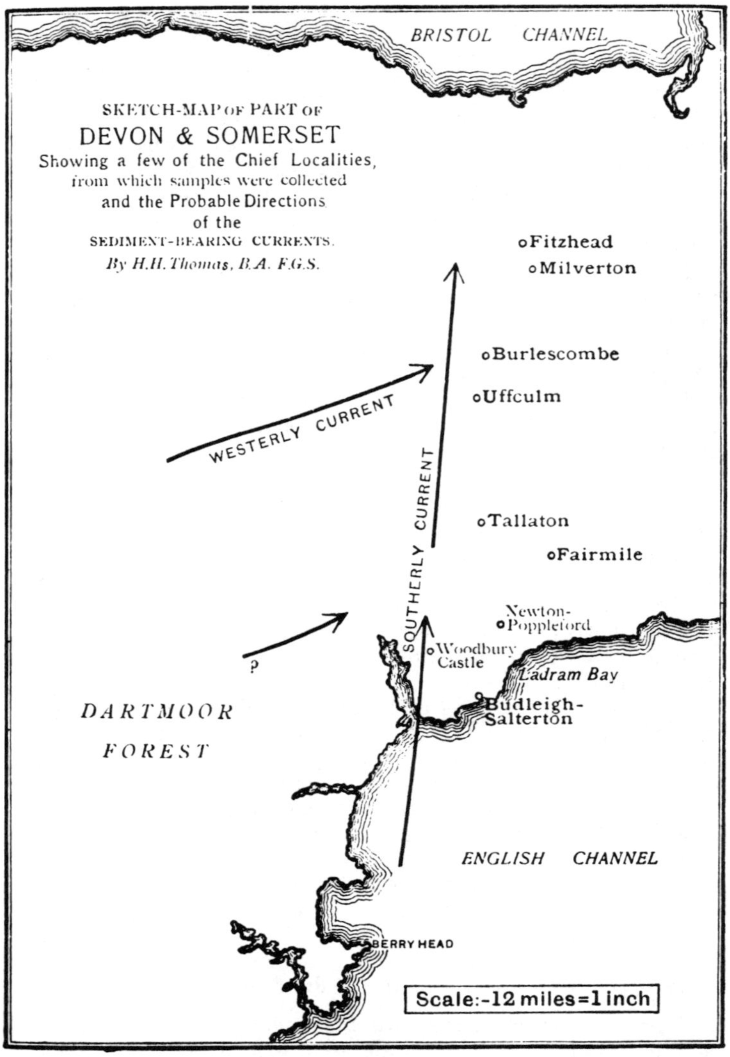
BRISTOL CHANNEL
SKETCH-MAP OF PART OF
DEVON & SOMERSET
Showing a few of the Chief Localities,
from which samples were collected
and the Probable Directions
of the
SEDIMENT-BEARING CURRENTS.
By H.H. Thomas, B.A. F.G.S.
Fitzhead
Milverton
Burlescombe
Uffculm
WESTERLY CURRENT
SOUTHERLY CURRENT
Tallaton
Fairmile
Newton-Poppleford
Woodbury Castle
Ladram Bay
Budleigh-Salterton
?
DARTMOOR FOREST
ENGLISH CHANNEL
BERRY HEAD
Scale:-12 miles=1 inch

An examination of the Pliocene sands from St. Erth and St. Agnes, in Cornwall, which contain local minerals[1] such as andalusite, cyanite, cassiterite, and brown and blue tourmaline, shows that staurolite is extremely rare or absent; and in view of the essentially local origin of these sands, it should be present if derived from a local source. It seems, therefore, that we must look to some other area to supply us with staurolite. Evidence already given points to a southerly source, and the question arises what land-area existed at the time capable of supplying it.

In Triassic times, the Armorican massif is generally believed to have been high land. That it formerly extended much farther north-westward is also probable. Indeed, most geologists hold the view that the present Armorican peninsula was connected with the South-west of England.[2] In the schists and gneisses of Britanny[3] and of the North-west of France generally, rock-masses occur in which staurolite is fairly abundant. This area therefore, or its northerly extension in Triassic times, seems to be the probable source of the staurolite, and many of the other heavy minerals associated with it.

It seems likely that the brown tourmaline is partly of local origin, and partly derived from a more distant area. This would account for the different degree of rounding noted in grains of identical size. The idea of a distant origin is possibly strengthened by the fact that Mr. E. B. Wethered[4] seems to have obtained detrital tourmaline from rocks of Middle Devonian age, which were deposited prior to the intrusion of the tourmaline-bearing granites of the West of England.

The blue tourmaline was probably derived from the metamorphosed sedimentary rocks in which it is so common. It is abundant in the Permian Breccias of the centre of Devon. Much of the rutile has probably had the same source.

To summarize, therefore, I put forward the following points:—

(i) That the distribution of material, and individual mineral species, indicates a main southerly sediment-bearing current, joined by a westerly minor current, which makes itself apparent about 20 miles north of the coast.

(ii) That a mass of material was furnished by a highly metamorphosed area, and that this differed widely from any now exposed in the South-west of England.

(iii) That the probable source of much of the material was the Armorican massif of Triassic times, and the high land of Devon and Cornwall to the west, feeding the southerly and westerly currents respectively.

In conclusion I wish to express my thanks to Mr. Ralph Morgan, of Exeter, for his kindness in supplying me with many samples of New Red Sands from Devon; to Prof. Sollas, for the kindly interest

[1] J. J. H. Teall, 'Brit. Petrogr.' 1888, pp. 386–87.
[2] A. de Lapparent, 'Traité de Géologie' 4th ed. (1900) p. 1004 (map).
[3] A. Lacroix, 'Minéralogie de la France' vol. i (1893–95) p. 6.
[4] Quart. Journ. Geol. Soc. vol. xlviii (1892) p. 383.

which he has taken in my work; and to Mr. C. J. Bayzand, of the University Museum, Oxford, for preparing the drawings for the plates which illustrate this paper.

EXPLANATION OF PLATES XXXI & XXXII.

PLATE XXXI.

Figs. 1, 2, 5, 6, 7, & 8. Zircons showing various forms. × 150 to 200 diameters.
Fig. 3. 'Capped zircon' (from a drawing and photograph by Mr. J. B. Scrivenor). × about 200.
4. Zircon with basal plane (001). From Burlescombe. × about 150.
9. Pink stumpy zircon or xenotime (?). × about 150.
10. Simple crystal of rutile. × about 200.
Figs. 11 & 12. Polysynthetic twins of rutile. × about 150.
Fig. 13. Knee-shaped twin of rutile. × about 150.
Figs. 14–16. Crystals of tourmaline (brown). × about 200.
Fig. 17. Basal section of tourmaline, perpendicular to the optic axis. × about 200.

PLATE XXXII.

Figs. 1–5. Tabular crystals of anatase, showing the large base and narrow pyramids. × about 225 diameters.
Fig. 6. Crystal of brookite from the Upper Bunter of Ladram Bay. × 250.
7. Grains of staurolite, from a microphotograph. × 50.
8. Group of rudely radiating blue tourmaline, from the Permian Breccias (from a microphotograph). × about 150.

[*Editor's Note:* The Discussion has been omitted. In footnote 1 on page 628 of the article the complete reference is: Ussher, W. A. E., 1902, *Geology of the Country Around Exeter (Explanation of Sheet 345),* Memoirs of the Geological Survey of England and Wales, Wyman & Sons, Ltd, London, 122p.]

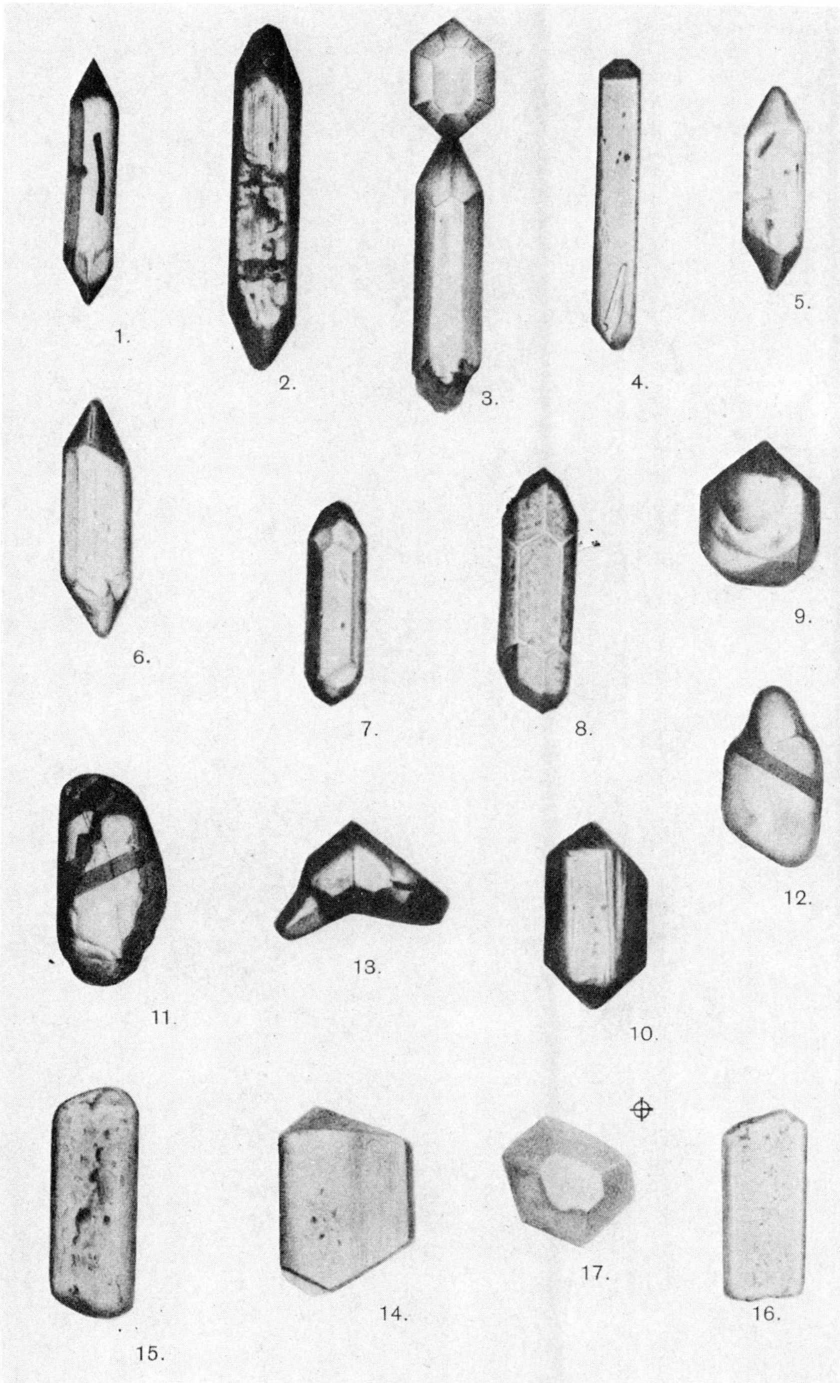

MINERALS FROM THE BUNTER PEBBLE-BED.

C. J. Bayzand, del. *Bemrose, Collo.*

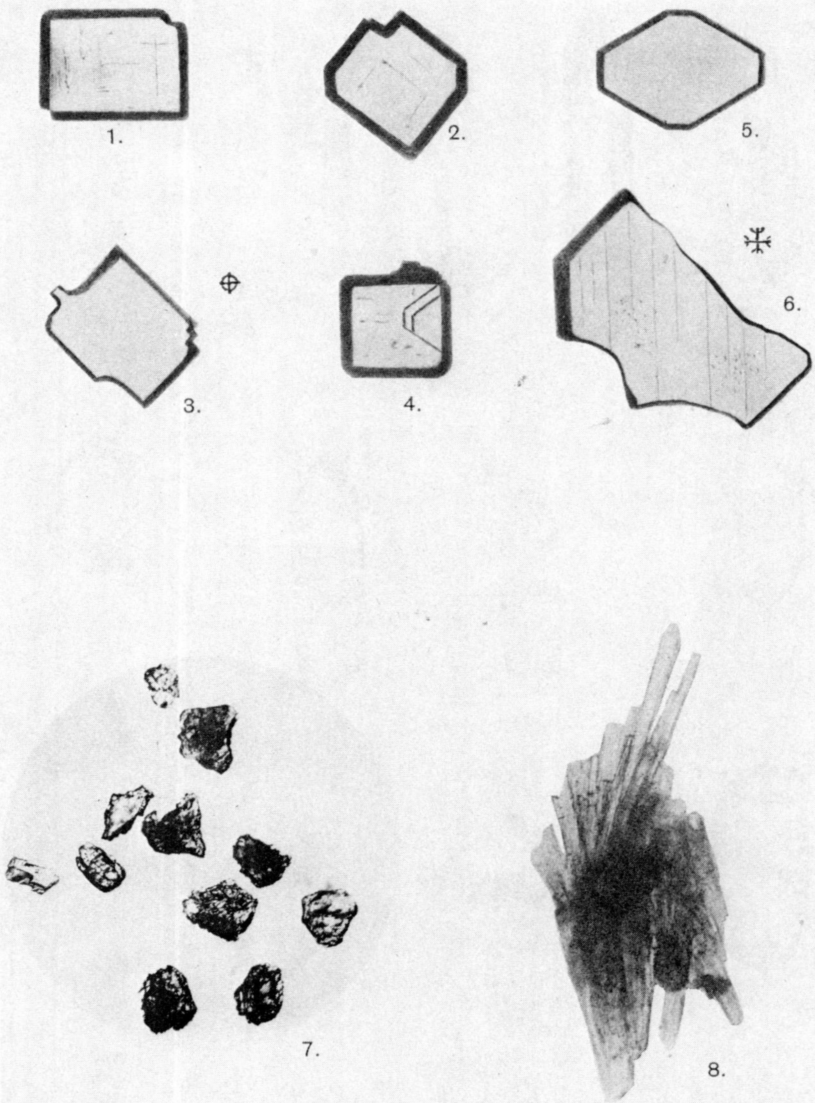

MINERALS FROM THE BUNTER PEBBLE-BED.

C. J. Bayzand, del. *Bemrose, Collo.*

4

Reprinted from *Edinburgh Geol. Soc. Trans.* **11:**138–164 (1923)

The Principles that Regulate the Distribution of Particles of Heavy Minerals in Sedimentary Rocks, as illustrated by the Sandstones of the North-East of Scotland. By WILLIAM MACKIE, M.A., M.D.

(Read 21st January, 1920.)

THE interest that attaches to the particles of heavy minerals found in sedimentary rocks turns chiefly on the indications they afford of the character of the rock masses that have yielded the debris that has gone to the making of the particular sediment in question. These can generally be indicated broadly by referring the minerals isolated each to the particular kind of rock in which it naturally occurs. To be able to refer a particular sediment to a definite local origin by the finding of a mineral of unique characteristics is a less frequent achievement. But they have other interests, *e.g.*, in relation to the light they may throw on the conditions of deposit of the sediments of which they form a part; in regard to the changes they may themselves have undergone since they were first included in the sediments; and in regard to the principles that regulate their distribution in the strata of the various formations. It is more particularly in reference to the last aspect that I beg to submit the conclusions embraced in this paper, derived as they are from a study of the mineral particles of all the sedimentary formations of the North-East of Scotland, and illustrated by such examples from present-day deposits and conditions as seem apposite to the issue raised.

THE PRINCIPLES REGULATING THE PRESENCE OR ABSENCE OF MINERALS IN A SEDIMENT.

Naturally the chief principle that governs the presence of a given mineral in a sediment is the fact of its presence in the rock mass or the rock debris which has gone to make up the sediment in question. With the exception of certain heavy minerals which occasionally occur as infiltration products, and of one or two others which result from the decomposition of minerals originally present—and both classes are easily recognised for what they are—no new minerals belonging to what are usually known as the group of heavy minerals are to be found in the sediments other than those directly derived from the original rock debris, unless a certain amount of metamorphism has taken place in the sediments. This, in the present case, is strictly excluded. On the other hand there exist quite a number of reasons why a mineral originally present in the rock debris should fail to find a place in the particular sediment derived from it, and even granting that it may have reached the sediment and been incorporated therein, why it should not now be found in it.

Size and Specific Gravity of Heavy Mineral Particles in relation to Transporting Agencies.

A mineral may be present in the original rock debris and absent from the corresponding sediments in consequence of conditions affecting the distribution of the heavy minerals under the influence of the transporting agencies, whether of water or wind. Minerals of high specific gravity or occurring in particles of some size may be left behind in a river channel or on a land surface by the transporting agencies, or they may come to rest in a deposit in virtue of their size or specific gravity at a distance from the general mass of the debris originally associated with them. Again the chemical composition of some minerals may be such as lead to their decomposition or solution in transit, while other species again that have reached and been included in the sediments may totally disappear by decomposition or solution within the strata in the period intervening between their inclusion in the sediments and the present time. Numerous examples of the disappearance of minerals have been noted under all these conditions in the course of this study, and further reference will be made to particular cases as they arise.

Experiment as to Effect of Specific Gravity on Distribution of Heavy Minerals.

As bearing on the question of distribution of mineral particles as affected by their specific gravity, the following experiment on a small scale may be described :—

About a quarter of an ounce of *moist* sand relatively rich both in regard to species and individuals in heavy minerals (the specimen used was scraped from a concentration layer on the surface of the sand of the sea-shore of the Moray Firth) was placed on the end of a broad micro-glass slip. This was held under the drip of an ordinary household tap at about an inch below it, the slip being held at a gentle inclination, which can be varied at discretion according to results. The current carries first of all down to the further end of the slip all, or mostly all, the lighter debris—that is the quartz, felspar, etc.—which form a light-coloured zone. By continued action of the current, which is carefully regulated by increasing or diminishing the inclination of the plate, the other minerals arrange themselves in more or less definite zones across the plate, each zone showing the tint of the prevailing mineral in it. Thus at the end nearest the tap there appears a narrow zone (*a*), in which many colourless particles mixed with some dark minerals appear, then comes a broader zone (*b*), of a decidedly dark colour, then a still broader zone (*c*), of a purple colour, then an irregular bunch (*d*), of dirty greenish black, and finally the straw-coloured band (*e*), already described. These bands on closer examination are found to be characterised by an excess in each case of the following minerals :—

(*a*) Zircons and some iron ore. (*b*) Iron ores. (*c*) Garnet. (*d*) Hornblende and other ferro-magnesian minerals. (*e*) Ordinary quartz and felspar.

It will be seen that the general arrangement here is roughly in the order of the specific gravities of the minerals, the heaviest being carried the shortest distance ; the next lower in specific gravity are carried a little further and so on, the particles of lowest specific gravity being carried furthest. There is thus effected in a broad way a separation of the different minerals according to their specific gravities. The conditions of this experiment have their counterpart in nature, wherever rock debris is subjected to the action of running water, and, as will readily be granted, with corresponding though, in the main, perhaps less definite results.

Concentration of Heavy Minerals in River Channels.

Applying the results of the foregoing experiment to the case of river action, we may premise that wherever currents play on the products of rock disintegration, there will always tend to be a separation of the lighter materials from the heavier. The specific gravity of the material carried away will be proportional to the force of the stream. The greater the force of the current the higher will be the specific gravity of the material left behind. In a stream the force of the current will vary both in a transverse and longitudinal direction in the various parts of its course, and, while there will be a general concentration of the heavier minerals in the debris left in the channel of the stream, this will be richer in some places than in others. The proportions of the different minerals will also tend to vary according to their specific gravity, the minerals of higher specific gravity being more abundant where the current has been strongest, provided always that it has not been strong enough to carry even the heaviest away.

Selective Action of Running Water.

In this way a selective action takes place under the varying currents of a stream. Thus, if among the materials in a stream channel there are such minerals as hornblende, apatite, zircon, garnet, iron ores, the hornblende and apatite may in a particular instance be practically all carried away by a current of just sufficient force to move these minerals, but which is insufficient to move the zircons, garnets and iron ores. The residual material will therefore be enriched in zircon, garnet and iron ores.

Broadly speaking, it may be said that deposits in river and stream courses are all over relatively rich in heavy minerals, though the degree of concentration will vary from point to point along their courses. If a general acceleration of the currents in a stream takes place, or if their courses become steeper, thus leading to torrential action, these residual concentrations will in whole or in part be swept out into the area of deposit whether it be lake or sea and will there give rise to a marginal deposit relatively rich in heavy minerals.

Of those examined the river sand richest in heavy minerals was from the Lour or Burn of Aberlour, which drains the granite area of Benrinnes and runs in its lower course through glacial deposits. It is periodically torrential and deposits of sand in its channel are rare. The specimen examined was rather coarse and was found in the shelter of a large boulder. It contained a very large amount of garnet, a less quantity, though still large, of staurolite, much iron ores, zircon, rutile, kyanite, etc., and some of the largest detrital monazites I have seen. Something on these lines has evidently taken place in Middle Old Red times as the deposits of that period (particularly the more basal ones) are relatively richer in heavy minerals than either the upper beds of the same division or indeed than any of the other sedimentary formations of the area. The proportion of minerals in the marginal beds of the area of deposit, following in this particular the deposits in the river channels, also varies. There is seen in the Middle Old Red a higher proportion of *garnet* and *staurolite* than in the later deposits. These minerals, in addition to their being of high specific gravity, also tend to break up into larger particles than any of the other heavy minerals, and for both reasons they appear in relatively greater proportion in the marginal deposits of the area, as they are the first to be dropped as the currents slacken on entering a lake or the sea. As a *per contra* the sediment occupying a more central position will be relatively deficient in these particular minerals.

Concentrations of Heavy Minerals within the Area of Deposit.

Concentrations of heavy minerals may occur within the area of deposit owing to the action of bottom currents during or *pari passu* with the deposition of the detritus. The concentrations, in consequence of the generally darker colour of the heavy minerals, give rise to what is known as *colour-banding*. These are apparently due to currents of moderate intensity, which, playing over the freshly deposited sediment of the lake or sea bottom, carry off the lighter material while leaving the heavier behind in a more or less definite stratum. This colour banding in sediments is well illustrated in Bishopmill quarry in the Upper Old Red Sandstones of the Elgin area. A number of these bands have been specially investigated in regard to their mineral content. Even when the bands are within an inch or two of one another they have been found to show varying proportions of the different minerals. Thus, while zircon and iron ores are generally abundant, some bands are found to show a high proportion of tourmaline, while that mineral may be almost or altogether absent from the next. This shows the selective action of currents of different strengths. It may be presumed that the particles of a sediment that have come to rest in a particular stratum are approximately equal in weight. This being so the mineral particles of lower specific gravity will have on the average the larger diameters. The gradually increasing pressure of the current on the greater cross section of the larger particles would reach a total pressure sufficient to move them on before it would do so in the case of the smaller particles of higher specific gravity. Now the particles of tourmaline are seen to be distinctly larger than the zircons and iron ores with which they are associated, hence the former are moved on by a current that is of insufficient strength to move the latter. Hence in such circumstances zircon and iron ores finally characterise the particular colour band to the exclusion of tourmaline. In the same way, on the assumption of the operation of currents of varying strength, one may predicate the separation of the different mineral species in virtue of their different specific gravities into definite zones or concentration layers each characterised by a particular mineral. The currents would of course be rarely so finely graded as to produce this ideal result, but it is theoretically possible. The separation of such a mineral as monazite or zircon in sea sand in a state of all but perfect purity may be cited as a case in point. Zircon, particularly in the finer sediments, *e.g.* of the Middle Old Red

Sandstones, tends to occur to the all but total exclusion of other minerals.

Note.—The relation of size of particle and its specific gravity, with regard to the pressure required to move it, may be symbolised thus :—A particle of diameter x would have as weight kx^3 sp.g. The pressure required to move or just lift it would be Px^2, when $Px^2=kx^3$ sp.g., that is $P=kx$ sp.g., *i.e.* the pressure required to move a particle varies directly as the product of its diameter into its specific gravity.

A section of a piece of quartzite exhibiting beautiful colour banding was found to show alternating lines of zircon and rutile to the exclusion of other minerals. The striking feature here was that the rutiles had recrystallised into new forms, while the zircons still retained their rounded outlines.

Concentration Layers of Heavy Minerals on Seashore. Description of a Typical Case.

Concentration layers of heavy minerals may be seen on the seashore. These under typical conditions are caused by a gentle to-and-fro movement, accompanied by as gentle a downward and backward movement of the water just below high tide limit which carries away the lighter leaving the heavier material. Though they may be temporarily formed at other points within tide limits, these are mostly only of a temporary character, being practically obliterated *pari passu* with their formation.

A widely extended concentration layer evidently formed about the limits of a spring tide on the sandy shore at Covesea supplied the material for the experiment described on p. 139. In the position indicated such a layer might last a long time without being obliterated ; it might become dry or even harden into a definite crust in consequence of the " running " of its constituent iron ores. In this case it could be lifted off the underlying sand in thin coherent scales. Wind-driven sand had cut into it in places and some distance into the sand layers below. At other places this drifting sand overlay the concentration layer in miniature dunes, the dark line of the concentration layer showing itself as dark contour lines around these particular accumulations. Another experiment was made in this connection. The question to be answered was, What would happen on this surface being gently submerged, as for instance by another springtide, the concentration layer being perfectly dry at the time of submergence ?

Selective Action of increased Surface Tension on Heavy Minerals.

By gently submerging a micro-slip on which was spread an even layer of the separated heavy minerals in a perfectly dry state, it was found that by reason of the increased surface tension a large proportion of the mineral particles was floated off. This contained an enhanced proportion of garnet. Here again a selective action took place, resulting in a concentration of garnet on the one side and of zircon and iron-ores in the residue. On repeating the process on the residual portion, a further concentration took place tending still further to increase the proportion of zircon in the residue. It is possible that similar conditions might occur in nature, but they are manifestly exceptional.

Action of Wind in relation to Heavy Minerals.

With regard to the action of wind as affecting the distribution of heavy minerals in deposits accumulated under its agency, some observations have been made, though they are very far from being exhaustive. Quite a number of extractions of the heavy minerals from specimens of Culbin sand, which may be considered as typically representing present-day conditions as regard wind-action, have been examined, but though they were found to exhibit marked differences both quantitatively and qualitatively, no definite inferences could be drawn from the results. A systematic study of these sands will be necessary before the principles regulating the distribution of heavy minerals in deposits accumulated under the agency of wind can be profitably summarised.

Concentrations of Heavy Minerals under Wind Action.

Evidence has been obtained, however, from the study of the smaller dunes of blown sand along the southern shore of the Moray Firth, that local concentrations of heavy minerals do occur under the action of the wind. These it has been observed are produced under moderate velocities of the wind, higher velocities tending to obliterate concentrations by the indiscriminate removal of both the light and heavy debris or preventing their formation. On one occasion, in the case of a small accumulation of sand in a corner which caused a wind eddy, it was observed that the carrying away of the lighter particles of sand had given rise to a residue which showed a deep purple colour from the high proportion of garnet which it contained.

On another occasion definite sand ripples were seen in process of formation on the surface of a dune under the action of a moderate breeze. It was observed that the troughs of the ripples were distinctly darker in colour than the crests from the higher percentage of heavy minerals they contained. This can be easily explained on theoretical grounds. A breeze of moderate strength would carry the grains of the lighter material further both in a horizontal and vertical direction than the heavier particles and thus have a tendency to effect a separation between them.

Changes in Minerals leading to their Disappearance in Transit.

A number of considerations in regard to the decomposition and solution of minerals while in transit from the parent rocks to the sedimentary deposits, whether in lake or sea, present themselves. These have an important bearing on the question of the presence or absence of certain minerals from particular deposits. Among minerals prone to decomposition, and hence to a tendency to disappear in transit, is *hornblende.* Though hornblende has been observed in quite a number of the collections of heavy minerals from the sediments of all the formations, thereby testifying to the contemporaneous denudation of hornblende-bearing rocks in all the periods, as an integral part of the aggregates of heavy minerals isolated it figures far less conspicuously than tourmaline, which, though an infinitely rarer, is a far more stable mineral. *Iron ores*, again, undergo oxidation and hydration on their surfaces but are usually sufficiently abundant to reach the area of sedimentation without marked reduction in their numbers. *Sphene* has been observed in decomposing rock debris with the characteristic fine crystallisation or etching of its surface, such as may be produced on fresh fragments of the mineral by heating them in acid sulphate of potassium to full fusion of the salt. Under natural conditions, however, it does not lose colour as it does under the treatment described.

Solution under surface conditions may be another reason for the elimination of certain minerals. *Fluor spar* and *apatite* fall to be considered under this head. The former mineral was found to be absent from specimens of sandstone at Spynie when collected from positions exposed to surface agencies, while it was found in abundance in specimens taken from some depth in the rock mass. *Apatite* is a mineral which shows varying results under different surface conditions. Its solubility in carbonated waters is well known. Its presence has been determined in river, sea and blown sand, and in soils of the present day, though in no great abundance. It has been found in many sedimentary rocks—sometimes in proportions so abundant as to be out of all measure as compared with those obtaining in sediments of the present day; while on the other hand it is entirely absent from many other sediments of former geological times. These differences may be probably due in part to original differences in the proportions of apatite in the rocks from which the sediments were derived, but it is just as likely that they are due to differences in the conditions of contemporaneous climate; to differences in the proportion of carbonic acid in the atmosphere and surface waters at the time of deposit; and probably as much to differences in the carbonic acid content of the waters circulating through the sediments since they were originally laid down.

Intrastratal Changes in Minerals.

Some of the changes commenced during transit may be continued and intensified with lapse of time under the influence of the intrastratal waters. *Hornblende*, if it has escaped during transit, is again attacked and the numbers of recognisable hornblende particles still further reduced. *Iron ores* tend to pass into solution, and in many cases these solutions are afterwards reprecipitated, giving rise to concretionary iron ores, which are often to be recognised by their irregular forms and roughly granular surfaces. When titaniferous iron ores decompose, their titanium dioxide tends to crystallise out as *anatase* in aggregates of fine plates or prisms of a steely grey or yellowish colour. These tiny crystals are often to be seen projecting from particles of undecomposed oxide, and even at times to be parasitic on other minerals, generally on minerals composed of or containing titanium dioxide, such as rutile or sphene.

Rutile is a secondary decomposition product frequently appearing in sedimentary rocks. If the iron ores are dissolved in fused acid sulphate of potassium, a number of spiny-looking aggregates will frequently be found in the residue, consisting of very fine interlacing dark brown plates, which are probably to be referred to rutile. At other times rutile in recognisable form is liberated from the iron-ores under the same process in definite crystals, often with sharp spear-like points, at other times in skeleton forms that could not possibly have resisted the abrasive action of the surface agencies, without being quickly broken up. From the frequency with which rutile particles appear with crystalline faces—oftener than the more resistant associated zircons—I am of opinion that worn rutiles may be rejuvenated in the sediments by the crystallisation of titanium dioxide around them, thus restoring their crystalline form, much in the same way as worn quartz grains may be restored by the crystallisation of secondary quartz around them. *Biotite* rich in titanium dioxide may also decompose in the sediments with the formation of definite crystals of rutile. This was seen in the case of pebbles of Rubislaw granite found in the conglomerate of Donmouth, which showed much more rutile than the parent rock in which rutile is so rare as to be counted an accidental constituent. But the change was seen in much greater perfection in a granite fragment included in the sandstone at Delgaty, Turriff, where the rutile had separated in large and brilliant crystals.

Sphene may go on to full decomposition in the strata, yielding anatase and it may be occasionally rutile. The other decomposition products of this mineral are sufficiently soluble to be speedily carried away and so to disappear. Sphene is a mineral so difficult to identify in sedimentary aggregates in consequence of surface changes, that I know of no list of heavy minerals isolated from sedimentary rocks that contains it. Cayeux, in describing *seriatim* the different minerals isolated from sedimentary rocks—and in particular the sedimentary rocks of France—does not even mention the mineral. Its recognition is facilitated by the fusion of the residues with acid potassium sulphate when the undecomposed sphenes appear with the characteristic etched surfaces, and their colour discharged or at least considerably modified. Sphene does occur in a number of rocks of the area, especially in the Kildrummie and Turriff outliers. In other places it has probably gone on to full decomposition and is now represented in the strata by rutile and more particularly anatase, which are widespread and often abundant.

Intrastratal Solution.

Apatite and *fluor* after reaching and being incorporated in the strata may thereafter disappear by simple solution in the intrastratal waters, solution being facilitated, as we have seen, by these waters being highly charged with carbonic acid.

Two other minerals that at first sight do not suggest susceptibility to the action of intrastratal waters are *garnet* and *staurolite*, both of which frequently appear in the sediments with beautifully etched surfaces, revealing their minute crystalline structure. The etched figures are very similar in these two minerals appearing as very definite small rectangular faces. Under certain conditions and with prolonged action, it is reasonable to conceive that particles of these minerals might eventually disappear by complete solution. The attenuated skeleton forms of these minerals which are found in some sediments—and this applies more particularly to garnet—are such as could not resist the operation of the surface agencies for any length of time and bear testimony to the possibility of such a result. If that be so, the inference that has been drawn from the absence of garnet in a set of sediments—namely that garnetiferous rocks were absent from the suite of rocks that went to make up the sediments in question—is altogether unwarrantable.

Staurolite, on the other hand, as the result of intrastratal changes, occasionally shows new outgrowths around an original grain of the mineral. These outgrowths are often of lighter colour than the body of the mineral, and may at times be altogether colourless. The new mineral areas occasionally show inclusions of fine dust like the parent mineral.

Other Mineral Peculiarities and their Effects.

Other minerals show peculiarities in regard to their chance of occurrence in any sediment which cannot be ascribed to decomposition, solution or other known physical or chemical characteristic. These may be broadly classified under the heading of methods of weathering. Take the following case:—

Calcium Fluoride occurs as a cementing substance to the extent of 25 per cent. to 30 per cent. in sandstones of Triassic age, which pass under the sea to the west of Hopeman. To ascertain how far this cementing substance was contributing to the sediments of the present seashore during the process of denudation of the parent rock, specimens of sea sand were taken directly from among the rocks containing the fluoride. Contrary to expectation it was found to be all but free from fluoride, while it was found exceptionally rich in barytes. Rocks containing barytes as a cement are known to exist on the shore four miles to the east at Covesea. Now both these minerals are highly insoluble, but the marked difference in quantity of the two minerals is probably to be attributed to difference in the methods of weathering of the two parent rocks. The fluoride-containing rocks in the tide-way are ground smooth under the action of the waves, the mineral going down into the sediment in a fine dust which tends to be washed further seaward and to be eventually probably in large part dissolved; while in the case of other rocks, the barytes cemented nodules weather out more or less entire and naturally, under the action of the waves, break up, liberating the cementing barytes in relatively large crystalline fragments. Another mineral whose behaviour is of interest, and at the same time quite exceptional, is *kyanite*. Along with staurolite, of which it is the usual associate, its presence has been ascertained in many surface sands—in river, sea, and wind-blown sands—derived practically from the same area as certain local sandstones. But up to the present only two fragments of the mineral have been identified in the sandstones—these were in the concentration layer from Newton and in Bishopmill, both of Upper Old Red Sandstone age. As to what the reason for this all but total disappearance of kyanite may be I cannot even hazard a guess. The *brittleness* of *garnet* along with the effect of its numerous gas inclusions, probably accounts for its occurrence in markedly angular grains and for their diminished numbers at other times. *Andalusite* and *sillimanite* are minerals not easy to identify, the former probably as much from the fact that it is often associated in aggregate with iron ores, the decomposition of which tends to render the mineral in many instances unrecognisable. Sillimanite appears rarely. Where identified it does not seem to have undergone any definite change. These three aluminous silicates require to be further studied from this point of view.

Effect of Cementing Substances on Heavy Minerals.

The general effect of cements—no matter what their character—whether quartz, calcite, barytes, fluor or other, is always in the direction of protection and preservation. Illustrations have been met with again and again in our study of the heavy minerals. Carbonate of lime preserves garnets, which, when extracted from sandstone highly charged with carbonate of lime, always appear with etched surfaces. It would appear that, so long as solutions charged with carbonate of lime circle round the garnets in the sediments, they probably tend to be dissolved, thus giving rise to etching of their surfaces. As soon, however, as calcite begins to be deposited around them in solid form all solution ceases, but they retain their etched surfaces. *Apatite*, a mineral particularly liable to solution, is practically absent from all the cementless rocks as at Cuttieshillock and Rosebrae. It was found at Bishopmill in a specially cemented band, when it was absent from all the other strata in that quarry. Ferric-hydrate acts as a preservative. Apatite is found in the Rosebrae division only where the rocks are red with ferric hydrate as at Oakbrae. Though apatite is abundant at Newton, it was found still more abundant in the red concentration parting from that quarry.

General Conclusions as to the Characters and Distribution of the Heavy Minerals in all the Formations.

With regard to the general graining of the lighter elements in relation to the graining of the heavy minerals in the corresponding sediments, it may be said that there is in the majority of instances a distinct correlation both as to the general size and rounding of the particles. Where the graining of the general particles is coarse, the heavy minerals though less in size are relatively coarse also. When the former are fine the latter are also relatively fine. Where again the former are rounded the latter also are rounded and rounded to a relatively greater degree, and where the former are angular, the latter are angular also—though a slight degree of rounding of the latter may exist. This last characteristic is a necessary corollary to their greater specific gravity.

Where exceptions to these general laws occur, some departure from the normal conditions of sedimentation may be expected. Thus at the footbridge across the Lossie at Kellas a mass of conglomerate has been faulted among the quartz-schists of the area. The rock fragments are markedly angular while the matrix is very fine and the heavy minerals correspondingly fine and rounded. The dropping of the angular rock fragments from a vertical rock face among fine sediments is suggested in explanation. Again in the north-west corner of the Upper Old Red Sandstone area of Elgin near Findrassie numerous pebbles, some of them of very large size, and many of them showing distinct *dreikanter* forms and other signs of wind action, occur along with rather fine-grained heavy minerals in a correspondingly fine-grained matrix. It is inferred that the pebbles were dropped from floating ice into relatively fine sediments. When the graining of the general sediment is rather coarse, while that of the heavy minerals is generally fine but varying markedly in size with occasional particles much larger than the average, the operation of wind agency may be suspected. These conditions obtain in the Gordonia beds at Cuttieshillock, in which another marked feature has been observed, probably also indicating the agency of wind. The heavy minerals have been found to diminish both in quantity and in general size when traced from the central position at Cuttieshillock, both towards the north-east to Quarrywood and eastwards to Crownhead, where some collections of heavy minerals were found to be almost negative. These observations are believed to indicate the operation of winds blowing prevalently from the west or south-west as at the present day. Similar observations have been made in the case of the sandstones of the coast range from

Burghead to Covesea and a similar cause is suggested in explanation.

Peripheral Deposition of Large and Heavy Grains of Minerals.

Heavy minerals of high specific gravities, or that are prone to persist as large grains, show a tendency to occur in relative abundance in peripheral positions in the various areas of deposit, and to be scanty or altogether absent from beds occupying a more central position. This is particularly noticeable in the case of *staurolite* and *garnet*, Staurolite has been observed to be most plentiful in beds occurring in proximity to schistose rocks in which staurolite is known to occur, *e.g.* in the Middle Old Red Sandstone rocks of Altash and Portessie and the Upper Old Red Sandstone of Scaat Craig. Some movement along the shore of the lake eastwards and westwards is indicated by the occurrence of the mineral at Donmouth in the Middle Old Red Sandstone and at Whitemyre, Boghole, on the Findhorn, Kingstep, and as far north as Balnagowan in Ross-shire, where there are no rocks in which staurolite is known to occur. On the other hand many aggregates of heavy minerals separated from the central beds have to be searched before a single fragment of the mineral is found. Only one fragment of staurolite has been seen at Bishopmill out of about thirty collections examined.

Much the same conclusions obtain with regard to garnet in the case of which, from its greater frequency of occurrence, results are even more emphatic than in the case of staurolite. In the basal beds of the Middle Old Red Sandstone they are particularly abundant. But the Middle Old Red rocks all over are much richer in garnet than either the Upper Old Red Sandstone or Trias, while the Trias contains relatively more than the central beds of the Upper Old Red Sandstone, and both of them more than the Gordonia beds of Cuttieshillock. Many collections rich in other minerals have been made from the Rosebrae division of the Upper Old Red Sandstone, in which not a single garnet has been seen. The result is perhaps not wholly attributable to the cause just stated. It has probably been aided, as will be seen later, by intrastratal changes, leading to a further reduction in the number of particles of garnet. The same distribution of this mineral has been observed in the Middle Old Red Sandstone rocks in the Turriff outlier. The conglomerate of Slack o' Causeway, Auchterless, shows numerous garnets in its matrix, while it contains many rock fragments, which also bear garnet. But the finer beds at Ardin and Delgaty only contain an occasional garnet and in many collections do not show even a single garnet.

Finer-Grained Beds Richer in Minerals occurring in Finer Grains.

The finer-grained beds in all formations tend to contain in enhanced proportions the minerals, such as zircon and rutile, which occur *ab initio* in finer grains. This is a natural corollary to the law of correlation of size of grain, but this tendency is accentuated in the case of these minerals by the fact that they are the most resistant to the operations of the denuding agencies. Hence they appear in greatest abundance among the debris that has been subjected to the greatest attrition—presumably now represented by sediments showing the finest state of division.

"Running" of Iron Ores in relation to Colouring of Sandstones.

Sandstones showing the darkest colour tints due to ferruginous cement have been found to show on the whole a much higher proportion of definite particles of iron ores than the lighter-coloured rocks even after the diffused hydrated-oxides have been removed with acid. This is particularly evident in the rocks of the Turriff outlier at Ardin and Delgaty, the darkest coloured sandstones of the area, where the particles of iron ore are superabundant. These may be compared with the sandstones of the Kildrummie outlier from Quarryhill, Rhynie, where the particles of iron ores are very much less abundant. The concentration layer from Newton, with 5·52 per cent. of heavy minerals, may be compared with the sandstones among which it was found. The natural inference is that red sandstones do not *always* owe their red colour to the contemporaneous precipitation of their hydrated oxides from the waters of the lake *pari passu* with the deposition of the sandstone, but that it is *sometimes* due rather to partial solution of the contained particles of oxide, followed by the diffusion of the solution with subsequent precipitation of the hydrates in place —in other words to what might be called the "running" of the constituent iron ores. This has undoubtedly taken place in some of the red beds in the Upper Old Red Sandstone, as at Oakbrae in the Rosebrae division, where the red colour is often seen to stop abruptly when traced along the same stratum. Light-coloured and dark red strata are there inter-bedded, and in places the red colour has spread irregularly to the contiguous light-coloured beds. The proportion of definite particles of iron ore in several comparative tests that were made in the case of the different coloured Oakbrae beds was always found to be higher in the red than in the light-coloured beds. The Permian sandstones of Dumfries were found to illustrate the same law.

Differences between the Several Formations.

The differences in the mineral particles in the several formations depend as much on differences in the conditions under which the different sediments have been accumulated as on differences in the rocks from which they were derived.

Rocks of the Middle Old Red Sandstone.

Like the grains of the lighter, the particles of the heavy minerals in the Middle Old Red Sandstone beds are almost always markedly angular. They are generally of large size, and occur in greater abundance than in the other formations. The torrential, and probably partly glacial, conditions which obtained in Middle Old Red times account for the marked angularity and abundance of the heavy minerals of these deposits. The prevailing mineral is garnet, which varies in quantity from bed to bed. It is very abundant in the beds at Donmouth and in the basal conglomerates of the Turriff outlier at Slack o' Causeway, Auchterless, as well as in the conglomerate of Corbie's Tongue, which is low down in the Kildrummie outlier. In the upper beds of the former at Ardin and Delgaty garnets are exceedingly rare, but in the Dryden Flags or upper beds of the latter they are exceedingly abundant. No garnets have been found in the fine-grained red beds to the west of Cullen or in the beds at Foyers, which are largely composed of an arkose derived from the Strath Errol granite. The mineral is for the most part almost colourless in small fragments ; brown garnet has rarely been observed, and the dark pink variety has been seen at Corbie's Tongue, Donmouth, and elsewhere.

Zircon is relatively scarce and occurs in idiomorphic or broken angular grains. Special spindle-shaped forms have been observed in the Donmouth sediments, agreeing with a form of the mineral which has been found typical of Rubislaw granite. Another form, similar to the zircons occurring in the syenites of Wardhouse and Bridge of Johnston, has been met with at Ardin in the Turriff outlier.

Tourmaline is relatively rare as compared with what obtains in the other formations, but is seldom altogether absent as at Cullen and Foyers. It occurs for the most part in irregular chips of a type which occurs chiefly in pegmatite veins, though the small idiomorphic or slightly abraded, variously-coloured prisms derived from the younger schistose rocks such as the black schists of Banffshire also appear. The latter have been seen as far west as Dalcroy, but it is doubtful whether the beds containing them should be classified with the Middle or the Upper Old Red Sandstone.

Iron ores are everywhere abundant, but the decomposition product of the titaniferous variety, anatase, is much less common than in the Upper Old Red Sandstone rocks ; it is seen in very large steely-grey to colourless plates at Culloden. Rutile has been recorded from every bed examined. It occurs in large, intensely dark, reddish-brown grains at Altash, Fochabers, along with much lighter irregular prisms enclosing quantities of black dust. The former variety is suggestive of a basal igneous rock mass as source. The latter has evidently been derived from the black schists of the adjoining area. Monazite has been found in every bed examined—not however in any abundance anywhere, but in numbers limited to one or two in a slide. The grains are almost invariably more or less rounded, though it has been met with showing idiomorphic contours particularly in the Donmouth sediments. The occurrence of this mineral in unabraded forms is significant. In a mineral of such softness it points conclusively to proximity of the parent rock. Though monazite has been found in small grains in other rocks such as schists and gneissose rocks, the mineral is typically derived from granite. Its widespread distribution in these rocks indicates the extensive denudation of granite in Middle Old Red times.

Staurolite has been found at Donmouth, Ardin, Portessie, at Cloddach and Kellas on the Lossie, Kildrummie, Cullen, and in the red beds to the north of Speybridge, Fochabers, but in greatest abundance at Altash, Fochabers, in close proximity to beds in the schists, which contain the mineral in quantity. Sillimanite appears at Donmouth; Quarryhill, Rhynie ; Kildrummie and Speybridge.

Hornblende is a rather rare mineral, but its presence has been ascertained at a few points in proximity to hornblende-bearing rocks, such as at Ardin, Rhynie, Kildrummie, Corbie's Tongue, Cloddach, Speybridge and Kilravock. A few plates of a colourless mineral of a lower refractive index than zircon, giving straight extinction and a positive reaction have been referred to topaz. It has been seen at a few places and its origin is probably referrable to granite. Epidote is a much commoner mineral in Middle Old Red Sandstone than in the Upper Old Red Sandstone or other rocks. The mineral evidently does not resist intensive attrition, and only from this point of view is its presence significant.

Fluorspar and barytes have been encountered in sparing quantities in a few places and in markedly angular grains, indicating that their presence is due to secondary infiltration. Fluor is noted as occurring at seven places, barytes only at one. Augite, hypersthene and enstatite, like hornblende, readily yield to surface and intrastratal decomposition, and rarely appear in recognisable form. Augite is recorded at two, hypersthene at four, and enstatite at three localities.

Andalusite probably occurs more frequently than recorded as its method of decomposition and association hinder recognition. Its presence has been determined with certainty at Donmouth and at three places in the Rhynie outlier—Quarryhill, Kildrummie and Smithston. It has also been recorded from Kilravock at some distance from rocks known to contain the mineral.

Apatite has been looked for at only a few places, but results so far were very different. It has been found exceptionally abundant at three localities in the Rhynie outlier, at Quarryhill, Druminnor, and in the ashy beds at the foot of Knock Caillachie. In the Turriff outlier apatite occurs in the basal conglomerates at Slack o' Causeway, but it has been searched for in vain in the finer sediments of Ardin and Delgaty. The mineral occurs in abundance at Donmouth, Altash, Fochabers and at Kellas in Elginshire. At the last locality the fine rounded grains of the mineral are covered with a red deposit of ferric hydrate or more likely perhaps of a mixture of that and basic phosphate of iron.

The difference between the outliers of Turriff and Rhynie in regard to the presence of apatite raised the question of similarity of climate and hence of the contemporaneity of the two sets of sediments. It may be remarked that in the case of the former locality the absence of the mineral does not appear to be due to intrastratal solution as the protecting ferruginous cement is much more plentiful in the Turriff than in the Rhynie outlier.

Rocks of the Upper Old Red Sandstone.

In the Upper Old Red Sandstone rocks again the conditions under which the beds were deposited have had a marked effect on the character and quantity of the heavy minerals present. From the high degree of rounding of the constituent particles —light as well as heavy—it is to be inferred that wind and water action co-operated to produce the observed results. The intensive attrition of the debris under both these agencies is responsible for the diminished proportion of some of the species of heavy minerals.

The proportion of heavy minerals is less than in the Middle Old Red Sandstone rocks. The individual particles are also smaller in size and are markedly rounded. The prevailing mineral is zircon, which everywhere predominates. In some of the beds it is almost the only heavy mineral. It occurs in well-rounded grains, though very occasionally idiomorphic crystals showing little rounding are seen.

Garnets occur in far fewer numbers proportionally than in the Middle Old Red Sandstone rocks. They are more numerous in the marginal beds of the area, but in the centre, particularly in the Rosebrae division, they are often not to be found at all. Several circumstances have contributed to this result. On the land surface of the time there would first of all be intensive attrition under the action of wind, accompanied by local segregations of garnet, which would probably remain on the land surface, while the lighter material blown into the river courses would eventually, under torrential rains, be carried down into the more or less temporary lakes of the time. Peripheral deposition of the remaining heavy minerals would result, and among them in particular would be most of the residual garnet. Finally there would be intrastratal solution which would account for the disappearance of the garnet particles that had escaped the operation of the other causes. Of the two last causes, there is much evidence in the sediments now. The higher proportion of garnet in the marginal deposits is evident at Scaat Craig and in beds on the Findhorn, at Whitemyre and Boghole; while the attenuated forms of the few garnets that are seen at Bishopmill with their solution-etched surfaces and skeleton forms, would have inevitably gone to pieces under the operation of the surface agencies. The etching of the surface of the garnets is a characteristic feature throughout the Upper Old Red Sandstone rocks and gives positive demonstration of intrastratal solution. In several places, though the garnets appear with these etched surfaces, they have been protected eventually from further solution by the deposition of carbonate of lime around them. This has been the case notably at Scaat Craig, where the highest proportion both of garnet and carbonate of lime occurs. In the Rosebrae division, and at Newton also to a less extent, there is evidence that carbonate of lime was once fairly widely present in discrete particles as a cement. It has been dissolved out during the intervening period, leaving in the sandstones dark spots impregnated with manganese and other peroxides, or areas without any cement whatever, in which case pitting of the sandstones takes place on exposed surfaces. With the carbonate of lime has gone in many places the last trace of garnet. Two characteristics of the garnets that survive in these beds are conspicuous—they are always markedly angular and they are often of relatively larger size than the particles of the other heavy minerals with which they are associated. The angularity, which is a feature of garnet everywhere, is to be attributed to the brittleness of the mineral. For its larger size in the particular position where it occurs no sufficient reason presents itself.

Tourmaline is usually abundant. It occurs in well-rounded grains, often in distinct more or less abraded prisms of various colours—brown, blue, green-yellow, pink, purple—or at times colourless. When the prisms are very fine they show no trace of wear and are often seen to enclose much fine dark dust which occupies a central position in the crystal, the borders remaining clear. These prisms have been derived from the younger or Banffshire schists and in particular from the Black Schist group in which similar tourmaline prisms have been observed. They are much more abundant on the eastern side of the area of deposit than the west, though they are frequent in concentration bands at Newton, while they have been seen as far west as Dalcroy in rocks usually classified as Middle Old Red Sandstone.

Rutile, both original and secondary, occurs throughout the series but is most abundant in the very fine-grained rocks such as at Oakbrae and in the beds which pass under the Trias at the bottom of the scarp at Lossiemouth. A few very dark and comparatively large fragments are seen at Scaat Craig, suggesting an origin from basic igneous rocks.

Sphene is rare, but one of its degradation products, anatase, is very abundant. Brookite has been observed at Findrassie Old Metal Quarry. Monazite has been found in every specimen examined with two exceptions—at Balnagowan in rocks similar to those of Newton and containing a similar suite of fossils, and in the very dark iron-stained, quartz-cemented rocks on the shore at Lossiemouth. The former are probably outside the monazite area, monazite being either exceedingly scarce or altogether absent from the granite masses *ex adverso* of the Balnagowan sandstones. The latter, from their flinty character, do not yield their heavy minerals readily on being broken up, and hence only a partial collection of them is made. Monazite is never plentiful—only a few small grains appearing in each collection. These are finely rounded and are on the whole smaller in size than the zircons with which they are associated.

Staurolite has been recorded from a number of localities, chiefly from peripheral positions in the area as at Scaat Craig, Kingsteps, Boghole, Whitemyre, Balnagowan, Dallas and

Alves. A single grain was observed at Bishopmill in the Rosebrae division out of about thirty collections from different horizons, some of them representing concentration bands containing many thousands of grains of other minerals.

Hornblende, augite, hypersthene and enstatite all appear at times. Hornblende is widely distributed, but is nowhere abundant. Both green and brown varieties are present, the former prevailing. Augite may be overlooked, or taken for tourmaline unless the reactions under polarised light are systematically taken. Specific identifications have been made in the instances recorded.

A number of minute angular bluish or greenish-grey particles showing pleochroism, but not the intense absorption of tourmaline, have been recorded as corundum. Clear plates referred to topaz appear a little more frequently than in the Middle Old Red Sandstone rocks. Epidote is recorded from only two localities, Balnagowan and Bishopmill, where a single grain was found out of a large number of collections. Like garnet, it probably falls a victim to intensive attrition.

Fluorite has been found in angular grains at a number of localities, but usually in small amount. In one specimen from Newton it was found in some quantity and of a distinct purple colour. In the beds below the Trias in the Scarp at Lossiemouth it is probably an infiltration product from the overlying Triassic beds in which the mineral occurs in some quantity. Barytes has been met with in the faulted mass at Park, Dallas, and in beds below the Gordonia beds of Cuttieshillock, in which there is always a sprinkling of fragments of barytes, as at the Leggat and Findrassie pebbly quarries. It has also been found in some quantity in the quarry on the north side of the road at Bishopmill. Green spinel or garnet has been seen at Whitemyre and Oakbrae. Small, absolutely colourless grains, generally hypidiomorphic and probably referable to celestine, were seen in numbers in one specimen from the pebbly metal quarry at Findrassie. They differ from barytes grains in being free from dust or inclusions.

Apatite, to judge from the limited number of instances in which it was looked for, appears to be very variable in quantity. It has been found in abundance in all beds in Newton and in particular in the concentration band, which was the subject of special study. It is almost, or altogether, absent from many beds of the Rosebrae division. At Bishopmill a prolonged search finally disclosed a few grains in a specially cemented band about a foot in thickness, which, no doubt owing to the cement it contained, showed itself as a glistening band along the quarry face. This would go to indicate that intrastratal solution may account for the meagre amount of apatite in this division.

Malachite and covellite, associated with much calcite and some rounded apatites, were found in a specimen from Kingsteps, Nairn.

THE GORDONIA BEDS.

A study of the heavy minerals of the Gordonia Beds of Cuttieshillock, Quarrywood and Crownhead discloses broad differences, particularly in their relative proportions, between these sandstones and the sandstones of the Upper Old Red rocks; while comparison with the suite of minerals yielded by the undoubtedly Triassic rocks of Findrassie, Spynie and Lossiemouth shows even greater differences, that is to say the Gordonia beds differ both from the Upper Old Red Sandstone and the Trias, but in greater degree from the Trias.

As wind is supposed to have been the chief agency in their accumulation, they are expected naturally to be deficient in garnet, and this they are, and to a greater degree even than the Rosebrae division of the Upper Old Red Sandstone. The grains are even more rounded than in that division. They show a finer polish of their surfaces and they are certainly more unequal in size. Zircon is abundant. Tourmaline is less abundant than in many beds of the Rosebrae division, and is less varied in character. The prism forms of the Upper Old Red Sandstone rocks are rare or absent. Monazite is present throughout. Hornblende, topaz and anatase were met with in all the beds, while augite, enstatite and corundum appear in a proportion of the specimens examined. Fluor is present at Cuttieshillock and barytes in all the beds, though in insignificant amount. Apatite, though carefully searched for, was not found.

THE TRIAS.

The Triassic rocks of the Elgin area differ *toto cœlo* from the other rocks of the area as to the relative proportions of their heavy minerals. This becomes quite evident at Findrassie, where specimens taken from opposite sides of the Old Findrassie quarry show remarkable differences between the Upper Old Red Sandstone and Triassic rocks. One of the marked differences is that the Triassic rocks nearly always contain a proportion of fluor, so that there is a higher proportion of light-coloured minerals, giving the specimens a lighter colour in the mass. The grains are sub-angular, and garnet is plentiful and as usual markedly angular. It shows the etched faces which are typically associated with carbonate of lime, which is present in discrete patches in the Findrassie, Spynie and Lossiemouth rocks, but not in those of the coast range which extends from Burghead to Covesea.

Zircon is always present ; tourmaline is less abundant, while the variously-coloured prism forms have all but vanished ; they have been seen, however, in numbers, in collections from Findrassie, just beyond the boundary with the Rosebrae division of the Upper Old Red Sandstone. This probably indicates some erosion of the latter beds during Triassic times.

Iron ores are probably in great part of secondary or concretionary origin. They are of irregular form, giving a "shreddy" appearance to the preparations. Rutile is present throughout. Sphene has not been seen. Monazite has been identified in a number of places, but does not appear to be so widespread as in the other formations. Hornblende is, if anything, more frequent, Fluor is present throughout, but varies in quantity from 30 per cent. in the Cummingston rocks down to one or two grains in a collection at Covesea. Barytes occurs in abundance at Covesea and in traces east of Burghead. Anatase occurs in all specimens examined. A green, highly refracting spinel in small angular fragments has been seen at Spynie. Some of the fragments showed fine parasitic anatases. Biotite has been met with at Findrassie.

A notable feature of the rocks at Lossiemouth and Spynie is the fact that they contain fluor in both rounded and angular grains associated with a few rounded apatites. The question of the origin of both *barytes* and *fluor* present as a cement in these deposits has often been raised. The former mineral is known to occur in some quantity in the quartzites to the south-east. While the presence of large blocks in the surface deposits to the west and south-west of Rothes, and the high proportion of flakes of barytes separated from the sand of a small stream on the south side of Hunt Hills would go to indicate the existence of other large veins in the quartz-schists in that direction. On the other hand the investigation of the heavy minerals of the granites of the North of Scotland discloses the abundance of fluor in such local granite masses as that of Abriachan and the granite of the Helmsdale coastal range. The granite of Peterhead, too, in which fluor is also relatively abundant, might not be considered too far afield to be a source of supply for this area. It may be conceived that during the denudation of these granites, the fluor would pass into the sediments in part in solid fragments, while the remainder would pass into solution, to be deposited afterwards around the grains of the sediment on concentration of the waters of the Triassic lake.

Apatite is nowhere abundant. A few grains have been seen in specimens from Spynie and Lossiemouth. In the rocks of the coast range only an occasional grain has been seen. It may be remarked that the examination of the minerals of the coast range would go to support the contention that they belong to a different formation, or at best a different division of the Trias, from the Findrassie, Spynie and Lossiemouth group of rocks. In several respects they approach more closely the Gordonia or Cuttieshillock beds.

Jurassic and other Rocks.

A few separations have been made of the heavy minerals in Jurassic rocks. Large blocks of sandstone closely resembling Upper Old Red Sandstone rocks in their macroscopic characters occur in an excavation in boulder clay at Teethill, two miles east of Elgin. These, however, were found to differ from rocks of Upper Old Red Sandstone age in showing exceedingly angular grains, an abundance of garnet, large dark rutiles and absence of the prism-derived tourmalines. They could not be classified with the Upper Old Red Sandstone, and have been referred to the Jurassic formation. Again specimens of the cornstone from New Elgin and from Main have been examined. They show abundance of heavy minerals, and are rich in garnet. The other minerals are fairly well rounded, but never come up to the degree of rounding of the Upper Old Red Sandstone beds. They were all rich in dark rutiles.

Illustrative examples were also examined from the Lower Old Red Sandstone of Central Scotland. They probably were not typical. A sample of ashy sandstone from near Perth showed a considerable amount of garnet, some tourmaline and a solitary monazite. A sample of the red Permian sandstone of Dumfries, showing fine colour-banding, gave a suite of heavy minerals in which iron ores bulked most largely. It was characterised by the absence of garnet, and the presence of one or two very minute monazites. Apart from the iron ores zircon was the most abundant mineral.

Origin of Deposits.

As regards the origin of the deposits, the character of the rock or rocks from which they have been derived can usually be indicated broadly by a reference to the suite of minerals isolated at each locality. The following may be given as an illustrative example :—

Heavy Minerals of the Donmouth Sediments.

Minerals Present.	Kind of Rock from which derived.
Zircon Tourmaline (pegmatite) Topaz Monazite (Garnet)	Granite.
Hornblende Enstatite Hypersthene Sphene	Basic Igneous Rocks.
Garnet Staurolite Andalusite Sillimânite	Metamorphic Aureole.

Examination of Pebbles.

Two classes of pebbles found in the sandstones in the course of this investigation have proved of interest—sandstones and granites. The latter are of greater interest in connection with the study of the granites, and only a limited number have been examined. Of sandstone pebbles four examples in all have been examined—three from the Middle Old Red Sandstone of Ardin and Delgaty, and one from the Newton division of the Upper Old Red Sandstone. All of them were very fine grained, and the heavy minerals they contained were equally fine and consisted chiefly of zircon. None of them contained garnet and only one—the Upper Old Red Sandstone specimen—contained monazite. They gave no indication as to the formation to which they originally belonged.

Of the granite pebbles, four were examined from the Middle Old Red Sandstone of Donmouth. All of them were referable to Rubislaw granite, thus showing the sediments there to be in the main a local deposit.

Twelve granite or granitoid pebbles from the Turriff Middle Old Red Sandstone outlier (ten of them from Slack o' Causeway, Auchterless) were also examined. Five of them proved to be sphene-bearing granites, which are not known to exist in the immediate neighbourhood, but four of them were probably referable to the Abriachan granite or the Helmsdale range, and the fifth had the characters seen in some large veins cutting the Corrennie mass. Of the others, one was a garnetiferous norite such as that of the Battlehill, Huntly ; one was

Benachie granite, and showed the characteristic drusy cavities; one was Corrennie granite; another very rich in biotite showed many zircons with curved faces as seen in foliated rocks, and contained a considerable amount of monazite in fine crystals. This mineral was not seen in any of the others. The last of the Slack o' Causeway specimens was a crumbling specimen of the foliated granite of Inchbae, as shown by the excessive amount of garnet it contained. The Ardin and Delgaty specimens were: one a typical pegmatite, the other a fine-grained rock showing abundant secondary rutile, derived from the abundant titaniferous biotite which it contained.

The results of the examination of the granite pebbles go to indicate a distant origin from localities 70 to 80 miles to the west, and from 20 miles to the south. These observations have an important bearing on the question of the conditions under which the Middle Old Red Sandstone sediments were laid down.

Other points in regard to the source of deposits have been indicated as they arose.

A Wide-spread and Far-travelled Zircon.[1]

Some observations with regard to a particular form of zircon deserve special notice. This is a deep purple-coloured mineral always occurring in highly-rounded forms—unless accidentally fractured in the process of breaking up the containing rock. If fractured at some intervening period of their history, the sharp edges of the fractures were found to have been smoothed down. Many of them are almost perfect spheres, and have been met with in all the formations investigated. They occur from Orkney to Pennegent in Yorkshire. They have been seen in their typical form in the grits associated with the clayslates of Aberdeenshire, and in the boulder-bed in the clayslates in the valley of the Kirkney Burn at Culdrain, Gartly, and in the Torridon sandstone. They also occur in marked abundance in the Gordonia beds of Cuttieshillock. That it had its origin outside the British Isles I am convinced. Its wanderings constitute the romance of this investigation. These, so far as yet observed, had better be put on record:—

Schistose Grits.—Culdrain, Bridge of Bogie, and Pourins, Rhynie.

Torridonian.—Island of Handa, etc.

Middle Old Red Sandstone.—Cloddach, Portessie, Cullen, Ardin, Bodnamuir, Craigbeg, Knoc Caillachie, Donmouth, Orkney.

[1] See also *The Source of the Purple Zircons in the Sedimentary Rocks of Scotland*, this volume, p. 200.

Upper Old Red Sandstone.—Whitemyre, Balnagowan, Cuttieshillock, Newton, Kingsteps, Cliff, Lossiemouth, Bishopmill, Rosebrae, Dallas, Rocky Park, Oakbrae, Pluscarden, Boghole, Alves, Scaat-Craig, Findrassie, Laverockloch.

Carboniferous.—Pennegent, etc.

Trias.—Cuttieshillock, Crownhead, Cummingston, Spynie, Findrassie, Lossiemouth.

Part II

TRANSPORTATION AND DEPOSITION OF HEAVY MINERALS

Editor's Comments on Papers 5 Through 9

Studies dealing with transportation and deposition of heavy minerals invariably cite two papers, one by Rubey (Paper 5), the other by Rittenhouse (Paper 6). William Walden Rubey (1898–1974) spent forty years of his geologic career with the U.S. Geological Survey; after his retirement from the survey, he was professor of geology at the University of California at Los Angeles. A member of the National Academy of Sciences and winner of the Penrose Medal (1963) and the Presidential Medal of Science (1965), he served as president of the Geological Society of America (1950) and on numerous other scientific councils and committees. His impressive bibliography includes twelve papers on stream dynamics, sixteen on structural geology, eight on sedimentology and diagenesis, and three on the origin of atmospheres and oceans.

Rubey (Paper 5) was the first to explain clearly the reasons for a given size-distribution of heavy minerals in a sandstone, although many of the principles he discussed had been outlined earlier by

Mackie (Paper 4). He used Stokes's Law to determine what sizes of grains of differing densities will settle out together; thus, he explained the concept of hydraulic equivalence although he never used the term.

Rubey pointed out that the mineralogy of sandstones derived from the same source can be very different if the grain sizes of the sandstones are significantly different. He also pointed out that heavy minerals are more abundant in the finer sizes because of their lower resistance to abrasion compared with quartz. He recommended comparisons of heavy-mineral fractions from comparable grain-size ranges in sandstones.

Gordon Rittenhouse (1910–1974) was an early practitioner of the "new petrology," described by F. J. Pettijohn as petrology that went beyond the simple description of a rock to include the rock's entire history. He worked for the Soil Conservation Service (1935–1943), U.S. Geological Survey (1943–1947), as associate professor of geology at the University of Cincinnati (1947–1951), and with Shell Oil Company (1951–1973). He was elected a fellow of the Geological Society of America in 1944.

The origin of the concept of hydraulic equivalence is commonly attributed to Rittenhouse, although he pointed out in a footnote (Paper 6, p. 1741) that the idea was not original with him. Paper 6 grew from a Soil Conservation Service study of arroyo cutting and soil erosion in the middle Rio Grande Valley of New Mexico. At the time there was no easy way to evaluate the relative importance of tributary streams and surrounding terrain as sediment sources. By comparing the heavy-mineral content in these streams and relating the variability to hydraulic equivalence, Rittenhouse was able to define the mineral composition of deposits regardless of their textures. He also suggested that discrepancies between observed relationships and settling equivalence can be explained by the absence of minerals of a given size from the overall sediment population.

Paper 6 offers the most extensive summary in English of the various problems associated with the transport and deposition of heavy minerals. It presents a detailed discussion of methods used to compare heavy-mineral compositions; the hydraulic ratio, the weight of a heavy mineral in a given size class divided by the light minerals in the hydraulically equivalent class, is recommended as a genetically sound method of representing heavy-mineral composition. The section on techniques and Tables 1 through 4 are not reproduced here due to space limitations.

Frank W. Stapor, Jr., a geologist with Exxon Production and Research in Houston, Texas, does not recommend Rittenhouse's method of determining hydraulic equivalence because the mineral populations are not treated as Gaussian distributions but are analyzed into intervals rigidly obtained by sieving (Paper 7). In his study of a coastal area in southern Florida, Stapor performed mineralogic segregations and then sieved each mineral separate, a procedure that differs significantly from that used by many previous workers. Furthermore, Stapor used only minerals of a constant and restricted density in his analyses, which helped to eliminate problems associated with varying densities of minerals.

In contrast to most workers, Stapor proposes two different but complementary modes of concentration of heavy minerals. In his field area there appears to be an inverse association of concentration, grain size, and energy level. He proposes that the heavy minerals are delivered to the shore as a single slug, then concentrated as follows: The higher energy level of open beaches removes coarser particles, leaving the finer particles behind. But on the sheltered, low-energy beaches, only finer particles are moved.

While many previous workers have discussed the concept of entrainment of heavy minerals (see Introduction), Rudy Lynn Slingerland (Paper 8) goes the furthest in developing a theory of the mechanism of entrainment. An assistant professor of geology at the Pennsylvania State University, he has developed a critical velocity equation and critical suspension velocity equation, which set boundaries on the hydraulic equivalent sizes of heavy and light minerals in sands. If a particular size has been established on a bed, like-sized particles will tend to be selected for deposition. Also, the heavier grains of a given size will test a bed more than will light grains and therefore have a better chance of filling a given position. Paper 8 is based on Slingerland's master's thesis work at Penn State.

In his study of inversely graded grain-flow deposits, Asbury Hilliard Sallenger, Jr., an oceanographer with the U.S. Geological Survey, introduces the concept of dispersive equivalence (Paper 9). This concept is derived from the dispersive pressure hypothesis of how inverse grading develops. He compares the hydraulic equivalence that should result from both differential entrainment and dispersive pressure in natural and experimental grain-flow deposits and concludes that the dispersive pressure hypothesis most convincingly explains the observed relationships. His study suggests there are at least three kinds of equivalences associated with sedimentary deposits. Sallenger's paper (Paper 9) received an honorable mention for the Best Paper in the *Journal of Sedimentary Petrology* for 1979.

5

Reprinted from *Jour. Sed. Petrology* **3**:3–29 (1933), by permission of the publisher

THE SIZE-DISTRIBUTION OF HEAVY MINERALS WITHIN A WATER-LAID SANDSTONE

WILLIAM W. RUBEY[1]
U. S. Geological Survey, Washington, D.C.

ABSTRACT

The factors that control the distribution of heavy minerals in natural sandstones are so numerous and complex that their separate effects are difficult or even impossible to untangle. However, a theoretical approach to the problem indicates something of the relative importance of each.

Factors such as differences in the density and hardness of the various minerals, differences in the original size of the various mineral grains in the source rock, the amount of abrasion that all grains have undergone during transportation, the different settling velocities of the various grains at the site of deposition, and the degree of sorting to which all grains were subjected there—all these factors seem competent to cause large variations in the relative abundance of various minerals in different samples of deposits that have been derived from exactly the same source rock and also in different size fractions of each of the samples. Certain of the factors, such as abrasion, tend to concentrate the heavier minerals in those samples that are predominantly finer grained; whereas other factors, such as sorting, tend to concentrate the heavier minerals in the finer grained portions of each sample, regardless of whether that sample is predominantly fine-grained or coarse-grained.

The selective effects upon different minerals of the various influences, as deduced from theoretical considerations, appear to be confirmed by the evidence of actual sand samples. These effects seem sufficiently large to merit attention when using heavy mineral suites as a basis for correlation or for inferences about the character of source rock. Unless all the samples to be compared are of approximately the same grain size and the same degree of sorting, it is suggested that, in order to make allowances for these effects, heavy mineral separates be taken from *two* selected size fractions of each sample—one, representing the same *actual* grain size, the other, the same *relative* grain size in each sample.

INTRODUCTION

Sedimentary petrologists are familiar with the fact that heavy minerals tend to be concentrated within the finer grained portions of sandstones and that this concentration is due largely to processes of sedimentation.[2] Yet it may be of interest to review the various factors that influence the size-distribution of different mineral grains and to point out the differences, both apparent and real, in the heavy mineral suites of different samples that may result, even among sands that have been derived from the same source rocks.

These effects were called very forcibly to the writer's attention by a comparison of the heavy minerals that occur in two Pleistocene terrace deposits along Mississippi and Illinois rivers in

[1] Published with the permission of the Director of the U. S. Geological Survey and the Chief of the State Geological Survey of Illinois.

[2] Mackie, William, The principles that regulate the distribution of particles of heavy minerals in sedimentary rocks, as illustrated by the sandstones of the northeast of Scotland: *Trans. Edinburgh Geol. Soc.*, vol. 11, pp. 138-164, 1923.

the Hardin and Brussels quadrangles, Ill.[8] Many of the samples collected from different rock formations in this area were subjected to various types of commercial tests by members of the State Highway Department of Illinois, the Department of Ceramics of the University of Illinois, and the Illinois Geological Survey. Samples from two stream deposits, one of Illinoian age and the other of Wisconsin age, were sieved and examined to determine the nature of rock particles and heavy minerals by H. B. Willman of the State Geological Survey. Inasmuch as reconnaissance studies outside the area of the Hardin and Brussels quadrangles had indicated that these two deposits may become important reference-beds in deciphering the river history of this general region, it seemed worth while to examine Willman's data for any suggestions of petrographic differences that might supplement the physiographic and stratigraphic criteria by which the two terrace deposits are distinguished. But, to the writer's surprise, these data were found to indicate that the heavy mineral assemblages depend, not upon whether a sample came from the Illinoian or the Wisconsin deposit, but upon the coarseness or fineness of the particular sample. For example, the coarsest grained and the finest grained samples that were examined both came from the Wisconsin deposit, yet these two samples were reported to contain the extreme differences in heavy mineral assemblages.

[8] Rubey, W. W., Geology of the Hardin and Brussels quadrangles, Ill.: *Illinois Geol. Survey, Bull.* (Awaiting publication.)

It seemed desirable to re-examine the samples and for this purpose the original heavy mineral separates were very kindly loaned to the writer by J. E. Lamar of the State Geological Survey. Re-examination completely confirmed Mr. Willman's reports in all essential matters. Epidote, kyanite, andalusite, rutile, and hypersthene were found to be increasingly abundant in the successively coarser grained samples, and magnetite, ilmenite, zircon, muscovite, and biotite increasingly abundant in the successively finer grained samples. Six other mineral species were found in the samples and, of these, the relative abundance of four seem also to be dependent upon the grain size of the sample. However, the evidence regarding the four additional minerals is not conclusive and, for the present purpose, it will suffice to point out the relationship suggested by the ten minerals whose occurrence clearly varies with the grain size of the samples. Of these ten minerals, the five that are clearly most abundant in the finer grained samples are either heavier or softer than the five minerals that occur most abundantly in the coarser grained samples. Furthermore, detailed comparisons of the samples seem to show that this same relationship also governs the distribution of fragments of lighter materials, such as quartz, chert, and dark and light igneous rocks. In other words, these samples indicate that, other things being equal, the grains of heavier and softer minerals tend to be concentrated in finer grained samples and the grains of lighter and harder minerals in coarser grained samples.

This empirical observation is, of course, far from conclusive; and the series of efforts made to test the validity of the generalization led to the preparation of the present paper. The writer wishes to acknowledge gratefully the helpful criticism received from many of his colleagues of the U. S. Geological Survey, particularly W. H. Bradley and M. I. Goldman, who suggested many needed revisions of the first draft of the manuscript.

EXPECTED DISTRIBUTION OF HEAVY MINERALS ACCORDING TO STOKES' LAW

The probable cause of the concentration of heavy minerals within the finer grained portions of sandstones can perhaps best be approached by discussing the distribution of heavy minerals that might be expected in the simplest imaginable case—some theoretical deposit of sand about which all the conditions of source, transportation, and deposition are known. Let us assume that the source rock from which this sandstone was derived is made up entirely of quartz, magnetite, and tourmaline and has such a homogeneous composition that the proportionate abundance of these three minerals remains essentially the same whatever the grain size.

If the particles produced by the breaking up of this source rock were very soon entrained by flowing water, carried a short distance, and then deposited, many of the selective effects on different minerals of weathering and abrasion and of different agents of transportation would be slight. If, in addition, the relative velocities with which quartz, magnetite, and tourmaline grains are carried and deposited in water are known, it becomes possible to estimate the size-distribution of the heavy mineral grains that would be carried along and deposited with the different sized quartz grains.

To be more specific, let us take any random sample from this sandstone deposit and determine by some method the complete size-distribution of the quartz grains that were deposited in it (See Fig. 1). Then, the relative abundance of each different size of the quartz grains in the deposit and the relative abundance of quartz and magnetite at the source being known, we need only the relative settling velocities of quartz and magnetite grains in order to determine the complete size-distribution of the magnetite grains in the deposit. This use of settling velocities to determine the sizes of grains of different minerals that would be deposited together involves no particular assumptions about the mode of transportation of the grains nor does it imply that all the grains in a given deposit have the same settling velocity. Obviously, small grains that fell from near the bottom of a muddy fluid might reach the bottom long before larger grains that started falling from near the surface of the fluid. The only assumption involved here is that, whatever the conditions may have been which permitted the deposition of quartz grains of a certain size, these conditions would also permit the deposition of magnetite grains that had the same settling velocity.

The velocities with which different particles fall through a fluid depend chiefly upon the density and volume of the particles and the density and viscosity of the fluid. It is true that other factors,[4] especially differences of shape, also influence settling velocities —as an extreme example, flat micaceous particles have a much larger surface area opposed to the currents of water than spherical particles of the same volume and density and, for that reason, the flat particles are picked up by gentle currents, fall slowly, and are deposited along with much smaller, more spherical particles. Nevertheless, the effect of the more important of the various factors can conveniently be summed up in Stokes' law. This law states that the settling velocities of essentially spherical particles vary as the square of the particle diameters, as the effective density of the particle in the fluid, and inversely as the viscosity of the fluid. Experimental studies[5] have shown that this law holds closely for particles with diameters between 0.0002 and about 0.2 millimeters (that is, for grains of clay, silt, and very fine sand). Larger particles depart more and more from Stokes' law until quartz grains with diameters greater than about 1.5 millimeters (that is, particles of very coarse sand and gravel) fall with velocities proportional to the square root instead of the square of the diameters.

If Stokes' law can be used for estimating settling velocities, then the relative sizes of a quartz grain and a magnetite grain that have the same settling velocity in water can be found.

According to Stokes' law, where,

$$v = \frac{g}{18} \cdot \frac{\rho_q - \rho_w}{\eta} d_q^2 = \frac{g}{18} \cdot \frac{\rho_m - \rho_w}{\eta} d_m^2$$

where

v = settling velocity of both grains, measured in centimeters per second

ρ_q = specific gravity of quartz grain = 2.66

ρ_m = specific gravity of magnetite grain = 5.18

ρ_w = specific gravity of water = 1.00

d_q = diameter of quartz grain, measured in centimeters

d_m = diameter of magnetite grain, measured in centimeters

η = coefficient of viscosity of the water (a constant under these conditions)

g = acceleration due to gravity

Then,

$$(2.66 - 1.00)\ d_q^2 = (5.18 - 1.00)\ d_m^2$$

and $d_m = \sqrt{\frac{2.66 - 1.00}{5.18 - 1.00}} d_q = 0.63 d_q$

That is, a magnetite grain only 0.63 mm. in diameter would fall through water with the same velocity as a quartz grain 1.00 mm. in diameter.

The effects of probable departures from Stokes' law are discussed more fully in another part of this paper. However, if we may here provisionally accept it as a basis for estimating the relative sizes of all magnetite and quartz grains that have the same settling velocity, it then becomes possible to determine the size-distribution of

[4] See also Rubey, W. W., Lithologic studies of fine-grained Upper Cretaceous sedimentary rocks of the Black Hills region: *U. S. Geol. Survey Prof. Paper* 165, pp. 17-18, 23-24, 26-27, 1930.

[5] For references see *Idem.*, p. 18; Rubey, W. W., Settling velocities of gravel, sand, and silt particles: *Amer. Jour. Sci.*, vol. 25, 5th ser., pp. 325-338, 1933; and p. 17 of the present paper.

the magnetite grains that, under these assumed conditions, would be deposited along with the quartz grains in a random sample of the sandstone. If it is known that the source rocks were furnishing some quartz grains of essentially all sizes and we find that the sample contains no quartz grains with diameters less than 0.15 mm. (Fig. 1), this would mean that any still finer grains supplied at the source must have been kept in suspension and swept past the site of deposition by the currents that laid down the sandstone. Similarly, if the sample contains no quartz grains with diameters more than 6.6 mm., this means that the currents were not swift enough to transport any of the still larger grains to the site of deposition. And furthermore, the fact that quartz grains with diameters of 1.0 mm. occur most abundantly in the sample would indicate that the complex of currents at and above the site of deposition were, on the average, best able to transport to and deposit there quartz grains of this particular size.

Inasmuch as the assumed source rock had a homogeneous composition, it follows from Stokes' law that the smallest magnetite grains laid down with these quartz grains would be 0.094 (that is, 0.63×0.150) mm. in diameter, the largest magnetite grains 4.16 (that is, 0.63×6.6) mm. in diameter and the most abundant magnetite grains 0.63 mm. in diameter. In other words, the magnetite grains would be concentrated in the finer grained portions of the sample. If the magnetite grains were known to be, say, 1/10 as abundant in the source rock as the quartz grains, the complete size-distribution or frequency curve of the magnetite grains could be worked out and represented as in Figure 1.

The same method of attack would give the complete size-distribution of the tourmaline grains. If the specific gravity of the tourmaline is 3.09, then under Stokes' law a tourmaline grain 0.89 $\left(= \sqrt{\frac{2.66 - 1.00}{3.09 - 1.00}}\right)$ mm. in diameter would fall through water with the same velocity as a quartz grain 1.00 mm. in diameter. Hence, in this sample, there would be no tourmaline grains less than 0.133 mm. in diameter and none more than 5.87 mm. in diameter, and those with diameters of 0.89 mm. would occur most abundantly. In other words, the tourmaline grains would be concentrated in portions of the sample intermediate in grain size between those of the quartz and magnetite concentrations. If the tourmaline grains occur in the source rock in, let us say, the same proportions as the magnetite grains, their frequency curve could be worked out in the same manner (Fig. 1).

In order to contrast clearly the frequency curves of the different minerals in Fig. 1, an unreasonably high proportion of magnetite and tourmaline to quartz has been assumed. To make the sample appear more nearly comparable with an actual one, a more reasonable $\frac{\text{quartz}}{\text{magnetite}}$ and $\frac{\text{quartz}}{\text{tourmaline}}$ ratio of 100 to 1 might have been taken. Furthermore, it is obvious that, had we chosen a somewhat finer grained sample but one with exactly the same type of sorting, the distribution or frequency curves of the quartz, magnetite, and tourmaline in it would have had precisely the same shapes and relative positions as those in the coarser sample but their actual positions would all be shifted the same distance toward the smaller grain sizes.

For purposes of later comparison, let us take such a finer grained sandstone with exactly the same type of sorting as the first sample but with $\frac{\text{quartz}}{\text{magnetite}}$ and $\frac{\text{quartz}}{\text{tourmaline}}$ ratios of 100 to 1 and with all grains only $\frac{1}{2.828}\left(=\frac{1}{(1.414)^3}\right)$ as large (that is, with the smallest quartz grains 0.053 instead of 0.150 mm. in diameter, the largest quartz

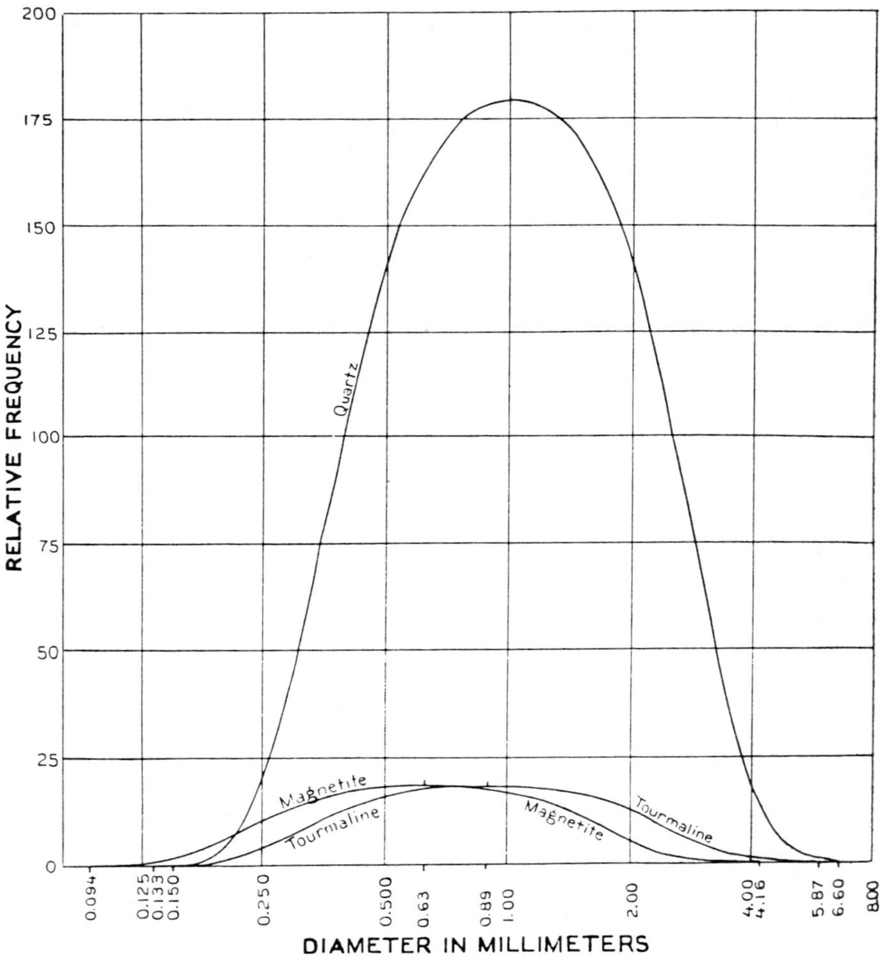

FIG. 1. Complete size-distribution or frequency curve of quartz, tourmaline, and magnetite grains within a well-sorted sandstone, according to Stokes' law.

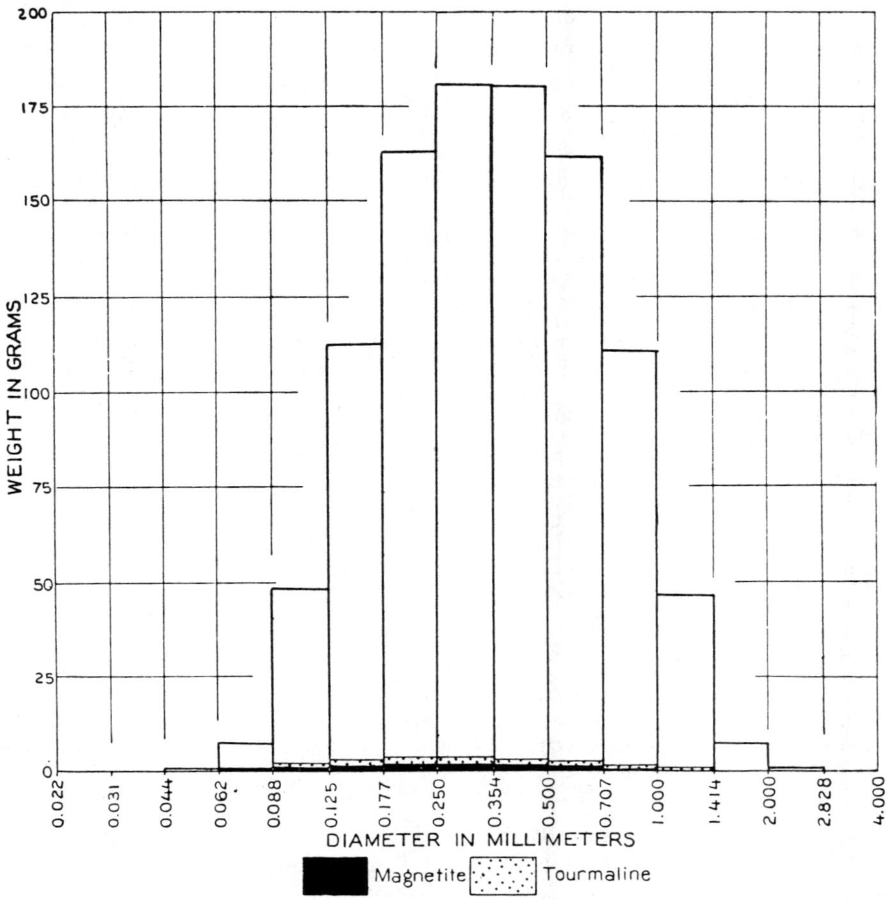

FIG. 2. Pyramidal diagram or histogram of entire sample showing distribution, between certain sieve limits, of quartz, tourmaline, and magnetite grains within a well-sorted medium grained sandstone, according to Stokes' law.

grains 2.33 instead of 6.6 mm. in diameter, and the most abundant quartz grains 0.354 instead of 1.000 mm. in diameter). Let us separate the grains in this second sample by means of a series of sieves with mesh sizes that increase by the ordinary sieve ratio of $\sqrt{2}$ or 1.414 and then weigh separately the portions retained on each sieve. When this is done, the data for the frequency curves of the three minerals can be replotted as one pyramidal diagram or histogram[6] which shows the distribution of 1,000 grams of quartz, 10 grams of tourmaline, and 10 grams of magnetite between the several diameters used as mesh sizes in the different sieves. (See Fig. 2.)

The histogram of this second sample

[6] Wentworth, C. K., Methods of mechanical analysis of sediments: *Iowa Univ. Studies*, vol. 11, pp. 46-50, 1926.

is that of a moderately well-sorted medium grained sandstone in which the particles with diameters between 0.354 and 0.250 mm. constitute the ***chief ingredient*** or ***maximum***[7] (or, stated in other terms, the median diameter of the sample is slightly less than 0.354 mm.). A third sample, still finer grained [with its particles only $\frac{1}{(1.414)^7}$ as large as those in the first sample (Fig. 1), and therefore only $\frac{1}{(1.414)^4}$ as large as those in the second sample (Fig. 2)] but with exactly the same degree of sorting, would obviously give a histogram of the same shape as Fig. 2 but displaced four sieve-units toward the smaller grain sizes. And a fourth sample, coarser grained [with its particles $(1.414)^2$ times as large as those in the first sample (Fig. 1), and therefore $(1.414)^5$ times as large as those in the second sample (Fig. 2)] but with exactly the same degree of sorting, would give the same shaped histogram displaced five sieve-units toward the coarser grain sizes.

TABLE 1. DISTRIBUTION BY WEIGHT, ACCORDING TO STOKES' LAW, OF MAGNETITE AND TOURMALINE GRAINS IN A SAMPLE OF WELL-SORTED MEDIUM GRAINED SANDSTONE. (Sorting same as that shown in Figs. 1 and 2.)

Limiting diameters of sieve unit in millimeters	Magnetite Weight in grams	Tourmaline* Weight in grams	Total "heavies" Weight in grams
0.031–0.044	0.005 −	None	0.005 −
0.044–0.062	0.16	0.005 −	0.160 + to 0.165 −
0.062–0.088	0.67	0.16	0.83
0.088–0.125	1.30	0.67	1.97
0.125–0.177	1.68	1.30	2.98
0.177–0.250	1.78	1.68	3.46
0.250–0.354	1.74	1.78	3.52
0.354–0.500	1.46	1.74	3.20
0.500–0.707	0.89	1.46	2.35
0.707–1.000	0.29	0.89	1.18
1.000–1.414	0.03	0.29	0.32
1.414–2.000	0.005 −	0.03	0.030 + to 0.035 −
2.000–2.828	None	0.005 −	0.005 −

* The value of 3.09, used for the specific gravity of the tourmaline, falls about midway in the reported range of tourmaline specific gravities and was chosen because it greatly reduces the labor of computation necessary to prepare this table. With its use, the quantity of tourmaline in one sieve unit is exactly the same as the quantity of magnetite in the next smaller sieve unit

Let us now remove by means of a suitable heavy liquid all the quartz grains from each of the 13 size fractions in the second or medium-grained sample (Fig. 2) and then determine the relative abundance of magnetite and tourmaline in each of the separate size fractions. For this particular sample, the frequency curve, the mineral densities, and Stokes' law give the quantities as shown in Table 1.

These quantities, plotted on an enlarged scale, are shown in the histogram of heavy minerals (Fig. 3B). A similar separation of quartz from the third sample (fine-grained) and the fourth sample (coarse-grained) would

[7] Udden, J. A., Mechanical composition of clastic sediments: *Geol. Soc. Amer., Bull.*, vol. 25, p. 658, 1914.

give the histograms shown in Fig. 3A and 3C.

If now we compare the heavy minerals found in some one sieve unit in each of these last three samples, we find that their relative abundance varies greatly.

The greater concentration of magnetite in the finer grained fractions of each sample means that, upon compar-

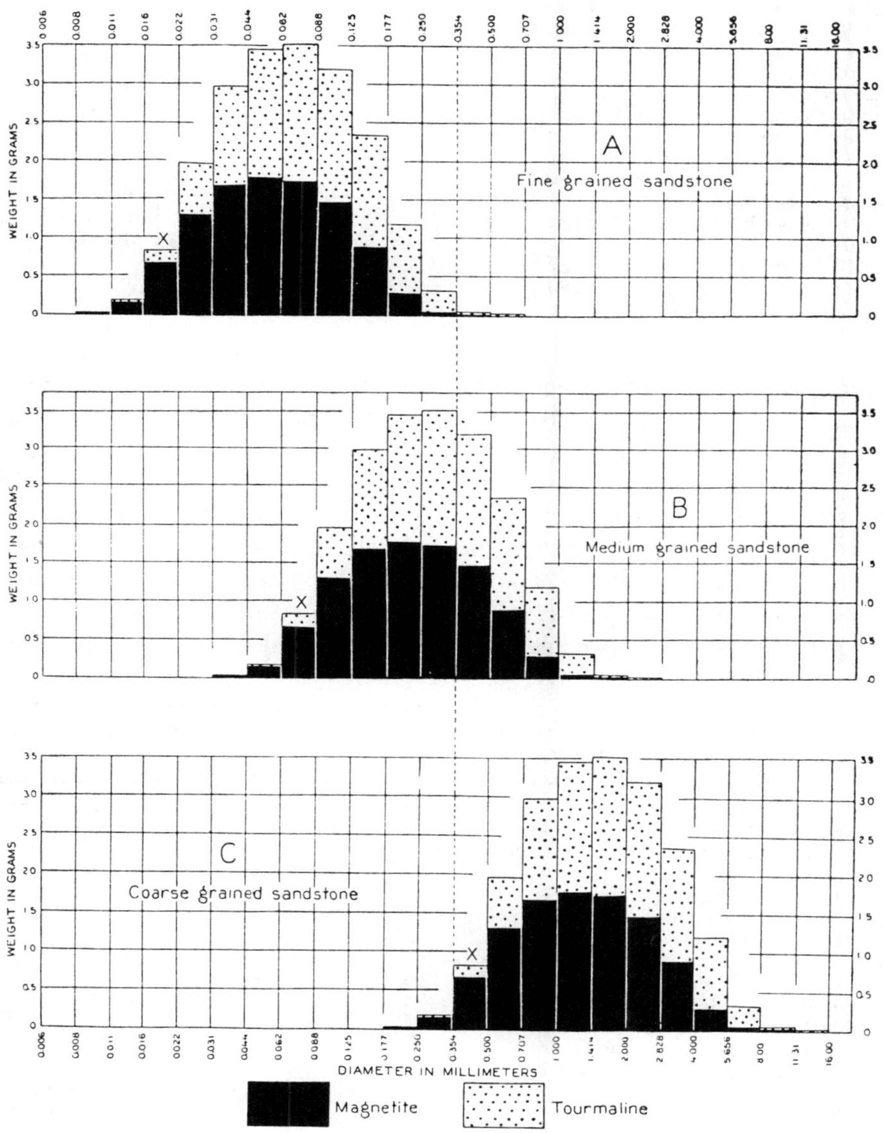

FIG. 3 A, B, and C. Pyramidal diagram or histogram of heavy minerals, showing distribution, between certain sieve limits, of tourmaline and magnetite grains within three samples of well-sorted sandstone, according to Stokes' law.

TABLE 2. RELATIVE ABUNDANCE, ACCORDING TO STOKES' LAW, OF DIFFERENT HEAVY MINERALS WITH DIAMETERS *Between 0.250 and 0.354 mm.*, IN THREE WELL SORTED SAMPLES (Sorting same as that shown in Figs. 1 and 2.)

Sample	Magnetite	Tourmaline	Total "heavies"
Fine-grained sandstone............. (Sample with minimum diameter of 0.013 mm., maximum diameter of 0.584 mm., and median diameter of 0.088 mm.—for quartz particles)	0.03 grams 9%	0.29 grams 91%	0.32 grams 100%
Medium-grained sandstone......... (Sample with minimum diameter of 0.053 mm., maximum diameter of 2.33 mm., and median diameter of 0.354 mm.—for quartz particles)	1.74 grams 49%	1.78 grams 51%	3.52 grams 100%
Coarse-grained sandstone........... (Sample with minimum diameter of 0.300 mm., maximum diameter of 13.2 mm., and median diameter of 2.00 mm.—for quartz particles)	0.16 grams 97+%	0.005− grams 3−%	0.160+ to 0.165− grams 100%

ing the relative abundance of the different heavy minerals of some particular grain size in different samples (that is, upon comparing the heavy minerals in the finer fractions of the coarse samples with those in coarser fractions of fine samples), magnetite seems to occur in greater abundance in the coarser samples, even though it is in fact equally abundant in all three samples.

The comparisons just made were of the relative abundance of magnetite and tourmaline in the same sieve unit of each sample. If, instead, we had followed another laboratory procedure and made the comparisons of different samples by considering all the heavy mineral grains smaller than some one sieve size, we would not have found so great a variation in relative abundance.

TABLE 3. RELATIVE ABUNDANCE, ACCORDING TO STOKES' LAW, OF DIFFERENT HEAVY MINERALS WITH DIAMETERS *Less than 0.354 mm.*, IN THREE WELL SORTED SAMPLES. (Sorting same as that shown in Figs. 1 and 2.)

Sample	Magnetite	Tourmaline	Total "heavies"
Fine-grained sandstone............. (Sample with minimum diameter of 0.013 mm., maximum diameter of 0.584 mm., and median diameter of 0.088 mm.—for quartz particles)	10.00 grams 50%	9.97 grams 50%	19.97 grams 100%
Medium-grained sandstone.......... (Sample with minimum diameter of 0.053 mm., maximum diameter of 2.33 mm., and median diameter of 0.354 mm.—for quartz particles)	7.33 grams 57%	5.59 grams 43%	12.92 grams 100%
Coarse-grained sandstone........... (Sample with minimum diameter of 0.300 mm., maximum diameter of 13.2 mm., and median diameter of 2.00 mm.—for quartz particles)	0.16 grams 97+%	0.005− grams 3−%	0.160+ to 0.165− grams 100%

With this laboratory procedure, magnetite would assuredly be found in all the samples but tourmaline might escape notice in the coarse-grained sandstone. It is important to note that both types of comparison would make the magnetite appear more abundant than the tourmaline in the coarser grained samples.

exposed to erosion at the time of sedimentation. Yet those who have estimated the relative abundance of different minerals in a number of samples probably will admit that minerals, like the magnetite and tourmaline in Table 2, which range from more than 90 per cent in some samples to less than 10 per cent in others, might sometimes

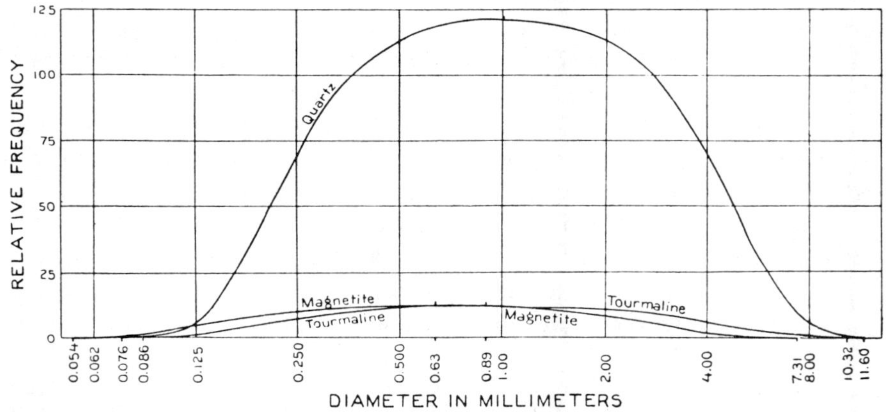

FIG. 4. Complete size-distribution or frequency curve of quartz, tourmaline, and magnetite grains within a poorly sorted sandstone, according to Stokes' law.

It is true that slight differences, such as some of those given above, in the relative abundance of heavy minerals in different samples are perhaps seldom used as a basis for the correlation of sandstones. But large differences in the relative abundance (and especially the mere presence or absence) of certain minerals in a rock are often used for correlation when more trustworthy criteria, like the occurrence of peculiar varieties, can not be found. Also, the general assemblages of heavy minerals in sedimentary rocks (that is, the presence and absence of certain species and the relative abundance of each) are sometimes used as indicators of the type or types of source rocks that were be reported, respectively, as "flood" or as "absent," especially when the effects of minor variations in technique of separation and the accidents of sampling for microscopic study are considered.[8] That is, two sandstones derived from the same source rocks and deposited at the same time might, un-

[8] Wentworth, C. K., The accuracy of mechanical analysis: *Amer. Jour. Sci.*, 5th ser., vol. 13, pp. 404-406, 1927.

Smithson, Frank, The reliability of frequency-estimations of heavy mineral suites: *Geol. Mag.*, vol. 67, pp. 134-136, 1930.

Ewing, C. J. C., A comparison of the methods of heavy mineral separation: *Geol. Mag.*, vol. 68, pp. 136-140, 1931.

Dryden, A. L., Jr., Accuracy in percentage representation of heavy mineral frequencies: *Proc. Nat. Acad. Sci.*, vol. 17, pp. 233-238, 1931.

der such circumstances, be interpreted as wholly unrelated to one another.

THE EFFECT OF DIFFERENT TYPES OF SORTING

Thus far, in order to eliminate some of the countless factors that make the interpretation of actual sandstones complicated and difficult, it has been assumed that all the samples compared have had exactly the same degree of sorting. Now, to determine the effect of the degree of sorting, let us take samples which, instead of being moderately well-sorted, are rather poorly sorted, with frequency curves of the shape shown in Fig. 4. From Stokes' law, the frequency curves of the magnetite and tourmaline associated with the quartz can be worked out; and, from these frequency curves, histograms can be constructed which show the distribution of 1,000 grams of quartz, 10 grams of tourmaline, and 10 grams of magnetite in poorly sorted fine-grained, medium-grained, and coarse-grained sandstones. Separating off the quartz as before, we can then make comparisons analogous to those made with the well sorted samples.

It is evident from these tables that the variation in apparent abundance of different heavy minerals in a series of poorly sorted samples is considerably less than it is in a series of well sorted samples of the same average grain size. This, of course, is the expected result, for, had there been no sorting whatever, the percentages of the different minerals would have remained exactly

TABLE 4. RELATIVE ABUNDANCE, ACCORDING TO STOKES' LAW, OF DIFFERENT HEAVY MINERALS WITH DIAMETERS *Between 0.250 and 0.354 mm.*, IN FIVE POORLY SORTED SAMPLES. (Sorting same as that shown in Fig. 4.)

Sample	Magnetite	Tourmaline	Total "heavies"
Very fine-grained sandstone (Sample with minimum diameter of 0.005 mm., maximum diameter of 0.73 mm., and median diameter of 0.062 mm.—for quartz particles)	0.08 grams 17%	0.39 grams 83%	0.47 grams 100%
Fine-grained sandstone (Sample with minimum diameter of 0.008 mm., maximum diameter of 1.03 mm., and median diameter of 0.088 mm.—for quartz particles)	0.39 grams 35%	0.73 grams 65%	1.12 grams 100%
Medium-grained sandstone (Sample with minimum diameter of 0.030 mm., maximum diameter of 4.10 mm., and median diameter of 0.354 mm.—for quartz particles)	1.20 grams 49%	1.24 grams 51%	2.44 grams 100%
Coarse-grained sandstone (Sample with minimum diameter of 0.172 mm., maximum diameter of 23.2 mm. and median diameter of 2.00 mm.—for quartz particles)	0.64 grams 71%	0.26 grams 29%	0.90 grams 100%
Very coarse-grained sandstone (Sample with minimum diameter of 0.243 mm., maximum diameter of 32.8 mm., and median diameter of 2.828 mm.—for quartz particles)	0.26 grams 87%	0.04 grams 13%	0.30 grams 100%

TABLE 5. RELATIVE ABUNDANCE, ACCORDING TO STOKES' LAW, OF DIFFERENT HEAVY MINERALS WITH DIAMETERS *Less than 0.354 mm.*, IN FIVE POORLY SORTED SAMPLES. (Sorting same as that shown in Fig. 4.)

Sample	Magnetite	Tourmaline	Total "heavies"
Very fine-grained sandstone......... (Sample with minimum diameter of 0.005 mm., maximum diameter of 0.73 mm., and median diameter of 0.062 mm.—for quartz particles)	10.00 grams 50%	9.91 grams 50%	19.91 grams 100%
Fine-grained sandstone............. (Sample with minimum diameter of 0.008 mm., maximum diameter of 1.03 mm., and median diameter of 0.088 mm.—for quartz particles)	9.91 grams 51%	9.52 grams 49%	19.43 grams 100%
Medium-grained sandstone.......... (Sample with minimum diameter of 0.030 mm., maximum diameter of 4.10 mm., and median diameter of 0.354 mm.—for quartz particles)	6.61 grams 55%	5.41 grams 45%	12.02 grams 100%
Coarse-grained sandstone........... (Sample with minimum diameter of 0.172 mm., maximum diameter of 23.2 mm., and median diameter of 2.00 mm.—for quartz particles)	0.94 grams 76%	0.30 grams 24%	1.24 grams 100%
Very coarse-grained sandstone....... (Sample with minimum diameter of 0.243 mm., maximum diameter of 32.8 mm., and median diameter of 2.828 mm.—for quartz particles)	0.30 grams 88%	0.04 grams 12%	0.34 grams 100%

the same in every size fraction of each sample. Yet it is interesting to note that, even in these poorly sorted samples, the percentage differences are still very large and, if samples somewhat finer and somewhat coarser grained (the "very fine-grained sandstone" and the "very coarse-grained sandstone" of Tables 4 and 5) than those shown in Tables 2 and 3 are compared, the differences are virtually the same as those in the series of well sorted samples.

It is also worthy of note that the apparent contrasts in mineral proportions are large enough to be significant when samples with the same average grain size but with different degrees of sorting are compared (Table 6). This, likewise, is the expected result for, had this same size fraction (or, for that matter, any other size fraction) been

TABLE 6. RELATIVE ABUNDANCE, ACCORDING TO STOKES' LAW, OF DIFFERENT HEAVY MINERALS WITH DIAMETERS *Between 0.250 and 0.354 mm.*, IN SAMPLES OF FINE-GRAINED SANDSTONE (median diameter about 0.088 mm. in both samples but maximum and minimum diameters not the same in both samples).

Sample	Magnetite	Tourmaline	Total "heavies"
Well-sorted..........	0.03 grams 9%	0.29 grams 91%	0.32 grams 100%
Poorly-sorted........	0.39 grams 35%	0.73 grams 65%	1.12 grams 100%

taken from an absolutely unsorted sample, it would have contained exactly equal amounts of magnetite and tourmaline.

These comparisons have shown, that, among grains of the same diameter from samples of the same average coarseness, the differences in relative abundance of various minerals increase with the degree of sorting. Hence, it is pertinent to inquire whether or not the degree of sorting taken to represent the "well-sorted" samples is so extreme that it is but rarely found in nature. The data for the histogram of any one of the "well-sorted" samples (Fig. 2 for example) can readily be recast into a form suitable for comparison with the collections of mechanical analyses published by Udden[9] and Wentworth.[10] Inspection of these analyses will show that, judged either by the relative height of the maximum grade above the rest of the sample or by the number of grade sizes, the "well-sorted" samples discussed in this paper are somewhat less sorted than many fluvial sands and much less sorted than most eolian and beach sands. Had more perfectly sorted samples been chosen for the above comparisons, the apparent differences in heavy mineral assemblages with varying grain size of the samples would have been much greater.

We have also assumed that the size-frequency curves of both the well-sorted and the poorly-sorted samples are smooth, simple curves which are symmetrical on either side of one chief constituent or maximum grade.[11] Frequency curves with marked irregularities and secondary maxima[12] would, of course, give rise to comparable irregularities in the relative abundance of different mineral species in certain size fractions; yet it is clear that irregularities of this sort in the frequency curves can, in the aggregate, neither increase nor decrease the contrasts in mineral proportions caused by differences in grain size or degree of sorting.

The effect of symmetry or asymmetry of the frequency curves perhaps requires an additional word. Sandstones with the perfectly symmetrical size-distribution assumed for our theoretical samples are no doubt rare in nature. Nevertheless, the distribution curves of many actual samples are approximately symmetrical, and departures from the assumed symmetry would not invalidate the conclusions reached in this paper. Of the two possible types of asymmetry, those in which more than half of the grains are larger than the chief ingredient appear to be somewhat the commoner. Asymmetrical distribution of this more common type would tend to increase the

$$\frac{\text{magnetite}}{\text{tourmaline}} \text{ ratio in the finer grained}$$

portions and decrease it in the coarser grained portions of a sample. Asymmetrical distribution of the opposite type (more than half of the grains

would tend to decrease the $\frac{\text{magnetite}}{\text{tourmaline}}$

[9] *Op. cit.*, pp. 693-727.

[10] Wentworth, C. K., The mechanical composition of sediments in graphic form: *Univ. Iowa Studies,* vol. 14, pp. 36-126, 1932.

[11] Udden, *Op. cit.*, p. 658.

[12] Udden, *Op. cit.*, pp. 736-737.

smaller than the chief constituent) ratio in the finer grained portions and increase it in the coarser grained portions of a sample. But, though the exact values of the ratios would be different for the various types of asymmetrical distribution, all types would give a concentration of the heavier minerals in the finer portions of a sample just as does a symmetrical distribution. Consequently, a comparison of the grains from different samples retained between some one pair of sieves would show a greater relative abundance of magnetite in the coarser grained samples, very much as does the symmetrical distribution represented in Figs. 3 A, B, and C. The effect of any asymmetry is, therefore, such that no important qualification is required of the conclusions that have so far been reached about the apparent variation of heavy mineral assemblages with varying grain size or degree of sorting of the samples.

THE EFFECT OF DEPARTURES FROM STOKES' LAW

For the purpose of estimating the relative settling velocities of the quartz, magnetite, and tourmaline grains, we have assumed that Stokes' law holds for all grain sizes. However, this law applies only so far as viscosity is the controlling factor in the resistance of the settling fluid. Theoretical considerations and experimental evidence indicate that for larger particles the resistance is dominated by the impact rather than the viscosity of the fluid and that very coarse sand and gravel fall with velocities which vary as the square root instead of the square of the diameters.[13]

If all particles in a sample were so large that their settling velocities followed this square root law, then the relative sizes of a quartz grain and a magnetite grain that have the same velocity in water could be found, for, using the same notation as that on p. 6.

$$v = \sqrt{2/3.g.\frac{\rho_q - \rho_w}{\rho_w} d_q}$$

$$= \sqrt{2/3.g.\frac{\rho_m - \rho_w}{\rho_w} d_m}$$

Then $(2.66 - 1.00) d_q = (5.18 - 1.00) d_m$ and $d_m = 0.397\, d_q$

That is, a large grain of quartz would fall through water at the same velocity as a magnetite grain with a diameter only $\frac{397}{1000}$ as large [instead of $\frac{63}{100}$, the expected ratio under Stokes' law]. And a tourmaline grain only $\frac{794}{1000}$ [instead of $\frac{89}{100}$]; the diameter of the quartz grain would fall with the same velocity. Hence, the frequency curves of the magnetite and tourmaline grains, though unchanged in shape, would be shifted much farther away from the frequency curve of the quartz grains than those shown in Figs. 1 and 4.

[13] Richards, R. H., Velocity of galena and quartz falling in water: *Trans. A.I.M.E.*, vol. 38, pp. 210-235, 1908.

Rubey, W. W., Settling velocities of gravel, sand, and silt particles: *Amer. Jour. Sci.*, vol. 25, 5th ser., pp. 325-338, 1933.

Furthermore, the magnetite curve would be shifted much farther than the tourmaline curve, thereby causing much greater differences in the $\frac{\text{magnetite}}{\text{tourmaline}}$ ratios of the coarse- and fine-grained portions of each sample.

If the Stokes' or square-of-the-diameter law holds for small grains (less than about 0.2 mm. in diameter) and the square-root-of-the-diameter law holds for large grains (more than about 1.5 mm. in diameter), then in a medium-grained sandstone, with many grains of intermediate sizes, the heavy minerals would be distributed very differently from those in a very coarse-grained sandstone or gravel or those in a very fine-grained sandstone or silt. In the sample of sandstone represented in Fig. 1, the smallest quartz grains (diameter of 0.150 mm.) would settle by the square-of-the-diameter law at approximately the same rate as magnetite grains with a diameter of 0.63×0.150 mm. or 0.094 mm., and the largest grains in this sample (diameter of 6.6 mm.) would settle by the square-root-of-the-diameter law at the same rate as magnetite grains with a diameter of 0.397×6.6 mm. or 2.62 mm. This would mean, not only that the magnetite would be finer grained than the quartz, but that the frequency curve of the magnetite would have a shape very different from that of the quartz curve because it would be condensed between narrower size limits. In other words, the magnetite grains would be much better sorted than the quartz grains with which they were deposited. This corollary seems to be confirmed by available data[14] on natural magnetite-bearing sands with particles that range between these two size limits. The degree of sorting of the tourmaline grains would, however, be affected much less than that of the magnetite grains because the two factors for tourmaline, 0.89 and 0.794, differ much less than those for magnetite. The net result of this more perfect sorting of the magnetite grains would be to increase still further the $\frac{\text{magnetite}}{\text{tourmaline}}$ ratio in the finer grained portions of a sample.

What we have just been saying is that the probable departures from Stokes' law would affect different samples differently, depending upon their prevailing grain sizes. In coarse sands and gravels, the frequency curve of the magnetite grains would be displaced much farther from the quartz curve than it would in very fine sands and silts, but the shape of the curve would remain the same. In medium-grained sands, however, the magnetite curve would be displaced only an intermediate distance but its shape would be changed greatly—the magnetite would be much better sorted than the quartz.

A general formula expressing the settling velocities of both large and small particles has been proposed elsewhere,[15] and it might perhaps be used to determine the relative diameters of quartz and magnetite grains, whether

[14] Udden, *Op. cit.*, pp. 733-734.
[15] Rubey, *Amer. Jour. Sci.*, vol. 25, pp. 325-338.

large, small, or medium sized, that have the same settling velocity.

$$v = \frac{\sqrt{2/3.g.\rho_w(\rho_q - \rho_w)d_q^3 + 36\eta^2} - 6\eta}{\rho_w d_q}$$

$$= \frac{\sqrt{2/3.g.\rho_w(\rho_q - \rho_w)d_m^3 + 36\eta^2} - 6\eta}{\rho_w d_m}$$

However, this general expression (here stated in the terminology used on p. 6) has not yet been tested adequately, and it seems to the writer that the computations necessary for such a determination would needlessly encumber the present paper. Suffice it here to state that graphic solutions based upon this general formula bear out the conclusion stated in the preceding paragraph. That is, the departures from Stokes' law tend to increase rather than decrease the irregularities in distribution of different mineral species within a sandstone; when these departures are taken into consideration, the heavier minerals still tend to be concentrated within the finer fractions of a sample; and, upon comparing some one grain size in samples of different coarseness (that is, upon comparing the relatively fine fractions of coarse samples with the relatively coarse fractions of fine samples), the coarser samples still appear to contain higher $\frac{\text{magnetite}}{\text{tourmaline}}$ ratios. In short, the estimation of relative settling velocities by means of Stokes' law does not exaggerate but, in fact, for the common size range of sand grains, it considerably minimizes the effect of water sorting upon the size-distribution of grains of different minerals.

THE EFFECT OF ABRASION DURING TRANSPORTATION

The sand grains in the samples we have been discussing were assumed to have traveled only a very short distance from their source to their site of deposition. But, had they been carried a considerable distance, the effects of mutual abrasion during transportation would have modified the results sufficiently to require attention.

It is doubtless true that most sand grains have, in their repeated cycles of transportation and deposition, been transported considerable distances and that they have therefore undergone appreciable abrasion. However, the principles that govern this abrasion and grinding down to smaller sizes are still too imperfectly known to permit quantitative estimates of the sort just attempted in dealing with settling velocities. To be sure, suggestions have been made that the abrasion of particles due to their transportation varies directly as the effective weight in water and the distance traveled[16] and inversely as the hardness of the particles.[17] These proposed relations may possibly apply quite strictly to rolling grains, for they have been found to hold closely for the largest pebbles in the gravel along the upper Rhine.[18] But moving water carries debris in various ways.[19] The larg-

[16] Sternberg, H., Cited by Barrell, Joseph, Marine and terrestrial conglomerates: *Geol. Soc. Amer. Bull.*, vol. 36, p. 327, 1925.

[17] Mackie, William, On the laws that govern the rounding of particles of sand: *Trans. Edinburgh Geol. Soc.*, vol. 7, pp. 298-311, 1897.

[18] Barrell, *op. cit.*, pp. 329-330.

[19] Gilbert, G. K., The transportation of debris by running water, *U. S. Geol. Survey Prof. Paper* 86, pp. 15, 26-34, 156, 223-224, 230, 1914.

est particles being carried at any one time are rolled along the bottom, the smallest ones are held in suspension, the particles of intermediate sizes travel by a process of leaping called saltation, and some particles of all sizes may move along the bottom by sliding or a kind of plastic flow. It seems very improbable that all the smaller particles being transported by suspension, saltation, and sliding should follow exactly the same law of abrasion as the rolling grains. Among these smaller particles, the distance of travel is not measured simply by the circumference and the number of revolutions of each grain, and some more complex relation probably holds.

For the present purpose, however, this qualification can perhaps be ignored and the generalization made that the abrasion of particles due to their transportation varies approximately as the volume, as the effective density (specific gravity — 1), as the distance of travel, and inversely as the true hardness of the different mineral grains. That is to say, the effect of transportation in any unit distance of travel would be to abrade most the largest grains of each mineral and to abrade least the smallest grains, thereby increasing the proportion of small grains. Different minerals, of course, would be affected differently by this process of comminution, for the heavier and softer minerals would suffer the greater loss in a unit distance of travel.

"True hardness" as here used means simply durability or resistance to wear. In general, this resistance of different minerals would be expected to increase as some function of their position or number in the ordinary Moh's scale of hardness, modified, however, in some minerals by such other factors as readiness of cleavage or rate of decomposition. The best readily available estimates of the resistance to abrasion of different minerals appear to be quantitative determinations of the resistance to scratching as measured with the sclerometer. Tests of this sort indicate[20] that a mineral with the hardness of tourmaline is about 2 or 2½ times as resistant to abrasion as a mineral with the hardness of magnetite. This lower resistance of the magnetite, together with its greater effective density in water (5.18 — 1.00 as against 3.09 — 1.00) would, therefore, lead us to expect that, in traveling a given distance, magnetite grains would be subjected to several times as much loss by abrasion as tourmaline grains of the same size.

At this point, it is well to call attention to another element of uncertainty which prevents us from making precise estimates of the amount of abrasion undergone by grains of different minerals. A unit distance of travel is not necessarily equivalent to a unit of time. The heavier grains (the larger ones and those of higher specific gravity) will, because of their greater mass and inertia, be transported less rapidly than the lighter grains.[21] Presumably the very smallest and lightest grains transported by a stream are accelerated during their long leaps until they attain a

[20] Jaggar, T. A., Jr., A microsclerometer, for determining the hardness of minerals: *Amer. Jour. Sci.*, vol. 4, 4th ser., p. 411, 1897.

[21] Gilbert, *op. cit.*, pp. 26, 29.

Daubrée, G. A., Géologie expérimentale, vol. 1, pp. 250-251, 1879; cited by Barrell, *op. cit.*, p. 327.

velocity essentially that of the water; whereas the very largest and heaviest grains (those that the stream is just able to transport) stop repeatedly and so have a very low average velocity; and the other particles of intermediate sizes and weights transported at the same time presumably travel at velocities intermediate between the velocity of the water and zero. That is, the heavier grains tend to lag behind the lighter ones. Therefore, since all grains do not travel with the same average velocity, the particular velocity of a given grain must be known in order to determine the amount of abrasion it undergoes in a unit of time.

The problem thus raised—the relative velocity of the water and the sand grains being transported—is of considerable importance, yet very little appears to be known about it. Manifestly, the actual velocity of the sediment in a stream should be known or else the load transported in a unit of time cannot be calculated accurately. But estimates of the total solids being eroded from drainage basins and of the total silt being carried into reservoirs have had to rest upon the rather improbable assumption that the sediment travels at exactly the same velocity as the water. Were it not for the fact that these estimates have also had to neglect the load dragged along the stream beds, the total sediment thus calculated would almost certainly be too great.

This uncertainty about the velocity of transportation of sand grains of different sizes and weights means that, although large heavy grains clearly suffer more loss by abrasion than small light grains in a unit distance of travel, it is not certain that the wear due solely to their own movement causes them to lose more in a unit of time. But there is an additional factor in abrasion, thus far neglected in this discussion, that tends to offset this uncertainty by increasing still further the wear per unit distance upon the heavier grains. This factor, which was clearly pointed out by Barrell,[22] is the abrasion of the slowly-moving heavy grains, not by their own movement, but by the continuous blast of rapidly-moving lighter grains always being swept past them. This process of wet-blasting appears to be of more than merely theoretical importance; if properly interpreted,[23] it accounts for the rapid disappearance of the quantities of pebbles and boulders being dumped into the Mississippi River by steep tributaries and caving banks of Pleistocene gravel.

The effect of this wet-blasting of the heavier, lagging grains thus augments the effect of the ordinary abrasion due to the particles' own movement. With continued abrasion, the number of small grains increases—partly because the larger grains are reduced in size and partly because small particles are broken off from the larger ones. And, inasmuch as abrasion of whatever type works more effectively upon the heavier and the softer minerals, it tends to increase the number of small magnetite grains much more than the number of small tourmaline grains.

This increased number of small mag-

[22] Barrell, *op. cit.*, pp. 330-331, 335-336, 338.

[23] Rubey, W. W., Geology of the Hardin and Brussels quadrangles, Illinois: *Illinois Geol. Survey Bull.* (Awaiting publication.)

netite grains, like the greater settling velocity of the magnetite grains, would clearly result in a greater concentration of magnetite in the finer grained portions of each separate deposit or sample of the sand. But, in one important respect, this increase of small magnetite grains would give a result quite different from that caused by settling velocities alone. Even though the number of magnetite and tourmaline grains of any given size may have been exactly equal at the source, continued abrasion might so alter the original ratio that, among the grains being swept along by the water and therefore available for deposition at favorable places, magnetite had become ten times as abundant as tourmaline in the small sizes and only one-tenth as abundant as tourmaline in the large sizes. That is to say, the small fragments resulting from abrasion would not necessarily or even probably be deposited along with the larger particles from which they were broken off; they would in large part be swept away by the currents at the first site of deposition and eventually dropped elsewhere. Thus the proportion of different heavy mineral grains would vary from one deposit to another even within the same stream. The heavier and softer minerals, having lost most by the abrasion, would be rare in the coarse deposits and abundant in the fine deposits.

This marked difference between the effects of settling and abrasion may be emphasized by stating the relationship in another way. Both processes tend to concentrate the heavy minerals in the finer grained portions of a given deposit. However, sorting, considered by itself, merely acts upon the mass of debris available to it in the load being transported. With a uniform distribution of sizes in the load, sorting causes the heavy minerals deposited in the finer fractions of a coarse sample to exceed the heavy minerals deposited in the coarser fractions of a fine sample; and, therefore, with standard methods of comparison, the coarsest grained samples appear to contain the most heavy minerals. The distance of transportation and the hardness of the different minerals play no part in their size-distribution as effected by sorting alone. The effects of abrasion, on the other hand, depend largely upon the distance of transportation and the hardness of the different minerals, and they tend to alter the proportion of different minerals in the various grain sizes in such a way as to oppose and eventually to overweigh the effect of sorting. Thus, as the result of continued abrasion, the heavy minerals deposited in the predominantly finer grained samples come to exceed the heavy minerals deposited in the predominantly coarser grained samples; and, whatever the basis of comparison of heavily abraded sandstones, the finer grained samples contain more of the heavier and the softer minerals.

THE EFFECT OF THE SIZE-DISTRIBUTION OF DIFFERENT MINERALS IN THE SOURCE ROCK

The question now very properly arises of how essential to our conclusions is the premise of a source rock of homogeneous composition, that is, one in which the relative abundance of quartz, magnetite, and tourmaline is approximately the same in each grain size. This assumption is obviously an artificial one chosen merely for its convenience in computation. If large magnetite grains were rare, that is, if the $\frac{\text{quartz}}{\text{magnetite}}$ ratio, instead of remaining 100 to 1 throughout the range of sizes, became much greater among the larger particles, the distribution of the two minerals would be very different among these coarser fractions of all samples. Still, if this upper limit of homogeneous size-distribution were 0.354 mm. or larger, the relative abundance of different minerals among the grains less than 0.354 mm. in diameter shown in Tables 2, 3, 4, 5, and 6 would remain unchanged. If, however, these departures from homogeneity had also affected the smaller grains, the proportions of the different minerals in various size fractions would have been quite different. The full effects of any departures from homogeneity can be calculated only if the complete size-distribution of each mineral in the source rocks is known.

The writer is aware of no common, or at any rate no generally recognized, relationships of crystal sizes that result in marked differences in the grain size of the various mineral species present in igneous and metamorphic rocks. Perhaps such relationships actually exist, for the normal order of crystallization in magmas and the selective recrystallization of minerals during metamorphism record differences of chemical composition, solubility, and specific volume which are, in turn, undoubtedly related to differences of specific gravity and hardness. But the net effect of these differences appears to be too complex to serve as a basis for predicting the dominant sizes of the crystals of different minerals. Consequently, among at least the smaller mineral grains that make up igneous and metamorphic rocks, the assumption of a homogeneous size-distribution now appears justifiable, because any irregularities seem about equally probable in either direction and hence quite as likely to increase as to decrease the variations of heavy mineral proportions here discussed.

However, in source rocks made up of glacial deposits and of other sedimentary rocks, the size-distribution is clearly subject to influences that make this assumption err systematically in certain directions. During prolonged glacial motion, particles of the softer minerals probably suffer most from abrasion, so that in glacial tills the softer minerals may tend to be most abundant among the fine grains and the harder minerals most abundant among the coarse grains. Likewise, if the source were a sedimentary rock such as a water-laid sandstone, its constituent particles would have been subjected previously to sorting and to more or less abrasion. As we have seen, both sorting and abrasion tend to concentrate the heavier minerals in the finer grained portions of the deposits, and the latter tends also to concentrate the softer minerals in these finer grained portions. Consequently, the initial effect of a sedimentary source rock is likely to be very similar to that of long continued transportation before deposition—the proportion of different minerals being carried by the stream and thus available for deposition would vary with the grain size. Any additional sorting or abrasion that precedes the second deposition only intensifies these initial variations in the size-distribution of the different minerals. It is perhaps of some interest to note in passing that this interpretation of the effect of a sedimentary source rock or of long continued abrasion upon the size-distribution of heavy minerals in a deposit offers a possible explanation of the occurrence of "flour" gold in very fine-grained deposits as contrasted to the more normal types of placers in coarse-grained deposits.

The precise effect of irregularities of the size-distribution of different minerals in the source rock, like that of abrasion during transportation, obviously depends upon different circumstances. Yet a rough approximation to the net effect of these two factors in ordinary deposits can perhaps be made by comparing the distribution of magnetite and quartz grains in some natural water-laid sands. Data for such a comparison are available in seven samples of magnetite-bearing quartz sand mechanically analyzed by Udden.[24] This comparison can be made most conveniently if Udden's data are replotted as cumulative[25] curves. Of the seven samples, two from the beach of Matagorda Bay, Olivia, Texas, in which all grains are less than ¼ mm. in diameter, follow closely the size-distribution of the two minerals expected from a homogeneous source rock and from Stokes' law of settling velocities. The distribution curves of the quartz and magnetite grains in these two samples have very similar shapes and are displaced so that the lower and upper size limits and the average diameter of the magnetite grains are approximately 63/100 as large as the equivalent sizes of the associated quartz grains. However, in the other five samples (one beach sand and four stream sands, in which essentially all grains are between ⅛ and 8 mm. in diameter) the distribution curves of the quartz and the magnetite grains are very different. In all five, the magnetite grains are much more perfectly sorted than the quartz grains and the curves are so displaced that the diameter ratios between approximately equivalent sizes of the magnetite and quartz grains decrease from about 0.85 in the finest sizes to about 0.15 in the largest sizes. This decrease of diameter ratios with increase of grain size is approximately what would be expected if Stokes' law gradually gives way to the square-root law between the silt and gravel sizes (pp. 17-19). But it is also precisely what would be expected as the result of derivation from a previously sorted sedimentary source rock or of considerable abrasion during transportation. Without much more information about each sample, it is impossible to say which of these possible causes has been most influential. For the present purpose, it is sufficient to note that the factors we have been discussing actually cause large variations in the size-distribution

[24] *Op. cit.*, pp. 727, 734-735.

[25] Wentworth, C. K., Methods of mechanical analysis of sediments: *Univ. Iowa Studies*, vol. 11, pp. 46-51, 1926; or, Rubey, W. W., Lithologic studies of fine-grained Upper Cretaceous sedimentary rocks of the Black Hills region: *U. S. Geol. Survey Prof. Paper* 165, p. 22, 1930.

of different minerals and that these variations are commonly such that they increase the contrast of heavy mineral proportions in coarse and fine samples.

THE EFFECT OF OTHER FACTORS

Certain other assumptions involved in the discussions of our theoretical sandstones need to be re-examined before any final conclusions can be drawn.

For one thing, the analysis has been restricted to water-laid sandstone. Other agents of deposition and transportation, notably the wind, would give quite different distributions of the heavy minerals. In fact, the very low density of air would lead us to expect that sorting by the wind would not cause nearly so pronounced a concentration of heavy minerals in the finer grain sizes as does sorting by water. Also, no effort has been made in this paper, beyond that involved indirectly in the choice of certain types of sorting for comparison, to contrast the effects of fluvial transportation and deposition with those of littoral transportation and deposition. Nevertheless, this neglect of the distinctly littoral processes, such as the zone of confused wave motion along beaches, appears to call for no significant qualifications of the foregoing results, because the laws of sorting and abrasion under these conditions are, after all, probably very similar to those under stream conditions.

However, the neglect of any consideration of weathering at the source and during transportation may be an important source of error. Decomposition would decrease the effective hardness and effective density of mineral grains; and, consequently, minerals like magnetite, which are relatively susceptible to weathering, would be more quickly ground down to finer sizes and eventually eliminated.

Then again, differences in the shape of different mineral grains have been ignored. It is true that, because of differences in the readiness of cleavage and the resistance to abrasion, the detrital grains of different minerals tend to have characteristic shapes. Nevertheless, so long as the comparisons do not include such micaceous minerals as biotite and muscovite, the neglect of shape differences is believed to introduce no very serious errors into the results.

One other assumption scarcely needs any special discussion. Had there been thirty different mineral species in the source rock instead of only three, the principles of sorting that govern their distribution in the deposits of sandstone would have been altered in no way. However, the rate of abrasion of each different mineral might have been affected somewhat, for this depends in part upon the dominant hardness of the other mineral grains being transported at the same time.[26]

Finally, the samples chosen for comparison range from fine-grained to coarse-grained sandstones. Other things being equal, the variation of the

$\frac{\text{magnetite}}{\text{tourmaline}}$ ratio among grains of a certain size fraction in different samples would increase with the contrast in average grain size of the samples. Had we compared only moderately

[26] Barrell, *op. cit.*, p. 336.

coarse and moderately fine sand, the variation of the heavy mineral ratios would have been less; but had our samples ranged from gravel to silt, as do the materials of the Wisconsin and Illinoian stream deposits, the study of which led to the preparation of this paper, the variation would have been much greater.

CONCLUSIONS AND RECOMMENDATIONS

Apparently the assumptions made in discussing these theoretical sand samples have, in the aggregate, tended to minimize rather than to exaggerate the effects of the various factors that influence the proportions of different heavy minerals in sandstones. Many of the assumptions seem to demand no important qualification of the results either one way or the other because probable departures appear about equally likely to increase or to decrease the contrasts that were found. On the other hand, several of the assumptions seem clearly to err on the side of distinctly minimizing the heavy mineral contrasts that might be expected. Hence, the conclusion appears justified that, even in sandstones derived from identical source rocks, the heavy mineral assemblages taken from certain size fractions of the different samples may—either apparently or actually—depend largely upon the grain size or degree of sorting of the different samples. Stated another way, the relative abundance of two minerals in any particular size fraction of different sand samples depends upon a great many factors, only one of which is an actual difference of source rock. In general, the contrast varies with the difference in the density and hardness of the two minerals, with the difference in degree of concentration of the minerals in certain sizes at the source, with the amount of abrasion during transportation, with the degree of sorting at the site of deposition, with differences in the average grain size of the samples compared, and (because of the different law of settling velocities of the larger grains) with the absolute grain size of the samples.

If this general conclusion is accepted, it follows that, in making comparisons of heavy mineral assemblages from different sandstones for purposes of correlation or of information about the character of source rocks, the probable dependence of the assemblages upon these various factors should be borne in mind and, if possible, allowed for.

Both theoretically and practically, the most satisfactory method of allowing for these various factors would be to restrict all comparisons to the heavy minerals taken from comparable portions of sandstones that have essentially the same grain size and the same degree of sorting. This precaution would eliminate all the variations discussed in this paper, except those due to differences in the amount of abrasion. This single exception might, of course, be an important one—two sandstones might be derived from the same source rock and be exactly alike in grain size and degree of sorting yet, if one had been transported far from its source and the other only a short distance, the two could have very different proportions of heavy minerals. However, the effects of transportation and abrasion cannot at present be estimated quantitatively; and any allowances made for these effects must rest upon regional paleogeographic considerations rather than upon the textural features of the samples. For many types of studies, perhaps this comparison of samples of the same grain size and same degree of sorting will be a sufficient precaution; but, with samples in which the grain size or the degree of sorting differs perceptibly, some other method of comparison must be sought.

If sorting were the only factor that affected the size-distribution of different minerals, it would clearly be advantageous to make the comparisons of the entire heavy mineral content of each sample. This separation of the entire heavy mineral content might be made from every one of the several size fractions into which each sample can be divided or, more directly, from each original unseparated sample itself. The first of these two methods would undoubtedly give the most complete information about the size-distribution of the heavy minerals in the samples, and for that reason it is to be preferred wherever conditions permit; but this method is subject to two practical disadvantages which decidedly limit its field of applicability. First, it is time-consuming, because it involves a great multiplication of laboratory work. Second, the use of the petrographic microscope for mineral identification sets a very effectual upper limit to the size of grains that can conveniently be studied after separation. Grains more than about ½ mm. in diameter call for a materially different technique of examination. Consequently, for all but the very finest grained sandstones, it is impracticable to make routine determinations of the heavy minerals in the coarsest grained size fractions of all samples. Separate determination[27] of the heavy minerals in each of the grain sizes less than some particular diameter still remains quite useful for certain purposes but, because of the practical difficulties involved, the determination of all the heavy minerals in each size fraction cannot readily be made in the study of a large number of samples.

The other possible method of comparing the entire heavy mineral content of each sample by making one heavy mineral separation from the entire sample encounters one of these same practical difficulties. With all but the very finest grained sandstones, this method yields grains too large to be examined conveniently under the microscope. It is also open to another and perhaps more fundamental objection in that it gives no data by which corrections can be applied for the possible effects of sorting, abrasion, or irregularities in the size-distribution of different minerals at the source.

The impracticability of comparing the entire heavy mineral content of each sample has naturally led sedimentary petrologists to the next step of comparing the heavy mineral grains from that portion of each sample which passes some particular sieve

[27] Trowbridge, A. C., and Mortimer, M. E., Correlation of oil sands by sedimentary analysis: *Econ. Geol.*, vol. 20, pp. 414-415, 419, 1925.

Graham, W. A. P., A textural and petrographic study of the Cambrian sandstones of Minnesota: *Jour. Geol.*, vol. 38, pp. 710-711, 1930.

mesh, so that, let us say, all heavy mineral grains with diameters less than 1/3 mm. in each sample are examined. This method has the great advantage of mechanical simplicity and speed, but it is open to serious objection if the samples to be compared vary appreciably in their grain size or their degree of sorting. As may be seen in Figure 3 and in Tables 3 and 5, the portions of different samples that will pass some arbitrary sieve mesh, such as 0.354 mm., may not be at all comparable in their heavy mineral content. The fraction so separated from a coarse-grained sandstone is only a very small and unrepresentative portion of the total sample. Figure 3 illustrates the disadvantages of this method where sorting is the only factor and where the proportions of the different heavy minerals depend solely upon the relative position of a size fraction within the general size-distribution curve of the sample. But it may also be clear from previous discussions in this paper that the heavy mineral separates obtained by this method are equally unsatisfactory for purposes of detailed comparison where abrasion and irregularities of size-distribution at the source are the controlling factors. Under these conditions, the relative abundance of the different heavy minerals vary not with the relative size but with the actual size of the grains. Hence, the separation of *all* heavy mineral grains less than some particular diameter tends to emphasize the variations due to abrasion and irregularities of size-distribution at the source, for the reason that the heavy minerals thus separated from the finer grained samples may include minerals concentrated only in the very finest sizes and so absent from the medium and coarser grained samples.

The practical difficulties that stand in the way of making routine comparisons of all the heavy mineral grains in each sample and the fundamental objections to comparing all "heavies" less than some particular size in each sample suggest that the best basis for comparison may be the heavy mineral crops separated from some one size fraction of each sample. This general method commends itself to many sedimentary petrologists, because it gives a maximum amount of information with a minimum expenditure of time.

Yet, even granting the convenience of comparing the heavy minerals of only one size fraction from each sample, there still remains the question of deciding what particular size fraction from each sample affords the best basis for comparison. Common practice apparently is to compare fractions of the same *actual* grain size from each sample, regardless of the average coarseness of the different samples. In two respects, this method of comparison is excellent: first, because it tends to eliminate the variations caused by abrasion and by irregular size-distribution in the source rock, and, second, because the color, pleochroism, and birefringence of the different grains are more nearly comparable when all grains under the microscope are of approximately the same size. However, this method, taken alone, is poorly adapted to discriminative comparisons of water-laid sandstones because, as we have seen (Tables 2, 4, and 6), it tends to emphasize those very differences in heavy mineral proportions that are caused by sorting in water.

An alternative method would be to compare the heavy minerals from fractions that represent the same *relative* position within the size-distribution curve of each sample. That is, the comparisons may be made of fractions with a size, say, midway between the minimum and average diameters of each sample (for example, the fractions marked X in each of the three samples illustrated in Fig. 3). Among samples that have approximately the same degree of sorting, such comparisons would eliminate essentially all the variations of heavy mineral proportions that are caused by different settling velocities.

Still, even this method of comparing fractions with the same *relative* grain size is no cure-all. Its merits are least precisely where comparisons of fractions of the same *actual* grain size are greatest, because it tends to emphasize those variations that are caused by abrasion and by irregularities in the size-distribution of the different minerals at the source. Thus, neither of these two methods, used alone, gives all the information needed for careful comparisons. Fortunately, however, the two methods supplement one another very well and, used together, they indicate closely and with a minimum expenditure of time the actual size-distribution of the various heavy minerals in a group of samples.

Summarizing then, it appears that no one particular size fraction from a series of sand samples with different degrees of coarseness and sorting is best suited for detailed comparisons of the heavy mineral content. In order to make allowances for the variations in heavy mineral proportions due to sorting, abrasion, and irregularities at the source, the heavy minerals from at least two size fractions of each sample should be examined. These two fractions should be so chosen that one of them has the same *actual* size limits in all of the samples compared and the other represents the same *relative* size within the distribution curves of the different samples.

[*Editor's Note:* In footnote 3 on page 4 of the article and in footnote 23 on page 21 of the article the complete reference is: Rubey, W. W., 1952, Geology and Mineral Resources of the Hardin and Brussels Quadrangles, Illinois, *U.S. Geol. Survey Prof. Paper 218,* 172p.]

6

Reprinted from pages 1725–1730 and 1739–1780 of *Geol. Soc. America Bull.* **54:**1725–1780 (1943), by permission of the Geological Society of America

TRANSPORTATION AND DEPOSITION OF HEAVY MINERALS

BY GORDON RITTENHOUSE

CONTENTS

ILLUSTRATIONS

ABSTRACT

An attempt to use heavy minerals to evaluate sediment sources in the Middle Rio Grande Valley showed the inadequacy of commonly used methods of representing heavy-mineral compositions. To develop a usable method, six samples from a transverse channel section were studied by quantitative methods. The size-distribution curves (by weight) of 12 heavy minerals were similar in general form to those of the light minerals, but displaced toward the finer sizes. Among samples the heavy-mineral composition varied systematically with differences in average size and sorting of the light minerals.

The size distribution of a heavy mineral in a fluvial deposit appears to depend on (1) its relative availability in each size grade in the stream load, (2) its equivalent hydraulic size, (3) the hydraulic conditions at the time and place of deposition, and (4) some factor or factors now unknown. In the Rio Grande, the equivalent hydraulic sizes of 16 heavy minerals vary from 1.0ϕ to .2ϕ.

The hydraulic ratio is recommended as a genetically sound method of representing the heavy-mineral composition of sediments. This ratio is the same in the stream load and bed deposits. Changes in the hydraulic ratio caused by wear, selective transportation, and other factors are discussed.

Other methods of representing heavy-mineral compositions are discussed and their limitations outlined. All methods are of limited applicability, but ratios between heavies of about the same specific gravity may be used most extensively. If the limitations are not observed, invalid conclusions may be reached.

INTRODUCTION

Although heavy minerals have been used extensively in attempts to solve a variety of geologic problems, comparatively little is known about the principles or laws that govern their occurrence in sedimentary deposits. Most papers about heavy minerals have been primarily descriptive in character, the particular species present and some data on their relative abundance by number being recorded. In a few investigations, notably those by Pettijohn (1931), Pettijohn and Ridge (1933), Pettijohn and Lundahl (1943), Russell (1937), and Smithson (1939), quantitative data have been obtained on suites of related samples. Some principles of heavy-mineral transportation and deposition have been inferred from the progressive change or lack of progressive change in these series. Perhaps the best paper yet written on the transportation and deposition of heavy minerals, however, was that by Rubey (1933b), who made a theoretical analysis of the relative importance of factors that might influence the size distribution of heavies in water-laid deposits. In contrast to other workers, Rubey used weight, rather than number, as the basis of his reasoning.

A feature common to previous studies has been the adoption of some arbitrary, empirical basis to compare the heavy-mineral composition in the samples examined.

The presence or absence of "diagnostic" mineral species, the relative abundance by number of the mineral species in the heavy-mineral fraction (expressed either as rough scales of abundance or as mutual percentages), and ratios between the number frequencies of selected heavy minerals have been used. Such descriptive values are determined for the whole sample, or for one or more selected size grades of the sample. In a very few studies, methods of comparison were selected after rather critical survey of the then existing data on the occurrence of heavy minerals in sedimentary deposits.

The validity of conclusions reached through the use of such empirical methods is entirely dependent on the validity of the stated or unstated assumptions underlying the conclusions. Basically, these assumptions concern the principles or laws that govern the occurrence of heavy minerals in sedimentary rocks. Consequently, tests of validity, and even recognition of some assumptions yet unknown, depend on a better understanding of fundamental laws governing heavy-mineral occurrence. Unfortunately, the body of comprehensive data needed to provide such understanding has not been available. Accumulation and interpretation of such comprehensive data is, in the writer's opinion, the most important problem now facing sedimentary petrologists who are investigating heavy minerals.

For some time the writer has been engaged in an extensive study of heavy minerals in the Rio Grande and its tributaries in the Middle Valley in New Mexico. In this area, the problems of arroyo trenching and accelerated soil erosion in relation to channel aggradation and other types of injurious sedimentation have been under investigation by the Soil Conservation Service and other governmental and private agencies. It has been found that sediment accumulates in the Middle Valley at a rate of about 12,000 acre-feet annually, that accumulation is more rapid in the channel and floodway than on the flood plain, and that about 75 per cent of the sediment accumulates in the 33 miles of valley below San Acacia, New Mexico. Such sediment deposits, accumulating over a period of years, have caused much damage. Channel and floodway capacity for discharge of flood water has been reduced ground-water tables have been raised, drainage has been impaired, and infertile sand has been deposited on pastures and cultivated lands.

One critical problem in the comprehensive study was the relative importance of different parts of the drainage area as sources of injurious sediment—mainly medium and fine sand. Without reliable data on sediment sources, it would be impossible to plan a sound, economical program of remedial measures. Because methods then in use were very costly, and because sufficiently accurate results would not have been available for some years, it was decided to try to determine the relative importance of sand sources by comparing the heavy-mineral content of tributary streams with that in the main stream above and below the tributary junctions.

After a series of samples had been collected, and the heavy-mineral content determined, it was apparent that heavy minerals could be used for this purpose. But it was also apparent that the empirical methods then in use for representing and comparing the heavy-mineral content of river sands were not suitable for evaluating sediment sources. (*See* section on Hydraulic Ratios.) A new method of representing mineral composition, based on the principles of sediment transportation and de-

position, was needed. The primary purpose of the investigation reported in this article was to develop such a method.[1]

Many factors control the occurrence of light and heavy minerals in river deposits, and their interrelationships are complex. The mineral composition of a stream is dependent on the kind, the amount, and the size of minerals in the source rocks; on the way in which the source rocks are disintegrated; on the absolute and relative rates of mechanical wear and chemical disintegration during transport; on the absolute and relative rates of transportation of different minerals and different sizes of the same mineral; and on the hydraulic conditions that existed at the place and during the time of deposition. At a transverse cross section of a river, however, all these factors except the last will be constants insofar as differences in mineral composition at different places in the cross section may be attributed to them. Their net effect will be reflected in the physical characteristics, availability, and kinds of minerals in the stream load at the cross section. Therefore, these factors and their complex interrelationships need not be considered in explaining why the mineral composition of bed deposits varies from place to place in a cross section. If the factors that do cause such variation can be recognized and their relative importance established, it should be possible to develop a genetically sound method of representing heavy-mineral compositions—a method that would define the mineral composition of the deposits regardless of their texture. The solution of other phases of the broad problem would also be greatly simplified.

This reasoning has been the basis of the investigation reported in this paper. A series of samples were collected from a transverse section of the Rio Grande near Bosque, New Mexico. These samples were sieved, and quantitative number and weight data were obtained for the heavy minerals in seven $\sqrt{2}$ size grades of each of six samples. The heavy minerals were found to have size distributions (by weight) similar in general to those of the light minerals in the same samples but were displaced toward the finer sizes. There seemed little doubt that the size distributions of heavy minerals vary systematically with differences in average size and degree of sorting of the deposits in which they were found.

To determine what factors might control the accumulation of light and heavy minerals at a transverse section, a theoretical analysis was made. The data from the six samples were then analyzed to find whether the actual mineral occurrence could be explained by these factors. It was not possible, in making the assumptions and analyzing the data, to work with each mineral individually because the hydraulic conditions that existed during the accumulation of the sediment and the absolute availability of various sizes of each mineral in the stream load were not known. It seemed probable, however, that under given hydraulic conditions grains of heavy minerals would, because of their greater specific gravity, tend to be deposited with quartz and other light minerals of somewhat larger size. If this were true, avail-

[1] The "hydraulic ratio" method, described in this article, has been used successfully in the Middle Valley and will be widely applicable to other areas where the location of sediment sources is of economic or scientific importance. In the Middle Valley below San Acacia, for example, it was found that the Rio Puerco, which drains about 21 per cent of the area, contributes about 50 per cent of the sediment, and that the Rio Salado, which drains only 6 per cent of the area, contributes about 20 per cent of the sediment. The importance of such data in planning a sediment-control program is obvious. The results of the source investigations will be published elsewhere (Rittenhouse, 1943b; 1943c).

ability could be expressed as relative availability of each heavy mineral with respect to the light minerals that would be deposited with it—*i.e.*, that would be of the same "hydraulic equivalent size." The need for absolute data on hydraulic conditions was also eliminated, because the hydraulic conditions, whatever they may have been, would have been the same for hydraulically equivalent heavy and light minerals that accumulated together.

The experimental data indicate that the distribution of heavy minerals in river deposits is largely dependent on (1) the hydraulic conditions at the place and during the time of deposition, (2) the relative availability of various sizes of each heavy mineral in the stream load, and (3) the hydraulic equivalent size of the heavies. Some additional factor or factors now unknown also affect the occurrence of heavy minerals of high density—*i.e.*, magnetite, ilmenite, and zircon. Heavies of low density are also affected but to a lesser extent.

Because, for low density minerals at least, the relative availability is constant at a transverse section, it may be used as a basis for representing and comparing heavy-mineral compositions. Since it is a ratio between amounts of hydraulically equivalent grains, it may be termed the "hydraulic ratio." For representing mineral compositions, it may be defined as "100 times the weight of a heavy mineral in a known range of sizes, divided by the weight of light minerals of hydraulic equivalent size". Theoretically, this ratio is the same in the stream load and in the bed deposits, and consequently is independent of average size and degree of sorting in the deposits, factors that limit the use of most empirical methods. Because the hydraulic ratio is based on both the number and weight of heavy minerals, it can be used to determine the quantitative importance of sand source areas.

The extent to which the hydraulic ratio will be changed by such factors as wear, weathering, and selective transportation is discussed. Enough data are available to show that the hydraulic ratio will be extremely valuable in helping to isolate and evaluate the other factors that determine the distribution of heavy minerals in stream deposits.

The Bosque data and the principles of heavy-mineral transportation and deposition derived therefrom have been used to outline, for various types of problems, the limitations of commonly used methods of representing heavy compositions. Although the applicability of all commonly used methods appears to be limited, ratios between minerals of about the same specific gravity may be used most widely. For quantitative sediment source evaluations for which both number and weight of heavies must be used, percentages by weight can be employed to a limited extent. Insofar as possible, the conclusions have been verified by applying statistical tests of significance to the Bosque data. An understanding of these limits of applicability is essential for petroleum and other economic geologists, stratigraphers, sedimentary petrologists, and others who may use heavy minerals.

ACKNOWLEDGMENTS

The writer is greatly indebted to Dr. H. A. Einstein, Dr. E. M. Thorp, and other members of the Research Division, Soil Conservation Service, and to Dr. F. J. Pettijohn, University of Chicago, for comments, helpful suggestions, and criticisms. **Part of the computing was done by assistants furnished by the National Youth Administration.**

[*Editor's Note:* Material including Tables 1 through 4 and Figure 1 has been omitted at this point.]

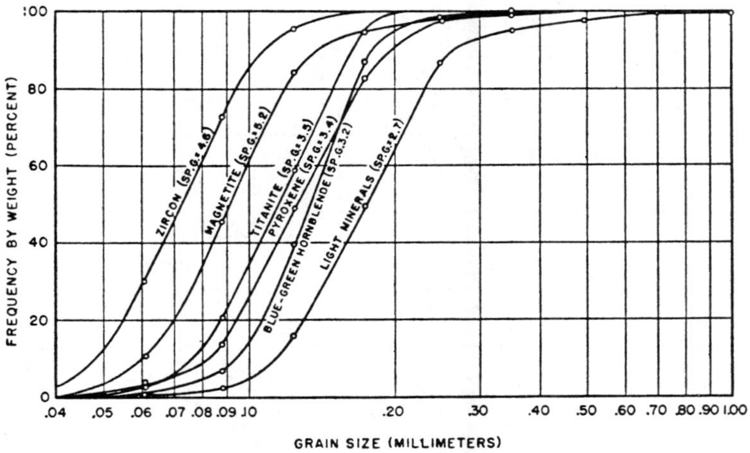

FIGURE 2.—*Size distribution by weight of light and selected heavy minerals*

Sample A

HEAVY-MINERAL SIZE DISTRIBUTIONS

SAME SAMPLE

The size distribution by weight, as determined for 12 heavy minerals in the 6 samples, is presented in Table 4. To illustrate differences in size distribution of various minerals within a single sample, the cumulative curves for the light minerals, and for blue-green hornblende, pyroxene, titanite, magnetite, and zircon in sample A have been plotted in Figure 2.

Several facts are apparent: (1) The size distributions of the heavy minerals form smooth cumulative curves of the same general shape as the cumulative curve of the light minerals; (2) all heavy minerals are finer in texture than the light minerals; (3) in general, increasing specific gravity is accompanied by increasing fineness of the heavy minerals; (4) zircon (sp. gr. = 4.6) is much finer than magnetite (sp. gr. = 5.2); and (5) difference in slope of the curves indicates that heavy minerals in the same sample have different sorting.

DIFFERENT SAMPLES

The size distribution of magnetite, ilmenite, blue-green hornblende, zircon, pyroxene, and all light minerals considered as a group has been plotted for samples A, D, F, and G in Figure 3. The heavy minerals of these four samples vary both in average size of grain and sorting in the four samples, and, in general, these differences are remarkably similar to those exhibited by the light minerals. Sample G,

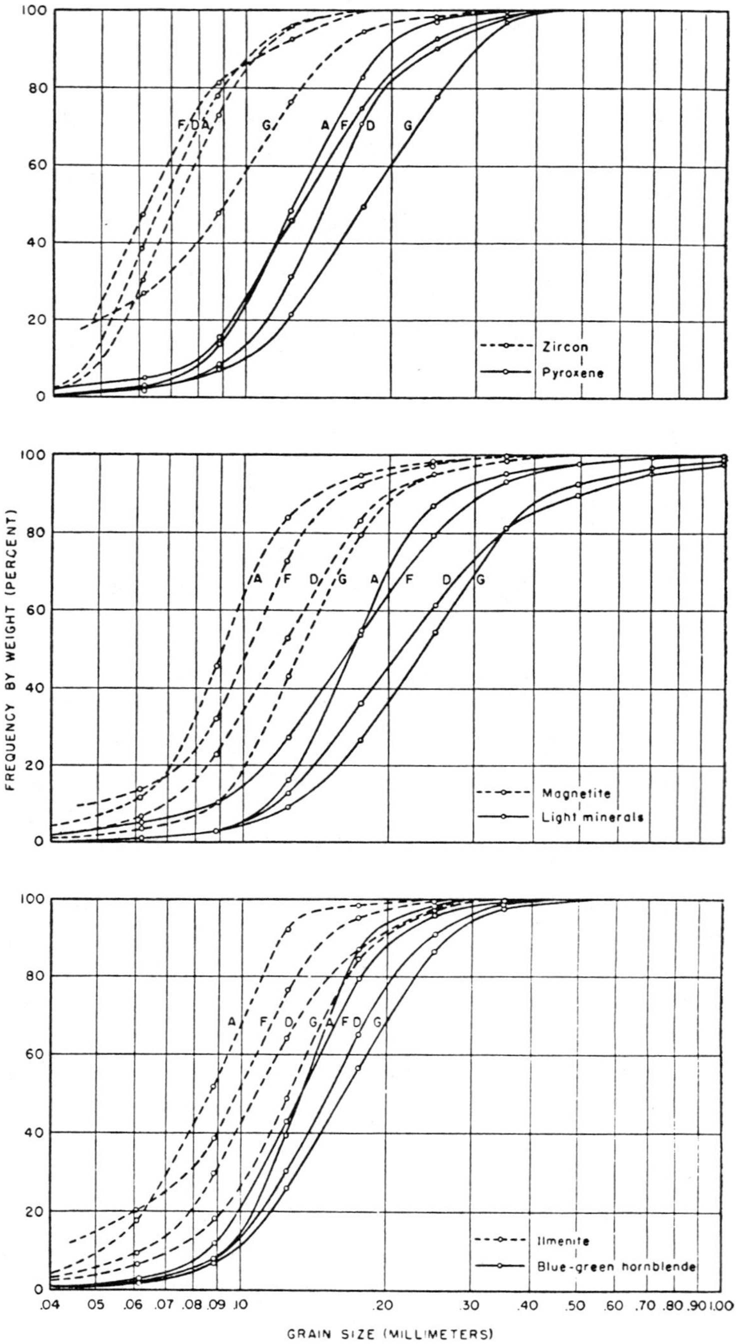

FIGURE 3.—*Size distribution by weight of light and selected heavy minerals*

Samples A, D, F, and G

which has the coarsest light minerals, also has the coarsest heavy minerals. Samples A and F, which have the finest light minerals, also have the finest heavies.

Differences in sorting of the light minerals, as exhibited by samples A and F, are accompanied by consistent differences between the size distributions of the various heavy minerals in these two samples. The heavy minerals of sample A, the better sorted of the two samples, are finer than the heavies in sample F. This difference in fineness is greatest for heavies of high specific gravity.

The similarity in relative position of the light and various heavy minerals in the four samples leaves little doubt that the heavy-mineral composition in these sand deposits not only varies with differences in average size and degree of sorting, but that the variations are systematic and should be relatable to some basic law or laws of sediment transportation and deposition. What factors may cause differences in size distribution within a single sample and systematic variations between samples?

FACTORS GOVERNING HEAVY-MINERAL OCCURRENCE

HYDRAULIC EQUIVALENT SIZES IN THE RIO GRANDE

For 50 miles or more above Bosque, New Mexico, no large tributaries contribute sediment directly to the Rio Grande channel. Consequently, as sand moves downstream through this reach of the river, there should be ample opportunity for thorough mixing of the sands brought to the Rio Grande by the tributaries above Albuquerque, New Mexico. The mineral composition of the sediment available for deposition at Bosque, therefore, should be essentially the same at all points across the river. If this is true, then the size distribution of the lights and heavies in the Bosque samples is the result of (1) the difference in hydraulic conditions in different parts of the transverse section at the time of deposition, and (2) the inherent character of the sand available for deposition. At each sampling point, some particular set of hydraulic conditions must have existed during the deposition of the sediment found there. Therefore, the difference in size distribution between minerals in each sample must be attributed to differences in inherent character of the minerals that are available for deposition. At different points in the cross section, the variation would result from both factors.

Sediment particles have a wide range of size, of specific gravity, of shape, and of other physical and chemical properties that affect their behavior under particular hydraulic conditions. Other factors including size being constant, two grains differing in density would not be picked up with the same frequency or bounced and rolled at the same speed. There would, however, be particles of different sizes that would settle or be transported along the bottom at the same rate. Differences in shape or other physical properties naturally would modify this simple postulated relationship between density and size.

This leads to the concept of hydraulic equivalence—namely, that whatever the hydraulic conditions may be that permit the deposition of a grain of particular physical properties, these conditions will also permit deposition of other grains of equivalent hydraulic value.[4] There is no specification in this general statement

[4] The concept of hydraulic equivalence is not original with the writer, having been stated previously in slightly different form by Rubey (1933b) and others.

as to what sizes, shapes, etc., of grains deposited together may be, or that the sizes, shapes, etc., that will be deposited together from suspension will be the same as those deposited from bed load. This concept has a great practical advantage in that the net effect of differences in physical characteristics of various kinds and sizes of grains may be evaluated, without requiring prior evaluation of size, shape, etc., which together determine the hydraulic equivalence for particular grains. The various subfactors can then be evaluated separately, if desired.

Because it is nearly always present and is commonly the most abundant mineral in stream sand, quartz would appear to be the best base to which the hydraulic equivalence of other minerals could be referred. Where quartz is associated with other light minerals of essentially the same specific gravity, however, it is much easier to use the light minerals as a group as the reference basis. This has been done in this investigation. Because sieves commonly have been used and are convenient for sizing sands, the hydraulic equivalence of heavies will be referred to that of the light minerals in terms of size as determined by sieving—*i.e.*, the light and heavy minerals will be of the same "equivalent hydraulic size".

At a given place and time in a stream, not only will there be inherent differences in physical properties between particles of different kinds, but there may also be differences in the availability for deposit of different sizes of each mineral. In addition, at different times or at different places in the stream, the *absolute* availability of any size of grains will vary because of differing hydraulic conditions. Thus, during high flows more and coarser sand will be transported than during low flows. Under varying hydraulic conditions, however, the *relative* availability of grains of any given equivalent hydraulic size should remain the same. If more light minerals are transported as the stream discharge increases, more heavies of hydraulic equivalent size will be transported; if coarser light minerals are transported, coarser heavies should also be transported. If this were not true, then the heavies and lights would not be of the same equivalent hydraulic size.

Similarly, when heavy minerals are deposited, light minerals of equivalent hydraulic size will be deposited in the proportion in which they are available in the stream load. Consequently, the relative availability should be a constant that is common to the stream load and to all deposits from that stream load, regardless of the differences in absolute amounts of the minerals of the same and other sizes that are deposited under the particular hydraulic conditions existing at different places of deposit.

The great difference in heavy-mineral size distribution that may be caused by relative availability is illustrated in Figure 4. The actual size distribution of light minerals in sample A has been used as a base to compute the size distribution of a heavy mineral that is equivalent in hydraulic size to light minerals one $\sqrt{2}$ size grade larger. Curve X assumes equal relative availability in all size grades; curve Y assumes relative availability increasing geometrically toward the finer sizes.

Relative availability may be expressed as a ratio between the amounts of two minerals by number, volume, or weight. The relation by weight, using light minerals throughout as one of the minerals, is easiest to obtain. In consequence, this hydraulic, relative-availability ratio, termed the "hydraulic ratio" for convenience, is defined in terms of weight. Because the weight of light minerals is great compared to that of most heavies, the ratio is multiplied by 100 to arbitrarily reduce the number of decimal places. Specifically, then, the hydraulic ratio of a mineral is "100 times the weight of that mineral in a known range of sizes, divided by the weight of light minerals of equivalent hydraulic size."

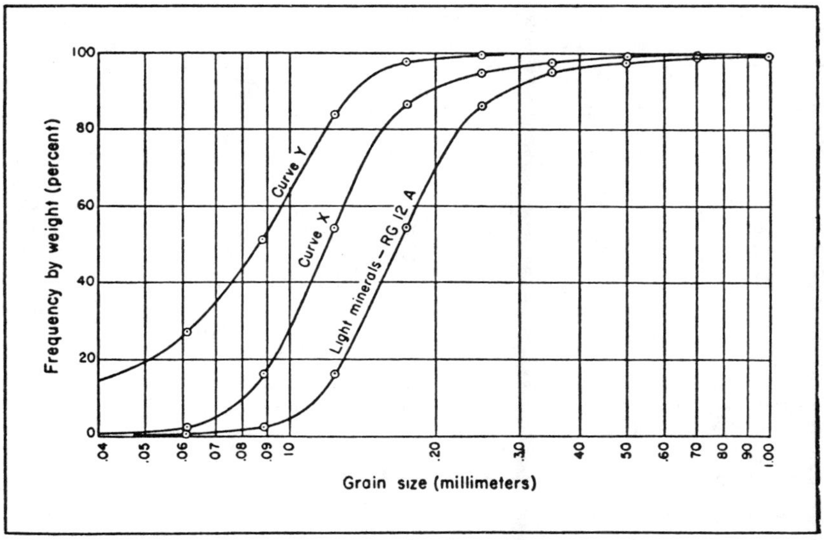

FIGURE 4.—*Hypothetical size distributions of magnetite with difference in relative availability*

Curve *X* assumes equal availability in all size grades, curve *Y* assumes availability increasing geometrically toward the finer sizes.

Three factors that affect the size distributions in the same sample or in samples from a transverse section across a stream may now be summarized: (1) the hydraulic conditions, which vary with time and position and are a composite of many interacting factors, (2) the hydraulic equivalent size, which is also the net effect of several factors, and (3) the relative availability for deposit of the different sizes of each mineral.

At any place in a river, the hydraulic conditions that govern sediment deposition change constantly. Short-time, almost instantaneous, fluctuations will be superimposed on long-time changes in average velocity. The size distributions of light and heavy minerals in the resulting deposit will reflect the net effect of these temporal changes. At any instant, however, all kinds and sizes of mineral grains that are accumulating will be subject to the same hydraulic conditions, whatever they may be. At other instants, this will also be true. Consequently, the differences in size distribution of different minerals in the same sample will be due to the second and third factors. If one factor is known, the other may be computed easily. If neither is known, however, neither can be computed from the size distributions in a single sample.

The size distribution of magnetite in sample A may be used to illustrate this point. It can first to be assumed that light minerals and magnetite of the same size are hydraulically equivalent, and the relative availability of various sizes of magnetite necessary to give the actual size distribution of magnetite can be computed. Similar computations can be made, assuming that magnetite grains are hydraulically equivalent to light minerals one, two, or three $\sqrt{2}$ size grades larger. The wide range in apparent relative availability is shown in Table 5. Obviously, if either the

TABLE 5.—*Relative availability of magnetite of various sizes in the stream load*

Computed from sample RG 12A, 1940, for a range of hydraulic equivalent sizes

Size grade	Hydraulic equivalent size (in $\sqrt{2}$ size grades)			
	0	1	2	3
Mm.				
>.351	0.04	0.21	0.51	0.18
.351–.250	.06	.51	.82	.63
.250–.175	.04	.39	1.01	.50
.175–.124	.10	.34	1.08	.87
.124–.088*	1.00	1.00	1.00	1.00
.088–.061	7.11	2.44	.76	.23
<.061	7.26	4.89	.59	.04

* For each hydraulic equivalent size postulated, the .124–.088 mm. size grade was used as a base to show the relative availability in the other size grades.

hydraulic equivalent size or the relative availability were known, it would be easy to compute the other.

In the transverse section of a stream, however, the relative availability for deposit of each size of each mineral theoretically should not vary greatly from place to place. Such other physical characteristics as specific gravity, shape, and roundness that determine the hydraulic value of the sediment particles should also be the same in different parts of the cross section. If both relative availability and equivalent hydraulic size are constant or essentially constant in a cross section, then the two factors can be determined simultaneously by comparing the mineral composition of several samples from different parts of the cross section. The process is similar, in general, to that in mathematics by which two equations are used to solve for two unknowns.

It is not known, however, that the hydraulic equivalent size would be the same for both large and small grains. In fact, it seems probable from theoretical considerations that the opposite would be true. For grains settling through a liquid, as from suspension, fine sand particles presumably would follow Stokes law, whereas coarse sand particles would not (Rubey, 1933a). Thus computation of the hydraulic equivalent size for all size grades at the same time would yield an average value that would be applicable to the entire sample as a whole but not to any particular size of sand. Therefore, single size grades or even smaller units should be compared. Notable differences in hydraulic equivalent size for coarse and fine grains will be shown by differences in the values computed for the different size grades or smaller units.

In actual practice, it is impossible to compute the hydraulic equivalent size for a single size—for example, for grains exactly .088 mm. in diameter. A range of sizes, from .124–.088 mm. diameter, for example, must be used. Since the weight of each kind of heavy mineral in each size grade is known (Table 4), size grades offer the most convenient subdivision of the samples for computing hydraulic equivalent sizes.

The procedure is as follows: The size-distribution curves of the light minerals of the six samples were plotted as large-scale cumulative curves, the relative boundaries

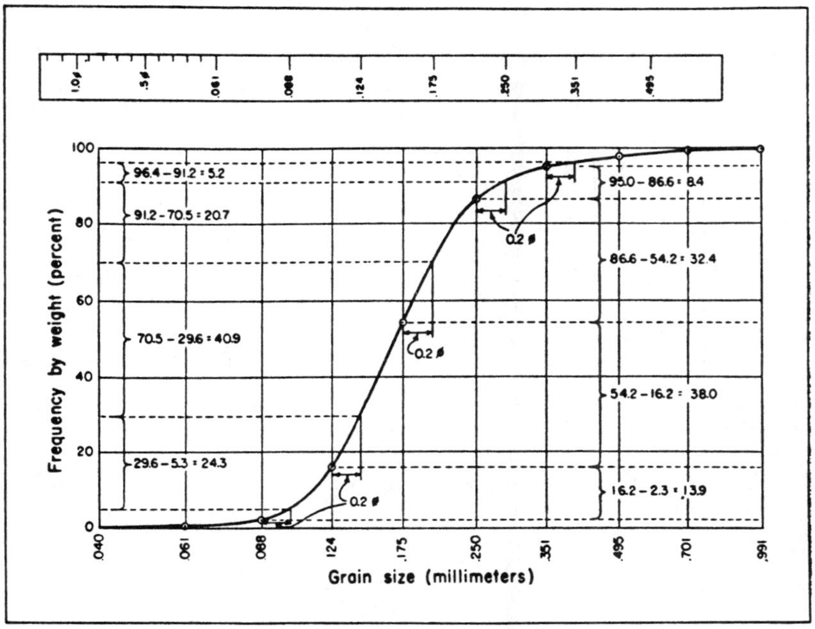

FIGURE 5.—*Diagram illustrating method of obtaining equivalent hydraulic percentages of a sample that are equivalent in hydraulic value to heavy minerals 0.2ϕ smaller*

Size grade boundaries are not those obtained by microprojection calibration

of the size grades being those determined by microprojection (Rittenhouse, 1942). For each size grade of each sample the percentage of light minerals, 0.0, 0.1, 0.2 . . . 1.0ϕ units[5] larger were read from the curves. Figure 5 illustrates this procedure. These percentages were multiplied by the total weight of light minerals in the test samples to give the weight of light minerals 0.0, 0.1, 0.2 . . . 1.0ϕ units larger than the heavy minerals of that size grade.

For each size grade of the six samples, the weight of each heavy mineral was divided by the weight of light minerals of the same size, then by the weight of light minerals 0.1ϕ larger, then by the weight of light minerals 0.2ϕ larger, etc. In this way,

[5] Krumbein (1936) originally suggested the use of a ϕ grade scale to simplify the computation of the moment measures of sediment distributions. In the ϕ scale, the unequal class intervals characteristic of geometric grade scales are eliminated by defining the class intervals in terms of their negative logarithms to the base 2. Thus, the interval between 1.0 and .5 mm. is 1ϕ, and the interval between .5 and .25 is also 1ϕ. For any $\sqrt{2}$ size grade, ϕ is 0.5. Since, in the ϕ notation any interval, such as 0.1ϕ or 0.3ϕ, is the same regardless of the absolute size of the material being compared, it is extremely convenient and useful for designating hydraulic equivalent size. For example, the statement that light minerals from .142–.102 mm. in diameter are equivalent hydraulically to hornblende from .124–.088 mm. in diameter and that light minerals from .201–.142 mm. in diameter are equivalent hydraulically to hornblende from .175–.124 mm. in diameter does not indicate immediately whether the hydraulic equivalent size for the coarser sand is smaller, larger, or the same as that for the finer sand. The statement that the hydraulic equivalent size is 0.2ϕ for both the .124–.088 and .175–.124 mm. size grades of the heavy does have an obvious significance.

hydraulic ratios (relative availabilities) were computed for a wide range of possible hydraulic equivalent sizes. Such a range of hydraulic ratios for ilmenite in samples A, D, F, and G is presented in Table 6.

If both the availability and the physical characteristics of ilmenite were the same at the points where samples A, D, F, and G were deposited, the hydraulic ratios should

TABLE 6.—*Hydraulic ratios for ilmenite in samples A, D, F, and G at Bosque, N. M. (RG 12)*

Computed for a range of hydraulic equivalent sizes

Sample or derived factor	Hydraulic equivalent size														
	.0ϕ	.1ϕ	.2ϕ	.3ϕ	.4ϕ	.5ϕ	.6ϕ	.7ϕ	.8ϕ	.9ϕ	1.0ϕ	1.1ϕ	1.2ϕ	1.3ϕ	1.4ϕ
A	3.4	2.0	1.14	0.77	0.56	0.54	0.54	0.54	0.54	0.54	0.55	0.70	1.05	1.54	2.1
D	2.2	1.8	1.15	.86	.69	.59	.57	.57	.56	.56	.57	.57	.62	.67	.7
F	1.5	1.1	.86	.72	.65	.65	.65	.65	.65	.65	.65	.70	.79	.95	1.1
G	3.7	2.7	2.06	1.57	1.17	1.05	.90	.76	.66	.59	.57	.56	.56	.58	.6
Mean	2.7	1.9	1.30	.98	.77	.71	.67	.63	.60	.59	.59	.63	.75	.93	1.1
Standard deviation	1.0	.7	.53	.40	.27	.23	.16	.10	.06	.05	.04	.08	.22	.42	.7
Coefficient of variation	38.	35.	41.	40.	36.	33.	24.	16.	10.	8.1	7.6	12.	29.	45.	61.

be exactly the same for one of the postulated hydraulic equivalent sizes in Table 6. For both low and high hydraulic equivalent sizes there is a wide variation in the hydraulic ratios of the four samples. In the center of the range, however, the agreement is good but not perfect. This is shown very well in Figure 6, in which the hydraulic ratios have been plotted against hydraulic equivalent size.

The coefficient of variation (standard deviation/arithmetic mean) provides a measure of the similarity between the hydraulic ratios in the four samples. The coefficient of variation for each hydraulic equivalent size of ilmenite is also presented in Table 6 and in graphic form in Figure 6. The lowest value is 7.6 per cent and occurs at 1.0ϕ. At higher values the coefficient of variation increases rapidly but at lower values the rate of increase is slow. The experimental results thus do not give an exact hydraulic equivalent size for ilmenite, but they certainly do indicate that it is probably between 0.8ϕ and 1.1ϕ, a fairly narrow range. Thus, for ilmenite of this size the probable range can be given as 0.8–1.1ϕ, and the best value as 1.0ϕ. In other words, ilmenite apparently is transported and deposited with light minerals about one Wentworth or two $\sqrt{2}$ size grades larger.

Similar computations, using the data from all 6 Bosque samples, were made for 4 size grades of ilmenite and 15 other heavy minerals. The best values are presented in the first four figure columns of Table 7. For convenience, the minerals have been listed in order of decreasing specific gravity. It will be noted that, in general, high specific gravity is accompanied by large hydraulic equivalent size, and low density by low hydraulic equivalent size. Of particular interest is zircon, which is notably finer in size than any of the other heavies (Table 4; Fig. 3) but has a hydraulic equivalent size in line with its specific gravity. Deficiency of coarse zircons is the explanation of this apparent discrepancy.

Three suites of suspended-load samples collected at Escondido, New Mexico, during the flood of June and July 1941 also provide information on hydraulic equivalent sizes. These samples were obtained from different parts of the transverse cross section and at different water depths. In the laboratory they were sieved, and the mineral composition of two or three size grades of each sample determined in the

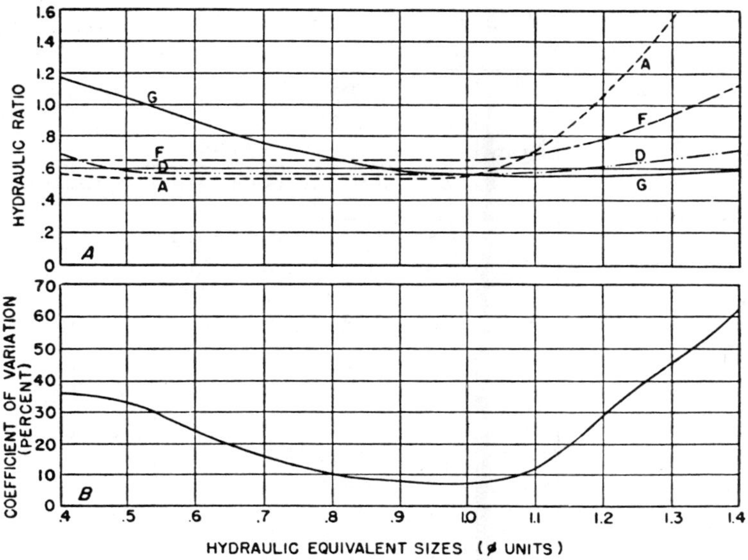

FIGURE 6.—*Hydraulic ratios and coefficient of variation*

A, Hydraulic ratios for ilmenite in samples A, D, F, and G, computed for a range of hydraulic equivalent sizes. *B*, Coefficient of variation for hydraulic ratios in the same samples, computed for the same range of hydraulic equivalent sizes.

same manner as for the Bosque samples. For each size grade hydraulic ratios were computed for a range of hydraulic equivalent sizes from 0.0ϕ to 1.0ϕ. Coefficients of variation were computed for all samples for enough 0.1ϕ units of the possible range to indicate the best value of the hydraulic equivalent size. These best values are presented in columns five and six of Table 7. This transit load data will be reported and analyzed in detail at a later date.

In general, the hydraulic equivalent sizes, as determined from the transit samples at Escondido, agree very well with those determined from the bed deposits at Bosque. Such agreement between values obtained by independent investigation of two different kinds of samples from widely separated points on the same stream indicates that the concept of hydraulic equivalence is a good one—*i.e.*, that on the average light minerals are transported and deposited with heavy minerals of a different size.

MAGNITUDE OF UNEXPLAINED VARIATIONS

In determining equivalent hydraulic size, the phi (ϕ) value at which the coefficient of variation was smallest was considered to be the true hydraulic equivalent size. Even at the best ϕ value, however, there remained some unexplained variation. For the example in Table 6, this was 7.6 per cent. For all other minerals studied, these unexplained variations, expressed as coefficients of variation of the hydraulic ratios (C_{hr}), are presented in the second figure column under each of the four size grades in Table 8.

To some unknown degree, these variations are caused by chance errors of sampling; the remainder, if any, represents variation due to some unknown factor or factors.

TABLE 7.—*Hydraulic equivalent sizes of various heavy minerals*

As determined from bed samples from Bosque and transit samples at Escondido, New Mexico

Mineral	Specific gravity	Bed samples (Bosque)				Transit samples (Escondido)		Best value all sizes
		Size Grade						
		.250–.175 mm.	.175–.124 mm.	.124–.088 mm.	.088–.061 mm.	.175–.124 mm.	.124–.088 mm.	
		φ units	*φ units*	*φ units*	*φ units*	*φ units*	*φ units*	*φ units*
Magnetite	5.2	.9	1.2	1.0	1.0	1.1	1.0	1.0
Ilmenite	4.7	.8	1.0	.9	1.0	1.1	.9	1.0
Zircon	4.6	......	.8	.9	.9	......	......	.9
Barite	4.5	.6	.4	.7	.2	......	......	.5
Garnet	3.8	.8	.8	.3	.6	.5	.8	.6
Kyanite	3.6	.3	.5	.0	......	......	......	.3
Titanite	3.5	.7	.4	.7	.3	.3	.7	.5
Pyroxene	3.4	.4	.7	.1	.3	.3	.3	.3
Hypersthene	3.4	.5	.3	.4	.2	.3	.3	.4
Diopside (Br)	3.4	.5	.2	.1	.5	......	......	.3
Diopside (C)	3.3	.3	.4	.6	.4	......	......	.4
Blue-green hornblende	3.2	.2	.0	.0	.1	.3	.3	.2
Green hornblende	3.2	.4	.1	.1	.2	.4	.2	.2
Brown hornblende	3.2	.3	.2	.0	.0	.3	.2	.2
Apatite	3.2	.6	.5	.2	.4	.4	.4	.4
Tourmaline	3.1	.0	.0	.3	.1	.3	.4	.2

Consequently, if the errors due to sampling are determined and compared with the total unexplained variations, the remaining unexplained variation, if any, will measure the extent to which factors, other than hydraulic conditions, hydraulic equivalent size, and relative availability govern the entrainment, transportation, and deposition of heavy minerals. In making the comparisons, the actual variations in hydraulic ratios for the average best value (column 7, Table 7) rather than the best values determined for individual size grades will be used.

The sampling errors of hydraulic ratios could not be predicted directly from any statistical equation known to the writer. Additional experimental work (Rittenhouse and Bertholf, 1942), however, showed that an equation, similar to that used by Dryden (1935) to predict the sampling errors of number frequencies, could also be used to predict the sampling errors of hydraulic ratios. The procedure was as follows: Eight splits were made from a composite sample, and the size distribution of each determined by sieving. The sand in the .124–.088 mm. size grade of each sample was split in two parts and the parts weighed. The heavies and lights were separated from one part by gravity settling, from the other part by centrifuge. After washing, drying, and weighing, the magnetite was removed and weighed. A random split of each of the 16 nonmagnetic fractions was mounted, and the mineral frequencies by number in each obtained by identifying 325 to 380 randomly selected grains.

The variation in number frequency of each mineral was computed for the eight gravity and the eight centrifuge fractions separately. These coefficients of variation were plotted against the percentages by number on logarithmic paper (Fig. 7).

TABLE 8.—*Mineralogical analysis, coefficients of variation*

(In per cent)

Mineral	Specific gravity	Size Grade											
		.250–.175 mm.			.175–.124 mm.			.124–.088 mm.			.088–.061 mm.		
		C_s*	C_{hr}†	C_w**	C_s	C_{hr}	C_w	C_s	C_{hr}	C_w	C_s	C_{hr}	C_w
Ilmenite	4.7	16.	62.	98.	10.	53.	103.	5.	23.	48.	4.	20.	67.
Zircon	4.6	......	......	......	57.	64.	82.	25.	44.	49.	12.	24.	58.
Barite	4.5	104.	83.	108.	70.	57.	58.	37.	58.	58.	30.	34.	32.
Garnet	3.8	41.	80.	143.	24.	59.	81.	22.	40.	29.	22.	17.	56.
Kyanite	3.6	79.	21.	35.	57.	48.	46.	49.	74.	55.	......	......	......
Titanite	3.5	49.	28.	65.	33.	28.	31.	25.	31.	35.	28.	50.	40.
Pyroxene	3.4	15.	13.	30.	13.	24.	28.	12.	14.	13.	15.	22.	35.
Hypersthene	3.4	28.	48.	68.	26.	20.	22.	22.	40.	58.	25.	40.	43.
Diopside (Br)	3.4	48.	74.	91.	31.	52.	51.	33.	65.	53.	36.	60.	47.
Diopside (C)	3.3	42.	32.	44.	31.	37.	43.	25.	34.	33.	30.	29.	44.
Blue-green hornblende	3.2	15.	38.	39.	11.	14.	11.	10.	21.	11.	15.	12.	18.
Green hornblende	3.2	22.	36.	45.	18.	19.	20.	14.	21.	24.	15.	6.	20.
Brown hornblende	3.2	48.	60.	56.	30.	28.	31.	23.	25.	19.	23.	28.	29.
Apatite	3.2	60.	71.	69.	34.	31.	42.	22.	21.	28.	21.	24.	35.
Tourmaline	3.1	70.	86.	73.	31.	46.	42.	23.	13.	22.	22.	25.	29.
Magnetite	5.2	......	29.	67.	......	39.	86.	......	23.	47.	......	32.	54.

* Coefficient of variation (theoretical) based on number of grains counted and frequency by number of each mineral species.

† Total coefficient of variation based on hydraulic ratios.

** Total coefficient of variation based on weight percentages.

For any mineral the sampling error (in number frequency) may be computed from the equation $C_s = \sqrt{npq}/np$, in which C is the coefficient of variation, n is the total number of grains counted, p is the number frequency of the mineral, and q is the sum of the number frequencies of all other minerals. For any n, the C for different values of p may be computed and a curve drawn to show the change in C with change in p. This was done separately for the average n of the eight gravity and the eight centrifuge samples. It was found that these curves fitted the points obtained experimentally from the gravity and centrifuge samples (Fig. 7). This was not unexpected, since chi-square tests of the same data (*see* section on Percentages by Number) showed the variation to be well within the limits to be expected in sampling. A similar close agreement between experimental points and theoretical values has been obtained for beach sands (Krumbein and Rasmussen, 1941). This means that error of reduction in sample size is so much larger than other laboratory errors that they may be neglected.[6]

Next, the number percentages were converted to weight per cent, the weights of

[6] Independent errors (c_1, c_2, etc.) are related to the total error (C) by the equation $C = \sqrt{c_1^2 + c_2^2}$ (Krumbein and Pettijohn, 1938, p. 266). Thus if one suberror is essentially the same as the total error, the other suberrors must be relatively small.

each heavy mineral were computed, and the hydraulic equivalent weights obtained graphically. From these the hydraulic ratios in the eight gravity and eight centrifuge samples were calculated. For each mineral the coefficient of variation of these hydraulic ratios was determined. It was found that the coefficients of variation of

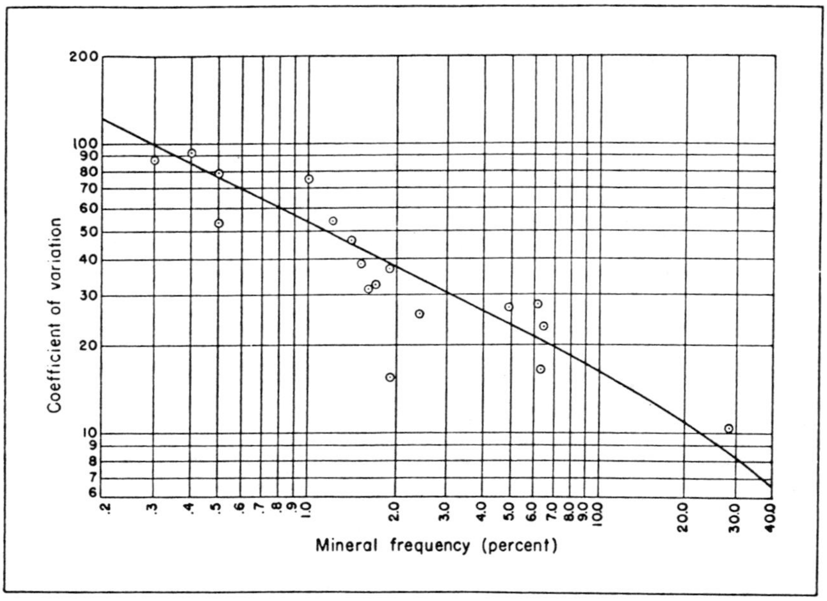

FIGURE 7.—*Experimentally determined and theoretical coefficients of variation of number frequencies*

Determined for the gravity separates of sample RG 8, 1938

the hydraulic ratios were essentially the same as the coefficients of variation of the percentages by number. Therefore, if the coefficient of variation is obtained for any mineral from (1) the total number of grains counted, and (2) the percentage by number, using the equation $C_s = \sqrt{npq}/np$, this coefficient of variation is also the coefficient of variation for the hydraulic ratio of that mineral. This is important since it provides a means by which the sampling errors of hydraulic ratios may be determined for any sample and permits tests of significance to be made between samples.

Coefficients of variation (C_s) so obtained may be compared with those (C_{hr}) obtained from the six Bosque samples, to determine the extent of unexplained variation In Table 8, the sampling variations are in the first column, and the experimentally determined variations are in the second column under each size grade.

For some minerals in each size grade C_{hr} is larger than C_s; for others the reverse is true. Thus, for 41 of the 58 comparisons (71 per cent), C_{hr} exceeds C_s; for 17, C_s exceeds C_{hr}. For the high-density minerals—ilmenite, zircon, barite, and garnet—the C_{hr} is greater for 12 of the 15 comparisons (80 per cent). For the minerals of low density, the C_{hr} is greater for 29 of 43 comparisons (67 per cent).

In interpreting these results, it must be remembered that the C_{hr} values are determined experimentally and consequently would be expected to scatter about any values derived from theoretical curves. Such scattering was well shown in the gravity and centrifuge experimental data (Fig. 7). About half the values would be expected to be larger and half smaller than those determined theoretically. Some deviation from this 50–50 ratio would be expected, but how much is not known. For the gravity samples the ratio was exactly 50–50; for the centrifuge samples it was 70–30, with most points being smaller than the theoretical; and for Krumbein and Rasmussen's data (1941) it was about 43–57.

TABLE 9.—*Percentage by which the coefficients of variation of hydraulic ratio (C_{hr}) and weight percentage (C_w) exceed the coefficient of variation of random sampling (C_s)*

(In per cent)

Kind of heavy minerals	Size Grade								Average of all size grades	
	.250–.175 mm.		.175–.124 mm.		.124–.088 mm.		.088–.061 mm.			
	C_{hr}	C_w	C_{hr}	C_w	C_{hr}	C_w	C_{hr}	C_w	C_{hr}	C_w
High density..........	94.*	255.*	141.	298.	144.	261.	121.	530.	127.	341.
Low density...........	1.	49.	15.	22.	36.	36.	13.	50.	16.	39.
All..................	20.	93.	49.	96.	64.	96.	43.	187.	45.	117.

* If C_s were 10 per cent, C_{hr} would be 19.4 per cent, and C_w would be 35.5 per cent.

For high-density heavies it appears unlikely that the deviations are due to chance. Not only are 80 per cent of the actual values larger than the theoretical, but many of them are several times as large. For ilmenite—the outstanding example—C_{hr} in each of the four sizes grades is more than four times as large as C_s. Apparently some factor or factors in addition to the three that were postulated affect the transportation and deposition of high-density heavies.

Although the deviations for low-density heavies might be due to chance, it seems more probable that there is some unexplained variation in excess of that to be expected as a result of sampling. If this is true, some factor or factors in addition to the three that were postulated also affect the transportation and deposition of heavies of low density, but to a lesser degree than they do high-density heavies.

From the Bosque data the importance of these additional unknown factors may be approximated.[7] Average percentages by which C_{hr} exceeds C_s are presented in the first figure column under each size grade in Table 9. For high-density heavies (ilmenite, zircon, barite, and garnet), the average C_{hr} was found to exceed C_s by 127 per cent. The range for individual size grades was from 94 to 144 per cent. For low-density heavies, the average C_{hr} exceeds C_s by only 16 per cent, ranging from 1 to 36 per cent in the individual size grades. If C_s of a high-density mineral were 20 per cent, for example, C_{hr} would be about 45 per cent (20×2.27); for a low-density mineral, C_{hr} would be 23 per cent (20×1.16). The increase of 25 and 3 per cent respectively for the high- and low-density minerals in this example represents the influence of a factor or factors other than the three outlined—*i.e.*, (1) hydraulic conditions, (2) hydraulic equivalent size, and (3) relative availability.

[7] For each mineral in each size grade C_{hr} was divided by C_s, or C_s divided by C_{hr}, the smaller value in each case being used as the divisor. (When C_{hr} is less than C_s, the ratio C_{hr}/C_s may vary from 1 to 0, but when C_{hr} is greater than C_s, the ratio C_{hr}/C_s may vary from 1 to ∞. Consequently, if the same factor were used throughout as the divisor, any average of a number of ratios would be too large.) The C_{hr}/C_s quotients were considered positive, and the C_s/C_{hr} quotients negative. Then $100\ (C_{hr}/C_s) - 100$ and $100\ (C_s/C_{hr}) + 100$ represent the percentages by which the larger variation exceeds the smaller. These percentages were summed by individual size grades and for all size grades together. Such sums were obtained for heavies of high density, for heavies of low density, and for all heavies. The sums were divided by the number of items in each to give an average percentage by which C_{hr} exceeds C_s.

The average C_s is dependent on the total number of grains counted in the nonmagnetic fraction. The average percentages (127 and 16) by which C_{hr} exceeds C_s were determined for a C_s based on an average count of 780 grains. If more or less grains were counted, C_s would decrease or increase, and the 127 and 16 per cent values

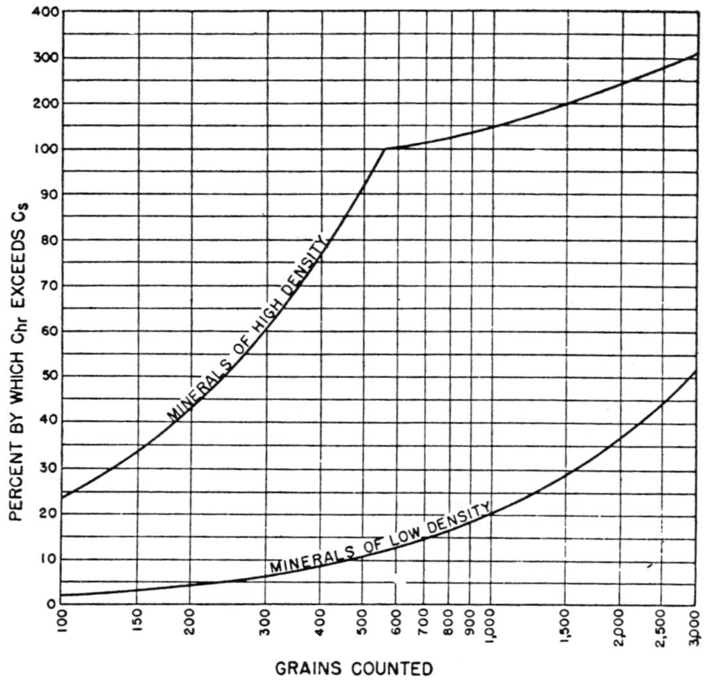

FIGURE 8.—*Percentage by which C_{hr} exceeds C_s (780 grain count)*

For counts of from 100 to 3000 grains

would no longer be applicable. Since the unexplained variations are independent of C_s, however, they may be evaluated from the equation $C_x = \sqrt{C_{hr}^2 - C_s^2}$, in which C_x is the average magnitude of the unexplained variations. These C_x values are presented in the first figure column under each size grade in Table 10. These C_x values are fixed and unlike the average excess percentages (Table 9) are independent of C_s, though they are presented in terms of percentages of a C_s from a 780 grain count. For high-density heavies, C_x is 204 per cent of the C_s for a 780 grain count; for low-density heavies it is 58 per cent.

With C_x evaluated, the C_{hr} for any mineral with any grain count can be estimated from the equation $C_{hr} = \sqrt{C_s^2 + C_x^2}$, in which C_s is based on the new grain count, and $C_x = 2.04$ [C_s (780 grain count)] for minerals of high density or $C_x = .58$ [C_s (780 grain count)] for minerals of low density. From this equation it may be seen that the relative importance of C_x in determining C_{hr} becomes smaller as C_s increases. Thus, for a count of 300 grains, a count commonly used in heavy mineral investigations, the average percentage by which C_{hr} would exceed C_s is about 61 per cent for high-density heavies and 6 per cent for low-density heavies. For counts of 100 to 3000 grains, the average percentage by which C_{hr} exceeds C_s (based on the C_x from the Bosque data) can be determined graphically for both high- and low-density heavies from Figure 8.

What are the factors that cause the unexplained variations? There are several possibilities, the relative importance of which are not known. First, in the Rio Grande some sand is transported in suspension, some by traction. The hydraulic equivalent sizes for these two modes of transportation may not be the same. If not, the hydraulic equivalent sizes that were determined experimentally from the Bosque samples represent some intermediate average hydraulic equivalent size for the cross section as a whole. But if the proportion of suspended and traction load was not the same in the six samples studied, some variation in the hydraulic ratios would result. Further studies in other streams, or under controlled conditions in flumes, should provide evidence on this point.

It is also possible that some variation in relative availability from place to place in the cross section may have resulted from inadequate mixing of the sand as it was transported down the Rio Grande. Although no major tributaries enter the Rio Grande for about 50 miles upstream, a relatively small tributary nearer Bosque might, under exceptional conditions, have contributed enough sediment to modify the mineral composition slightly.

Also, the variations for heavy minerals of low density might conceivably be due to errors of random sampling. It seems less probable, however, that the unexplained variations for heavies of high density could be due to sampling. Further investigations should indicate more accurately the effect of sampling errors.

VARYING HYDRAULIC CONDITIONS

To what extent do varying hydraulic conditions at different times or places affect heavy-mineral size distributions at a transverse cross section? Obviously, if the effect is minor, the additional time and effort involved in eliminating or reducing it is largely wasted. Weight percentages of heavy minerals in a size grade—*i.e.*, the weight of a heavy mineral divided by the weight of sediment of the same absolute size—should not eliminate the effect of varying hydraulic conditions. Consequently, a comparison of (1) the coefficients of variation of weight percentages (C_w), hydraulic ratios (C_{hr}), and the theoretical sampling errors (C_s), and (2) the mean weight percentages and hydraulic ratios will indicate the magnitude of effects due to varying hydraulic conditions.

Weight percentages were determined for each mineral in four size grades of the Bosque samples. The C_w values were computed and are presented in the third figure column under each size grade of Table 8. In 50 of the 58 comparisons (86 per cent) C_w exceeds C_s, in 1 they are the same, and in 7 C_s exceeds C_w. For high-density heavies, C_w exceeds C_s in 14 of 15 comparisons. For low-density heavies, C_w exceeds C_s in 36 of 43 comparisons (84 per cent). For both low- and high-density heavies, C_w exceeds C_s more times than C_{hr} exceeds C_s.

The average percentage by which C_w exceeds C_s was computed for each of the four size grades and for all size grades together. These percentages are presented in the second figure column under each size grade in Table 9. Except for low-density heavies in the .124–.088 mm. size grade, the average C_w values exceed the average C_{hr} values, in many cases being more than twice as large.

Since the variations caused by varying hydraulic conditions (C_y) are independent of those due to sampling and to other unexplained factors, they may be evaluated

TABLE 10.—*Errors due to unexplained factors (C_x) and varying hydraulic conditions (C_y) in terms of the coefficient of variation of random sampling (C_s) for a 780 grain count*

(In per cent)

Kinds of heavy minerals	Size Grade								Average of all size grades	
	.250–.175 mm.		.175–.124 mm.		.124–.088 mm.		.088–.061 mm.			
	C_x	C_y	C_x	C_y	C_x	C_y	C_x	C_y	C_x	C_y
High density..........	1.66*	2.97*	2.19	3.16	2.22	2.66	1.97	5.90	2.04	3.78
Low density...........	.14	1.09	.57	.41	.92	.00	.53	.99	.58	.77
All...................	.66	1.51	1.11	1.27	1 30	1.07	1.02	2.48	1.05	1.61

* If C_s were 10 per cent, C_x would be 16.6 per cent, C_y would be 29.7 per cent. Also $C_{hr} = \sqrt{C_s^2 + C_x^2} = 19.4$ per cent, and $C_w = \sqrt{C_s^2 + C_x^2 + C_y^2} = 35.5$ per cent.

from the equation $C_y = \sqrt{C_w^2 - C_s^2 - C_x^2}$. These C_y values, presented in the second figure column under each size grade in Table 10, are comparable to the C_x values in the same table. Variations due to hydraulic conditions, particularly for high-density minerals, are large and, therefore, well worth spending time and effort to eliminate. C_y may be expected to increase as the range in average size and degree of sorting of the light minerals increase, and to decrease relative to C_s as the number of grains counted is decreased. The C_y values determined from the Bosque data cannot be applied to other investigations in which the range and magnitude of hydraulic conditions are different.

Whatever C_x and C_y may be at any locality, they may be reduced, theoretically at least, by collecting two or more spot samples and combining them into a single composite sample. Such reduction would be inversely proportional to the square of the number of spot samples combined into the composite. This relationship would appear to hold, at least approximately, provided two important qualifications are made. First, if three spot samples are combined, the expected reduction in C_x and C_y will occur only for those size grades common to all three samples. There will be partial reduction in size grades common to two of them, and no reduction in the other size grades. Similarly, if some samples contribute more sediment than others to any size grade of the composite, the reduction in C_x and C_y will not be inversely proportional to the square of the number of samples but will depend on the relative amounts of sediment contributed by each. Second, the smaller C_y values obtained from composite samples will not apply to other sands that were deposited under a different range or magnitude of hydraulic conditions. Before C_y of a series of composite samples may be determined, the entire range and magnitude of hydraulic conditions under which all samples were deposited must be considered.

The C_w and C_y values measure the variation about the mean weight percentages. If weight percentages rather than hydraulic ratios are used to represent heavy-mineral compositions, however, the absolute mean value as well as the variation about the mean must be considered. Depending on the relative position of the size grade being studied to the size distribution of the light minerals, and on the hydraulic equivalent size of the heavy mineral, the weight percentages will be smaller or larger than the hydraulic ratios.

TABLE 11.—*Weight percentages and hydraulic ratios for various minerals in four size grades of the Bosque bed samples*

Mineral	Size Grade							
	.250–.175 mm.		.175–.124 mm.		.124–.088 mm.		.088–.061 mm.	
	Hydraulic ratio	Weight percentage	Hydraulic ratio	Weight percentage	Hydraulic ratio	Weight percentage	Hydraulic ratio	Weight percentage
Magnetite	.59	.15	.56	.48	.66	1.57	.41	3.30
Ilmenite	.34	.10	.41	.37	.58	1.40	.38	3.80
Zircon	......	......	.012	.011	.031	.074	.062	.56
Barite	.004	.003	.005	.006	.016	.031	.021	.096
Garnet	.019	.011	.029	.036	.038	.076	.024	.13
Kyanite	.003	.003	.007	.008	.008	.013	......	......
Titanite	.010	.007	.020	.023	.027	.052	.019	.086
Pyroxene	.073	.063	.101	.117	.114	.185	.101	.295
Hypersthene	.028	.022	.029	.033	.034	.066	.029	.107
Diopside (Br)	.010	.008	.022	.024	.018	.032	.013	.048
Diopside (C)	.011	.009	.019	.023	.030	.048	.023	.068
Blue-green hornblende	.070	.065	.132	.146	.172	.243	.135	.278
Green hornblende	.033	.030	.056	.063	.101	.146	.122	.253
Brown hornblende	.007	.007	.022	.025	.037	.053	.053	.110
Apatite	.005	.006	.016	.020	.034	.063	.044	.169
Tourmaline	.004	.004	.018	.020	.038	.056	.065	.131

This is well illustrated by Table 11, in which the average hydraulic ratios and the average weight percentages are given for four size grades of the Bosque samples. For the coarsest size grade, the weight percentages average about two thirds as large as the hydraulic ratios; for the .175–.124 mm. size grade, they are slightly larger for minerals of low density and smaller for minerals of high density; for the .124–.088 mm. size grade, they average about twice as large; and for the .088–.061 mm. size grade, they are about five times as large. In general, the minerals of low density show less difference then minerals of high density. This relationship would be expected to hold for other localities, though the particular size grades of best agreement would vary with the average texture of the deposit.

The difference in absolute value between hydraulic ratios and weight percentages will not be important if the size distribution of the light minerals is the same in all samples being compared, since the weight percentages will be in error the same relative amount in all of them. If the size distribution varies, however, quite misleading conclusions may be reached.

A common type of detailed heavy-mineral study, for example, is that in which the change in mineral composition is related to the distance of transport. If the average size of the deposits decreases with increasing distance from the source as is commonly the case, weight percentages may indicate trends that actually do not exist. This is shown in Figure 9A, in which the relative availabilities of magnetite and pyroxene actually remain the same in a series of samples, but the weight percentages change. The increase in weight percentage is greater for magnetite than for pyroxene, which

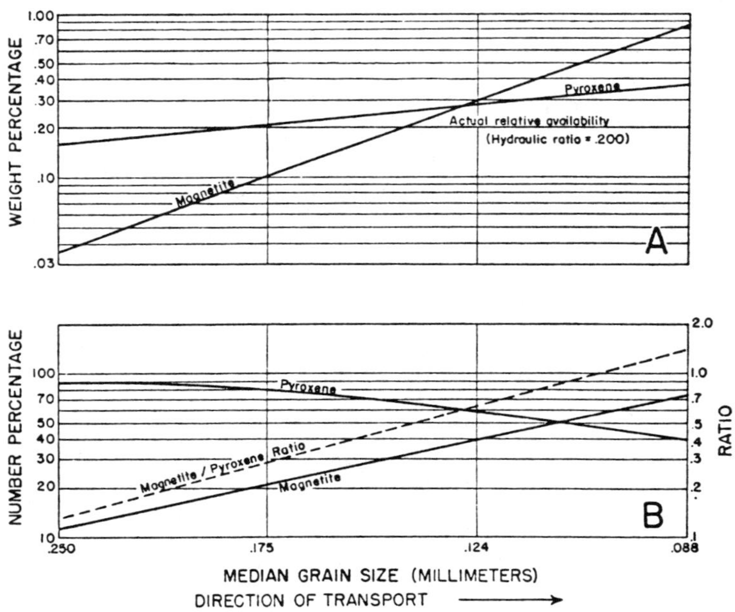

FIGURE 9.—*Changes with distance of transport*

A, Change in weight percentage of magnetite (1.0ϕ) and pyroxene (.3ϕ) with distance of transport, assuming an exponential decrease in the average grain size of the entire sample, no change in the actual relative availability, and the same sorting as sample G.

B, Change in mutual number percentages and in the magnetite/pyroxene ratio with distance of transport. Assumptions are the same as for *A*.

is of smaller hydraulic equivalent size. In Figure 9B, the same data have been presented as mutual number percentages of magnetite and pyroxene and as a ratio between magnetite and pyroxene. Several misinterpretations of the genetic significance of these number percentage and ratio data are possible.

When the size distributions of light minerals are plotted as cumulative curves on semilogarithmic paper, the central part of the curve, extending from about 25 to 75 per cent by weight, is relatively straight. For size grades located within this straight section, the weights of light minerals that are of the same absolute size and the same hydraulic size as the heavies will not differ greatly. For the .175–.124 mm. size grade in Figure 5, for example, the values are 38.0 and 40.9 per cent, and, consequently, the weight percentages and hydraulic ratios will be essentially the same. In contrast, for size grades in the curved sections, the absolute size and equivalent hydraulic size of light minerals will contain quite different weights of light minerals, and the weight percentages and hydraulic ratios will differ materially. For the .124–.088 mm. size grade in Figure 5, for example, the values are 13.9 and 24.3 per cent. The weight percentages would be almost twice the hydraulic ratios.

In considering weight percentages in samples of differing average size or sorting, it would be expected that those size grades located on or near any straight sections that are common to all samples would have weight percentages most nearly equal to the hydraulic ratios and would also show lower coefficients of variation between the weight percentages. This is well shown by the Bosque data in Tables 8 and 10 and by Figure 9A. The greater the differences in size distribution, the less chance there will be for straight sections of the curves to be common to all samples, and, consequently, the more need there will be to use hydraulic ratios rather than weight percentages to represent heavy-mineral compositions.

CHANGE IN HYDRAULIC EQUIVALENT SIZE WITH SIZE

In Table 7 there is some indication that the hydraulic equivalent size for the coarser size grades is larger than for the finer size grades. The Bosque samples provide data for further testing of this point. In using size grades to compute hydraulic ratios for a range of possible hydraulic equivalent sizes, it was assumed that the hydraulic equivalent size of light minerals, as expressed in ϕ units, was the same for both the lower and the upper limit of each size grade. For example, it was assumed that if pyroxene .124 mm. in diameter was hydraulically equivalent to light minerals 0.3ϕ larger, then pyroxene .175 mm. in diameter was also hydraulically equivalent to light minerals 0.3ϕ units larger. But, if hydraulic equivalent size increases with grain size, the coarser heavies would actually be hydraulically equivalent to somewhat larger lights—*i.e.*, pyroxene .175 mm. in diameter would be equivalent to light minerals more than 0.3ϕ units larger.

As the range of heavy mineral sizes to be used is reduced below a $\sqrt{2}$ sieve interval, the difference in size of light minerals that would be hydraulically equivalent to the smallest and largest heavies would also be reduced. When the range in size of heavies is reduced to zero, the difference in hydraulic equivalent size of light minerals would also be zero.

From the cumulative size-distribution curves of the heavy minerals, the percentage of heavies or lights may be obtained graphically for any given range of sizes. From these percentages, the weight of heavies in the selected size range is obtained by multiplying the percentage by the total weight of heavies in the sample. As the range is reduced, however, the errors involved in determining the percentages increase. Consequently, there is a practical lower limit to the range sizes that can be used.

In the present study, the range selected was one-fifth of a $\sqrt{2}$ size grade. Since a $\sqrt{2}$ interval is 0.5ϕ units, the selected range was 0.1ϕ unit (0.1 of a Wentworth size grade). The procedure was as follows: The size distributions of selected heavy minerals in samples A, B, C, F, D, and G were plotted as large-scale cumulative curves, using the size grade limits as determined by sieve calibration (Rittenhouse, 1942). Each heavy-mineral curve was then subdivided into 0.1ϕ units and the percent by weight in each 0.1ϕ unit was read from the curves. These weight percentages were multiplied by the total weight of heavy mineral in the sample to give the weight of that heavy mineral in each 0.1ϕ unit of the size distribution. The size distribution of the light minerals in samples A, B, C, D, F, and G were plotted to the same scale and divided into 0.1ϕ units that coincided with those for the heavies. The percent by weight and the weight of light minerals in each 0.1ϕ unit were thus obtained.

The data are similar in all respects to those used for determining hydraulic equivlent size by size grades, except that the range of sizes for each determination is 0.1ϕ

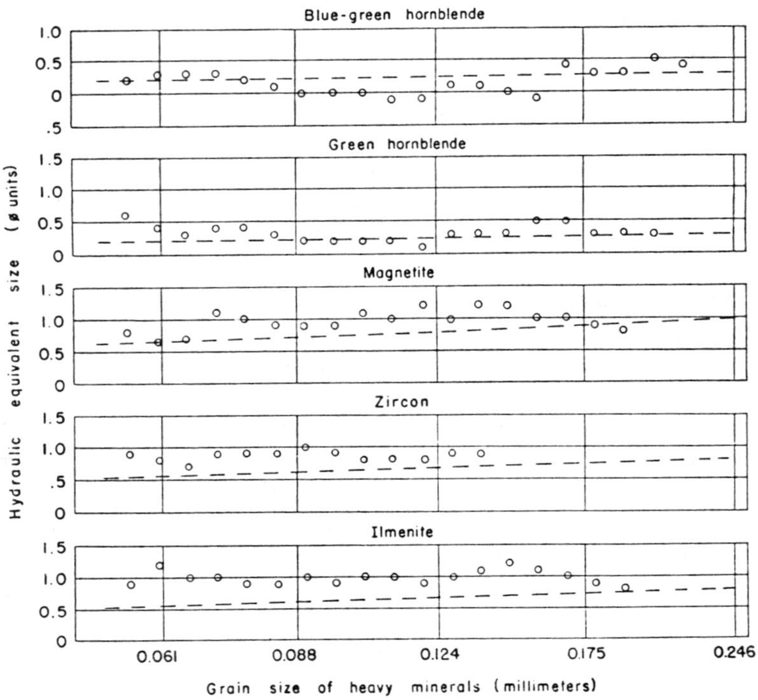

FIGURE 10.—*Relation between grain size and hydraulic equivalent size of selected heavy minerals from samples A, B, C, D, F, and G*

Dashed lines represent settling velocities of spheres having the same specific gravity as the mineral

instead of about 0.5ϕ. Consequently, the procedure in obtaining the hydraulic equivalent size for the new and smaller ranges is essentially the same. The weight of heavy mineral in the same 0.1ϕ unit of each of the six samples was divided first by the weight of light minerals 0.1ϕ larger, then by the weight of light minerals 0.2ϕ larger, etc. In this way hydraulic ratios were computed for a wide range of possible hydraulic equivalent sizes. Coefficients of variation of the hydraulic ratios were obtained for each of the possible hydraulic equivalent sizes, the lowest coefficient of variation in the range representing the least variation in the hydraulic ratios, and, consequently, the best available estimate of the correct hydraulic equivalent size.

Similar computations were made for other 0.1ϕ subdivisions of the heavy-mineral size distribution curves. The best values for each mineral have been plotted in Figure 10. For comparison, hydraulic equivalent sizes, as computed from Rubey's general settling formula for spheres (1933a), are shown by dashed lines in the same figure.

Several facts are apparent from examination of these curves. (1) The curves are not straight lines, but there is not a progressive increase in hydraulic equivalent size from fine to coarse; (2) the best values for all sizes of each mineral lie within limits of about 0.4ϕ—i.e., for green hornblende between 0.1ϕ and 0.5ϕ, for magnetite between 0.8ϕ and 1.2ϕ, etc.; (3) the heavies of high specific gravity have larger hydraulic equivalent sizes than heavies of low specific gravity; (4) the general shape of the curves is the same for all heavies, but the lowest hydraulic equivalent sizes of the high-density heavies are displaced toward the fine sizes relative to those for heavies of low density; (5) the smallest hydraulic equivalent sizes as determined for the Bosque samples are about the same as those computed for spheres settling through water, but larger hydraulic equivalent sizes (in the coarser and very fine sizes) are considerably larger than those computed from the formula.

Analysis of the curves show that either the large equivalent sizes in the finest sizes should be smaller or the smaller equivalent sizes in the immediately coarser sizes should be larger. As the curves are presented in Figure 10, heavies in the finest sizes are equivalent hydraulically to the same light minerals as the heavies in the slightly coarser sizes. This seems impossible.

To check the position of the best value points, data from a duplicate analysis of sample F were obtained. Substitution of the duplicate F for the original F causes no essential change in the shape or position of the best value curves of Figure 10. Exclusion of data from samples B and C generally does not change the position of the points in the coarser sizes but does lower the lower points for some minerals. This would seem to indicate that at least part of the curvilinear shape of the curves is caused by experimental or sampling errors. The Bosque samples thus do not give definite evidence concerning change of hydraulic equivalent size with size, other than to indicate that such change, if present, is probably of small magnitude in the size grades that are important at Bosque.

HYDRAULIC EQUIVALENT SIZE AND GRAIN SHAPE

In using hydraulic ratios to compare heavy-mineral compositions, weights of heavy minerals rather than their frequency by number are used. Because most of the minerals cannot be conveniently segregated from one another and weighed separately, the weights must be determined by an indirect method. The frequency by number is first obtained and then corrected for specific gravity to given weight percentages. These weight percentages are then multiplied by the weight of the nonmagnetic heavy-mineral fraction. In the specific-gravity correction, however, it is assumed that the average volume of the different heavies is essentially the same. However, if number frequencies were determined for the entire range of sizes in an unsieved sample, such minerals as zircon and ilmenite that, because of high specific gravity and/or relative availability, are concentrated in the finer sizes would have disproportionately high number frequencies, and the volume assumption would be greatly in error.

To reduce such errors, sieves are used to size the heavy and light minerals. However, sieves separate grains on the basis of shape as well as size, and consequently the reduction of one error introduces another factor that must be evaluated—namely, the relation between shape and the size of grain passed by sieves. It is pertinent to inquire how this relation affects the equivalent hydraulic size.

The relation of shape to size of grain passing sieves has been discussed previously by the writer (Rittenhouse, 1943a). It was found that in a sieve separate the average intermediate diameter increases as flatness increases. Thus, in the same sieve separate, the flat grains of large intermediate diameter will be associated with more spherical grains of smaller intermediate diameter. The flatter or more elongated grains would be expected to settle through water at a slower rate than spherical grains and then be of smaller hydraulic equivalent size. Actually, however, the flatter grains appear to have larger volumes and consequently greater weights than would spheres retained or passed by the same sieves. Thus some compensation in settling rates would be expected.

That such compensation exists and that, for a fairly wide range of shapes, it is extremely good, was shown by a comparison of the intermediate grain diameters as determined by microprojection and the settling rates of grains from the same size grades (Rittenhouse, 1943b). Although the intermediate diameters of 30 to 50 per cent of grains are larger than the rated openings of the sieve through which the grains have passed, the irregularly shaped grains settle at just about the same rate as spheres that would be passed or retained by the same sieves. These measurements were made on grains that had an average flatness ratio (short diameter/intermediate diameter) of about 0.74.

Because most of the heavy minerals have flatness ratios of about the same order of magnitude, it seems probable that most of the heavy minerals would settle at about the same rates as spheres of the same material. If this is true, then differences in shape caused by sieving would have no effect on the hydraulic equivalent sizes. The flatness ratio of mica, and perhaps blue-green hornblende, are sufficiently different, however, so that their settling velocities and their hydraulic equivalent sizes might be less than expected from their specific gravities. Data are not available for mica, since only part is in the heavy-mineral separate. For blue-green hornblende, which appears to have a hydraulic equivalent size slightly less than 0.2ϕ in comparison with brown and green hornblende, which are slightly more than 0.2ϕ (Table 7), shape may be important.

Actually, the compensation effect introduced by sieving simplifies interpretation of the relation between shape and hydraulic equivalent size, at least insofar as the settling rates are concerned. From shape calibration curves (Rittenhouse, 1943a), it is possible to estimate what the hydraulic equivalent sizes might be if they were based on the measurement of intermediate diameters rather than on sieve units. For minerals that are flatter than the light minerals, the hydraulic equivalent size would be greater by about $.07\phi$ units for each .10 difference in flatness. For minerals that are less flat than the lights, the hydraulic equivalent size would be less by about $.07\phi$ units for each .10 difference in flatness.

Although some experimental data are available on the relation between grain shape and transportation of pebbles as bedload (Einstein, 1937; Ho, 1939), the writer knows of no similar data on bedload movement of sand. Theoretical considerations would indicate that shape is not important in bedload transportation of fine sand. At low-water stages, the bed sediments characteristically move downstream in ripples, the sediment in each ripple apparently moving as a unit, regardless of difference in size or grain shape. In such transportation, grain shape would be an important factor only during the initial "picking up" of the sediment that forms the ripple. With increasing water velocity, the size of the ripples would increase, but the unity of each ripple should not be affected. With further increase of velocity, suspended-load movement would become predominant, and the settling velocity would be the important factor. Difference in grain shape resulting from sieving would thus not appear to affect the hydraulic equivalent sizes.

RELATIVE AVAILABILITY

The hydraulic ratio measures directly the relative availability of heavy and light minerals of equivalent hydraulic value in the stream load. Consequently, comparison of the hydraulic ratios for different sizes also shows how relative availability changes with size. The average hydraulic ratios for 16 minerals in 4 size grades in the Bosque samples have been presented in Table 11.

For all minerals except hypersthene, the relative availability is lower in the coarsest size grade than it is in some finer size grade. Examination of other data (Rittenhouse, 1943b; 1943c) indicates that this is usual for most minerals in the Rio Grande and its major tributaries. The most probable cause appears to be a deficiency of large-sized heavy minerals in the source rocks.

In general, the minerals at Bosque appear to be divisible into three groups with respect to the relation between availability and size. In one group, which includes zircon, barite, tourmaline, apatite, brown hornblende, and green hornblende, relative availability increases with decreasing size over the entire range presented in Table 11. In a second group, which includes magnetite, ilmenite, diopside, titanite, kyanite, garnet, and blue-green hornblende, the relative availability increases with decreasing size to the .175–.124 mm. or .124–.088 mm. size grade and then decreases. In the third group, which includes pyroxene and hypersthene, the relative availability is essentially the same in all four size grades.

Data on the relative availability are furnished also by the <.061 size grade, but for most minerals its interpretation is not clear. The <.061 mm. size grade includes a wide range of small sizes, and the hydraulic ratios are average values for this entire range. For zircon, hypersthene, pyroxene, titanite, garnet, and apatite, the hydraulic ratios are notably larger than those in size grades >.061 mm. This can only mean that for some sizes in this grade these minerals are available in larger amounts than in the coarser size grades. Whether the availability increases progressively with decreasing size to the finest sizes present or whether there is an increase in availability and then a decrease cannot be determined. For the other minerals, the hydraulic ratios in the <.061 mm. size grade are about the same as in the .088–.061 mm. size grade. This also could be interpreted in several ways—*i.e.*, to mean fairly constant relative availability in all sizes <.061 mm., an increased availability and then a decreased availability, or a decreased availability and then an increased availability.

In spite of the difficulty of interpreting the data for the <.061 mm. size grade, the other four size grades yield information that partially explain some of the apparent discrepancies in position and slope of the heavy-mineral size-distribution curves. The relatively low availability of zircon in the coarse size grades in comparison with magnetite and ilmenite, which are both of higher specific gravity, appears to account for the much finer average size of the zircon. The low relative availability of blue-green hornblende in both the coarse and fine sizes appears to explain why it is better sorted than pyroxene, whose relative availability is essentially the same in the four size grades for which the best data are available (Fig. 2).

SIZE DISTRIBUTION AT DIFFERENT PLACES IN THE STREAM

At different places in the same stream such factors as slope, width, depth, and roughness of the channel will not be the same. Consequently, hydraulic conditions will vary, not only with time and stage, but with place. Commonly, progressive downstream change in the hydraulic conditions will be reflected by progressive downstream decrease in the average grain size of the channel deposits. These differing hydraulic conditions will be accompanied by differing size distribution of the heavy minerals, similar to the differences that occur at any one transverse section of the stream. No change in the hydraulic ratio results.

At different places in the same stream, the relative availability and the physical characteristics of the minerals may be different, and this may result in changed equivalent hydraulic size and/or hydraulic ratios. Such changes may be progressive, as when various factors constantly act on the sediment during its downstream transportation, or may be sudden, as when new debris is introduced by tributaries. The changes in relative availability and/or physical characteristics may be caused by (1) selective wear or breakage, (2) selective sorting that does not follow the hydraulic equivalent size concept, (3) selective weathering or authigenic growth, (4) progressive change in source material, (5) floodplain contamination, and (6) tributary contamination. The change produced by each of these and the probable importance of such changes will be discussed briefly.

(1) Selective wear or breakage. All grains, as they are transported downstream, will be reduced in size, but the rate at which such reduction occurs will be selective. Presumably soft minerals will be reduced in size and will become rounded more rapidly than will hard minerals of the same original shape. Minerals with good cleavage will tend to split and thus change their sphericity or roundness or both at a different rate from minerals that lack good cleavage. Size of grain also is probably a factor in the absolute rate of wear of different minerals. The net effect of all these factors will be a differential change in the relative availability and the physical characteristics (mainly shape and roundness) of the sediment in different parts of the stream. These changes may be accompanied by changes in the hydraulic equivalent size and in the value of the hydraulic ratio.

Assuming no change in hydraulic equivalent size, the change in hydraulic ratio of any size grade will depend on the net loss or gain resulting from addition of grains from coarser size grades, loss of grains to finer sand grades, and loss of very fine particles (silt and clay). Thus when a heavy is reduced in size more rapidly than the light minerals of equal hydraulic value, the hydraulic ratios in the coarser size grade will be smaller, those in the intermediate size grades will show moderate gains or losses, and those in the finest grades will increase. When the differential loss of very fine particles is great, the hydraulic ratios may decrease in all size grades. When the light minerals are reduced in size more rapidly than the heavy, the hydraulic ratios for the coarse size grades will increase, those for the intermediate size grades will not change greatly, and those for the fine size grades will decrease. Great differential loss of very fine particles may increase the hydraulic ratios in all size grades. Decrease in the hydraulic equivalent size of minerals that wear more rapidly than the light minerals will reduce, increase, or have little effect on the hydraulic ratios depending on the size distribution of the deposit and the size distribution in the stream load.

Most available evidence indicates that wear is a relatively slow process. In their Mississippi River studies Russell and Taylor (1937) found that selective destruction of "less resistant" minerals appears to be of slight consequence between Cairo, Illinois, and New Orleans, Louisiana, a distance of 950 miles. In the Rio Grande, many grains show little evidence of either wear or breakage. At Bosque, New Mexico, the original crystal form of the easily abraded minerals apatite and hypersthene commonly has been only slightly modified. At the same locality, many grains of hornblende and pyroxene are characteristically sharp cornered, and some grains of these minerals have needlelike projections. Preservation of such features after a minimum of 60 miles transportation (from above Alameda, New Mexico) must indicate very slow wear.

Theil's (1940) tumbling barrel experiments with quartz, feldspar, garnet, tourmaline, hornblende, and apatite show that differential wear occurs but is very slow. Apatite was reduced in size most rapidly, quartz least rapidly. Even after travel of 5000 miles, however, about 15 per cent of the apatite of the coarsest size (2–1 mm.) still remained in that size grade. Thiel's experiments apparently simulated bedload transportation. In streams like the Rio Grande, where suspended-load transportation of fine sand is important, the absolute and differential rates of wear may be expected to be much slower. In mountain streams of high gradient, or streams in which crushing of sand by gravel is possible, differential wear may be more rapid. In such streams cleavage and brittleness may be more important factors than hardness.

(2) Selective sorting that does not follow the hydraulic equivalent size concept. One question that will probably be raised by some readers concerns the occurrence of heavy-mineral concentrates in stream sands and their relation to the concept of hydraulic equivalence. Genetically, concentrates of heavy minerals may form in two ways. Some concentrates result from selective erosion, others from segregation during transportation. If sands containing magnetite or other dark-colored heavy minerals are placed in a stream (as by undercutting of a bank), a dark heavy-mineral concentrate will form at the points of current impingement, the light-colored minerals having been selectively removed. If the current velocities are low, ripples composed only of light-colored minerals will form a few feet downstream. Diversion of the current at this time leaves an "erosion" concentrate.

If erosion continues, however, both dark and light minerals will be removed. The ripples downstream will be composed of both dark and light minerals, the dark minerals being concentrated at or near the crest—apparently forming a ripple on a ripple. Each ripple then appears to move downstream as a unit, neither gaining nor losing dark or light material. Apparently no selective action is involved in such downstream transportation.

When the ripple goes over the edge of a hole or a small delta in the channel, dark and light minerals form separate layers in the foreset beds. Such "deposition" concentrates may also form on bars, where the reduced water depth and velocity apparently results in a rather uniform distribution of both dark and light minerals in thin horizontal beds. Provided enough layers are sampled, the concentration of dark and light heavies from such deposits will be the same as that in the ripples.

In bedload transportation, downstream change in the hydraulic ratios resulting from selective sorting that does not follow the hydraulic equivalent size concept appear to result from erosion or "picking up" factors. Although less is known about concentration that occurs during higher stages when the finer size grades are carried in suspension, picking up rather than deposition would appear to be of most importance. Because such selective action permits the heavies to lag behind the light minerals whose hydraulic size otherwise is the same, the deposits would contain too many heavies and too few lights. The hydraulic ratios as determined from the bed deposits would thus be too high. Presumably, high density heavies would be effected more than those of low density. This may account for part of the difference between the hydraulic ratios in transit and bed deposits at Escondido, New Mexico.

If such selective action occurred in all streams to the same relative extent—*i.e.*, the hydraulic ratios in each stream being 10 or 20 or 30 per cent too high, it would have no effect whatsoever on the comparison of the deposits. Only when the selective action is not the same will the comparisons be in error.

In the selective action described above, differences in specific gravity would appear to be of first importance. Studies by Russell and Taylor (1937) and Pettijohn and Lundahl (1943) indicate that grain shape also may be significant. They found that the average sphericity of the light minerals decreased with distance from the source. Presumably, such a shape change would also occur for heavy minerals. If so, the hydraulic equivalent size and the hydraulic ratio may increase, stay the same, or decrease, depending upon the relative rate of sphericity change for the light and heavy minerals.

(3) Selective weathering or authigenic growth. Quartz is much more stable chemically than many of the heavy minerals in stream sands. Apatite, for example, is much more soluble than quartz, and the amphiboles (hornblende), pyroxenes, and ilmenite are more easily altered. Thus, differential weathering is a factor in changing the relative availability and physical characteristics of minerals during their downstream transportation. The quantitative importance of this factor over distances of a few hundred miles had not been evaluated. Thiel's (1940) tumbling barrel experiments showed that measurable quantities of apatite, hornblende, and feldspar were dissolved in 5000 miles of transportation, but the experiments did not indicate how much of this solution was from the original grains and how much from the many tiny chips removed by attrition.

Field evidence on the relation of weathering to distance of transportation is meager. In the Rio Grande some grains of apatite and garnet show solution cavities, and all gradations between ilmenite and its alteration product leuxocene occur. However, the proportion of such grains does not appear to change with increasing distance from the source. Changes in size or physical characteristics (mainly shape) by weathering probably is a very slow process in the Rio Grande. Russell reached essentially the same conclusion in his Mississippi River investigation. Downstream change in hydraulic ratios from weathering will probably be very small.

Concerning authigenic or secondary growth of minerals during downstream transportation, very little is known. Grains of pyroxenes showing authigenic terminations are fairly common in Rio Grande sands, but there appears to be no change in the number of terminated grains or in the size of the terminations with distance of transport. It seems most probable that the terminations were developed before the grains left the source area and that they have not been destroyed during downstream transportation.

(4) Progressive change in source material. If the rates of erosion increase in part or all of a drainage basin, subsoils are exposed by sheet erosion, and C-horizons by gullying. To the extent that the mineralogical composition of the several soil zones differ, the mineral composition of the sediment delivered to the streams would change progressively. The extent of the resulting downstream change in mineral composition would depend on the relative importance of two factors—namely, the rate of change in composition in the source area, and the average rate of downstream transportation. If different sizes of debris are not transported at the same average rate, the downstream change in composition may not be the same for different sizes of the same sediment.

Differential change in the rate of contribution from various parts of the drainage area will also tend to produce progressive downstream changes in mineral composition. Such differential rates of contribution are probably the rule rather than the exception. Here again, the change will depend on the rate of change in the source area and the average rate of downstream transportation. In the Rio Grande channel between San Acacia and San Marcial, New Mexico, the heavy-mineral composition has been changed greatly since about 1870 by increased contribution of sand from the Rio Salado (Rittenhouse, 1943c).

(5) Flood-plain contamination. In many streams, the lateral migration or widening of the channel by bank erosion is an important process. In lateral migration, erosion on the outside of bends is wholly or partially compensated by deposition on the inside of bends. Although no increase in sediment load necessarily results, the mineral composition may be greatly modified. Channel widening, in contrast, will increase the sediment load as well as modify the mineral composition. Similar increase in sediment load and mineral composition may be caused by trenching or by local erosion of the bedrock. The extent to which the hydraulic ratios will be modified by such floodplain contamination depends on the amount of sediment added and its mineral composition. If the shape of the minerals in the new sediment differs from that in the old, some change in hydraulic equivalent size may also occur.

(6) Tributary contamination. Sediment introduced by tributaries immediately modifies the hydraulic ratios in the main stream. The extent of change will depend on the difference in mineral composition of the tributary and main stream sands and on the relative amounts of sand from each. Hydraulic ratios of some minerals may

be increased; ratios of others may be decreased; and ratios of the remaining minerals may not change. In the Rio Grande, all these changes have been noted (Rittenhouse, 1943b; 1943c). Notable differences in physical characteristics of the sediment may affect the average hydraulic equivalent size, as may notable differences in the mode of transportation (bedload or suspended load) of the tributary sediment.

METHODS IN COMMON USE

GENERAL CONSIDERATIONS

In the preceding sections of this paper, the data on the occurrence of heavy minerals in six samples from the Rio Grande at Bosque have been presented, several factors governing the occurrence of heavies in river deposits have been outlined, and the hydraulic ratio has been presented as a new and more generally useful basis of comparing heavies in river deposits. Other investigators have used or recommended other methods of comparing heavy minerals. In general, these other methods are simpler and involve less expenditure of time per sample worked than do hydraulic ratios. Consequently, they should be used whenever they are applicable. The Bosque data and the principles of heavy-mineral occurrence that have been outlined provide a means of determining how useful these other methods may be and the extent to which they may be applicable to various types of investigations. Insofar as possible[8] the conclusions reached will be verified by statistical tests of significance.

Some investigators use the presence or absence of one or more "diagnostic" minerals. Other investigators determine, by estimation or counting, the relative frequency by number of the minerals in the heavy-mineral fraction. The mineralogy may be determined for the whole sample, for a single size grade, or for combinations of several size grades. One common practice is to compare these number frequencies, either directly, if the same number of grains have been counted in all samples, or as mutual percentages. In another method (Pettijohn, 1931) one mineral (garnet) is made a base to which the relative proportion of other minerals is compared. Cogan (1935) proposed a somewhat similar scheme but used the ratio of each component to that in a "reference" sample.

Weight percentages of the entire heavy-mineral fraction and of magnetite have also been used to a limited extent. Presumably, weights of mineral species other than magnetite would also have been used, had a practical method of segregating and weighing the different species been available.

Most heavy-mineral investigations have one or more of the following objectives: (1) to describe the mineral composition of a particular deposit; (2) to establish the similarity or dissimilarity of samples, generally for the purpose of geologic correlation; (3) to determine the change in mineral composition within a series of related samples so that factors related to the occurrence of heavies may be discovered and evaluated; (4) to locate the sources or to evaluate the relative importance of various sources of a deposit; (5) to provide data from which the past history of a deposit

[8] Because the Bosque samples were sieved and the minerals in each size grade examined separately, the data are not well adapted to testing those methods based on examination of unsieved samples. By using reasonable assumptions, however, it has been possible to compute the mineral composition in the whole sample from that in the various parts and to test the methods with these data. Because of the assumptions and because application of tests of significance to such results is difficult, the verification of conclusions regarding unsieved samples is less complete than desired.

may be interpreted; and (6) to find and aid the exploitation of economically useful minerals. The usefulness of any method for attaining these objectives depends to a considerable extent on the accuracy needed. For some problems, methods giving qualitative data may be satisfactory; for others qualitative data may be inadequate or misleading. Consequently, guiding principles rather than hard and fast rules will be stressed in the discussion to follow.

TABLE 12.—*Mutual percentages by number of eight selected minerals in actual and postulated samples at Bosque*

Sample	Size Grade	Ilmenite	Pyroxene	Hypersthene	Blue-green hornblende	Green hornblende	Zircon	Apatite	Garnet
	Mm.								
A	.250–.175	8.6	31.6	11.5	34.6	11.5	....	1.6	.7
G	do	35.0	17.4	8.2	18.1	11.3	.6	2.1	7.2
X	do	26.7	23.5	8.9	23.2	11.3	....	1.9	4.2
A	.088–.061	70.2	5.3	2.1	4.3	3.9	9.2	2.4	2.6
A	All	66.3	5.5	2.7	5.5	3.8	9.8	3.4	3.0
G	do	62.4	7.4	3.7	7.0	5.4	8.2	3.2	2.8

DIAGNOSTIC MINERALS

In using diagnostic minerals, the entire sample or one or more size grades may be examined, and the presence or absence of selected mineral species noted. The presence or absence of certain species may, in some cases, be sufficient to complete objective 2 and partially solve objectives 1, 4, 5, and 6.

If a mineral occurs only in small grains, however, it must be used with caution as a diagnostic mineral, particularly if a size grade rather than the entire sample is examined. At Bosque, zircon is such a mineral, being absent from the coarser size grades in all samples except G (Table 3). If entire samples are examined, such minerals as zircon would be absent only when the samples contained no sediment of fine size. Similarly, such minerals may be used if a size grade common to all samples is studied.

But the presence or absence of zircon, or minerals like it, is of questionable diagnostic value if single size grades having the same relative position with respect to the size distribution of the samples are used. This is well shown in Table 12 in which sample X[9] has no zircon, and sample A has over 9 per cent zircon. In each sample, the size grade selected for comparison is half way between the modal grade and the finest material in the size distribution.

SCALES OF ABUNDANCE

From estimates or actual counts, some workers classify grains in order of abundance, using such terms as abundant (A), common (C), scarce (S), and rare (R), or

[9] The textural range of the Bosque samples was not great enough to illustrate this point. Since the weight or number frequencies of any sample of known or postulated size distribution may be computed from the hydraulic ratios (Table 11) and the hydraulic equivalent sizes (Table 7), the mineral composition of a sample having the same sorting as sample G but an average grain size one Wentworth size grade larger was calculated. Percentages for all samples in Table 12 are mutual percentages for the minerals shown.

an arbitrary scale of numbers. The entire sample or selected size grades may be examined. To illustrate the discussion of this method by use of the Bosque data, the following classification will be adopted:—VA, >60 per cent; A, 30–60 per cent; FA, 15–30 per cent; VC, 7.5–15 per cent; C, 3–7.5 per cent; FC, 2–3 per cent; S, 1–2 per cent; R, 0–1 per cent; 0, absent. Of the Bosque samples, A and G differ most in

TABLE 13.—*Abundance* of eight selected minerals in actual and postulated samples at Bosque*

Sample	Size Grade	Ilmenite	Pyroxene	Hyper-sthene	Blue-green horn-blende	Green horn-blende	Zircon	Apatite	Garnet
	Mm.								
A	.250–.175	VC	A	VC	A	VC	O	S	R
G	do	A	FA	VC	FA	VC	R	FC	C
X	do	FA	FA	VC	FA	VC	O	S	C
A	.088–.061	VA	C	FC	C	C	VC	FC	FC
A	All	VA	C	FC	C	C	VC	C	FC
G	do	VA	C	C	C	C	VC	C	FC

* Abundance symbols represent percentages as follows:—VA, >60; A, 60–30; FA, 30–15; VC, 15–7.5; C, 7.5–3; FC, 3–2; S, 2–1; R, 1–0; and O, Absent.

texture and should show the greatest difference in relative abundance. The abundance data are presented in Table 13.

For the entire samples, the abundance differs only slightly in samples A and G, and the similarity of the two samples is apparent. For a size grade common to all samples, or for a size grade having the same relative position with respect to the size distribution of the samples, however, the differences are so great that a decision on their similarity or dissimilarity becomes a matter of personal opinion. In samples having a greater range in average size and degree of sorting, the differences in abundance should be greater than those in Table 13, both for entire samples and for selected size grades.

For some qualitative work required by objectives 1, 2, 4, 5, and 6, the use of abundance scales, particularly those based on examination of entire samples, appears to be justified. When samples differ by a $\sqrt{2}$ size grade or more in average grain size, however, differences in abundance, particularly for single size grades, may be misleading.

The abundance method is based entirely on number percentages in the heavy-mineral separate or some selected part of it. The percentage of heavy minerals by weight is not considered. Consequently, deposits that are similar if compared on the basis of numbers actually may not be from the same source or have the same history. Misleading conclusions concerning the relative importance of various sources may also be obtained if the weight of heavies is not considered.

PERCENTAGES BY NUMBER

General statement.—In many investigations, the percentages by number of various minerals in the heavy-mineral separate are determined by counting, and the percentages in different samples are compared to determine the degree of similarity or difference. If the percentages are identical, there is no question regarding the statistical similarity of the samples, though it is recognized (Dryden, 1935) that geological formations of different age or history might have identical or very similar mineral frequencies. If the percentages differ, as they usually do, the attempt to describe and interpret the meaning of various degrees of difference, particularly when a large number of samples are involved, is awkward, difficult, and a matter of personal judgment, unless some objective measure of similarity or difference is available (Krumbein and Pettijohn, 1938). This has led to the application of statistical measures to the comparison of mineral compositions, measures that are based on the theory of probability.

When statistical methods are used, however, it is assumed that the deposits to be compared are each homogeneous with respect to mineral composition—*i.e.*, that differences in mineral composition in a set of duplicate samples from the same deposit would be caused only by chance errors of random sampling. Such factors as variation in hydraulic conditions at different times and places of deposit may influence the mineral composition within a single deposit sufficiently to invalidate this assumption. Unfortunately, the magnitude of such nonsampling variations cannot be approximated on the basis of probability. Consequently, if an objective measure of similarity or difference is to be used, nonsampling variations must be evaluated experimentally or eliminated through choice of a suitable method of comparison.

Same size grade.—In the same size grades of different samples at Bosque, the number percentages of heavy minerals varies. For some minerals such differences are large, as for ilmenite for which, in the .175–.124 mm. size grade of the six samples, the range is from 6.0 to 35.8 per cent (Table 3). For other minerals the differences are small, as for brown hornblende which, in the same size grade, has a range of from 4.6 to 6.0 per cent. To determine whether these differences might be due to errors of random sampling, the frequency of occurrence of 11 minerals[10] in the 6 samples was compared by the chi-square method (Eisenhart, 1935; Snedecor, 1940). Chi-squares for the .250–.175 mm., the .175–.124 mm., the .124–.088 mm., the .088–.061 mm., and the $<.061$ mm. size grades were, respectively, 125, 207, 219, 184, and 186 (df = 50). Random sampling should give a chi-square value greater than about 75 only 1 time in 100. The differences between samples are highly significant for all size grades compared.

There was still the possibility that the highly significant differences in mineral composition might be due largely to the influence of one or two of the six samples. To eliminate this possibility, all possible combinations of four samples in the .175–.124 mm., the .124–.088 mm., and the .088–.061 mm. size grades were compared by the chi-square method. For none of these combinations was chi-square less than that to be expected 1 time in 100 as a result of random sampling.

These highly significant differences in heavy-mineral composition are not caused by nonrandom errors in the laboratory analysis of the samples. For sample F and a duplicate split from the same composite, the differences in mineral composition, when expressed as number frequencies, did not exceed those to be expected from ran-

[10] Ilmenite, pyroxene, hypersthene, blue-green, green, and brown hornblende, tourmaline, zircon, apatite, garnet, and diopside (C).

dom sampling for the .250–.175 mm., the .175–.124 mm., the .124–.088 mm., the .088–.061 mm., or the <.061 mm. size grades. In another test, the mineral composition in eight splits from the .124–.088 mm. size grade of a Rio Grande sample from Albuquerque, New Mexico, did not differ significantly (Rittenhouse and Bertholf, 1942). For beach sands, the nonsignificance of laboratory errors has been demonstrated by Krumbein and Rasmussen (1941).

Since the highly significant chi-squares are not caused by nonrandom errors in laboratory analysis, they must reflect highly significant differences in heavy-mineral composition (when expressed as number percentages) in the sand deposits at a transverse cross section of a river. If varying hydraulic conditions and other factors so modify the mineral composition of the closely spaced and texturally similar Bosque samples, it is difficult to see how statistical measures based on a size grade common to all samples can be used widely to compare the mineral composition of sediments. In most investigations, the textural range will be greater than that at Bosque, and consequently the mineral composition will differ even more.

It should be emphasized that the differences in mineral composition resulting from varying hydraulic conditions cannot be eliminated by counting more grains per sample, because the samples as brought to the laboratory are dissimilar. Nor, insofar as the comparison of samples from different localities is concerned, can the differences be reduced or eliminated by combining two or more spot samples into a composite, as recommended by Krumbein and Rasmussen (1941). It is true that such composites more nearly approximate the average mineral composition for the average hydraulic conditions existing at each sampling locality. But inherent differences in composition due to deposition under the different magnitudes or ranges of hydraulic conditions at various localities will persist. To determine experimentally the variation due to hydraulic conditions, the *entire* range and magnitude of hydraulic conditions under which *all* samples to be compared were deposited must be considered. Rarely will the necessary range be present. Probably comparison of number percentages of heavies from a single size grade will not be more useful than scales of abundance in reaching any of the six objectives.

Averaging size grades.—Russell (1936), recognizing that the heavy-mineral composition of sands varied with average size and sorting, averaged the composition of two size grades. He found that the average of a size grade common to all samples and a size grade "midway between the grade containing the maximum amount of material and the finest material" gave the best basis of comparison. Similar comparisons were made for samples A, D, F, and G of the Bosque suite, using the .088–.061 mm. size grade as that common to all, and the .175–.124 mm. (samples D and G), or the .124–.088 mm. grade (samples A and F) for the other size grade. A chi-square test indicates that the differences in mineral composition of these averages are highly significant.[11] This would indicate that averaging two size grades does not eliminate inherent differences in mineral composition at different parts of a trans-

[11] Combining size grades complicates statistical analysis, since the number of grains counted is an important factor in the tests. If the total number of grains counted in both size grades is used, the chi-square will be twice as large as if the arithmetic mean is used. In the test above, the differences were highly significant when the arithmetic mean was used. The differences would also have been highly significant if the total number of grains in both size grades had been used.

verse river section. Consequently, if statistical measures cannot be applied, the method is little better than that in which a single size grade is used.

Entire sample.—The number percentages of heavy minerals in all size grades of the six Bosque samples were calculated[12] and are presented in Table 14. A comparison of the percentages in this table with those in Table 3 indicates a lesser variation in

TABLE 14.—*Percentage by number of heavy minerals at Bosque*
As computed for entire samples

Sample	Ilmenite	Pyroxene	Hyper-sthene	Blue-green horn-blende	Green horn-blende	Brown horn-blende	Tour-maline	Zircon	Apatite	Garnet	Diop-side (C)	Others
A	46.9	3.9	1.9	3.9	2.7	.8	1.6	6.9	2.4	2.1	.9	26.0
B	34.7	4.9	1.0	6.7	4.4	2.1	2.0	7.6	1.8	1.3	.7	32.8
C	31.0	5.3	1.4	5.8	4.7	1.6	2.5	7.6	3.2	1.6	1.5	34.0
D	37.6	4.8	1.5	4.6	3.6	1.6	1.5	8.9	3.6	1.5	1.2	29.8
F	32.4	3.1	1.8	4.4	3.8	1.6	1.5	6.9	2.6	2.2	.9	38.8
G	44.6	5.3	2.6	5.0	3.9	1.6	1.9	5.9	2.3	2.0	.8	24.0

number percentages for the entire samples than for selected size grades. For ilmenite in the entire sample, the range is from 32.4 to 46.9 per cent, a difference of 14.5 per cent, whereas for the .175–.124 mm., the .124–.088 mm., the .088–.061 mm., and the <.061 mm. size grades the differences are about 30, 28, 25, and 28 per cent respectively.

Applying chi-square tests to this calculated data requires some assumption about the number of grains counted in each sample. Obviously, it is not correct to total the number of grains counted in the various size grades, because the three finest size grades are much more important than the others in determining the number percentages. However, the number must be greater than that in the finest size grade (which is most important in determining the number percentages). If this minimum number is used as a limiting approximation, a chi-square test indicates that the chances are about 1 in 20 that the differences in mineral composition in the samples at Bosque could be due to chance errors of sampling. For twice this minimum number of grains—a more reasonable assumption—the chances would be much less than 1 in 100 that the differences are due to sampling. Thus, it seems unlikely that statistical methods of comparison can be applied to number percentages of entire samples, and very unlikely that they may be applied where the difference in texture between samples exceeds the $\sqrt{2}$ size grade range existing at Bosque.

Where a series of samples changes progressively in texture, as along a beach or a

[12] It was assumed that the number of grains per unit weight of heavies was inversely proportional to the cube of the grain diameter (1.00, 2.83, 8.00, etc., for $\sqrt{2}$ size grades). For each sample the weights of nonmagnetic heavies were multiplied by these factors and the product for each size grade divided by the sum of the products for all size grades to give a weighting factor for each size grade in the sample. The number percentages in each size grade were then multiplied by the weighting factor for that grade. For each mineral the sum of these products for all grades was divided by the sum for all minerals to give the number percentage for the entire sample. Incidently, the three finest grades have a weight of 88 per cent or more in determining the final percentages. Therefore, if a sample is divided into a coarse and a fine part, and the heavies from only the finer part examined, the number frequencies so obtained will be essentially the same as those in the entire sample. Use of such partially sized materials will facilitate laboratory and petrographic work.

river, progressive changes in number frequencies may reflect only the textural differences at the various sites of deposit and not progressive sorting, wear, or other changes in mineral composition. This will be true regardless of whether the entire samples or selected size grades in them have been examined. In Figure 9B, the progressive changes in the number percentages of magnetite and pyroxene that might accompany progressive change in average grain size have been plotted. To attain objective 3, the effect of hydraulic conditions must be eliminated before the importance of other fundamental characteristics may be determined.

Coefficient of determination.—In the preceding discussion of number frequencies, no mention was made of a statistical measure designated the "coefficient of determination" (r^2) by Dryden (1935). This coefficient ranges from 1 to 0. When r^2 is 1, the percentages in the two samples are identical (or inversely proportional); when r^2 is 0, no species are common to both samples. The Bosque data have yielded some information on the use and the limits of applicability of this measure.

Using the number percentages for the entire samples, very high r^2 values were found. For the same size grades, r^2 was very high (.99) for some pairs of samples but was as low as .10 for other pairs (A and G, .175–.124 mm. size grade). If r^2 may range from .99 to .10 in the same transverse cross section of a river, it is difficult to see how this measure can be of value in comparing mineral compositions from different formations or parts of the same formation, especially where the textural range is great. If the entire samples rather than selected size grades are examined, however, the measure may be useful, at least for a textural range comparable to that at Bosque.

It should be noted, however, that values of r^2 cannot be correlated with significance or nonsignificance as determined by the chi-square method. For the .124–.088 mm. size grade of samples A and G, r^2 is .99 though the chi-square indicates highly significant differences between the samples. In contrast, for the .250–.175 mm. size grade of samples D and G, r^2 is .87, though the chi-square method indicates that the samples do not differ significantly.

It also appears that when there is a flood of one mineral in the two samples being compared, r^2 will be high regardless of the relative proportions of the other minerals in the samples. For example, in a pair of samples containing five minerals that have number percentages of 75 and 50, 10 and 30, 0 and 5, 15 and 0, and 0 and 15, r^2 is .64. If in a second pair of samples, the number percentages were 25 and 20, 12 and 20, 15 and 15, 21 and 20, and 27 and 25, r^2 would be .44. Probably most geologists, however, would consider the second pair of samples more alike than the first pair. With this the writer would agree. Incidently, if the number percentages had been 40 and 0, 30 and 10, 20 and 20, 10 and 30, and 0 and 40 for the five minerals, r^2 would have been 1.00. While it is not the writer's purpose to discuss the statistical basis of Dryden's coefficient of determination, further critical study seems necessary before r^2 is accepted as a basis for comparing mineral compositions.

RATIOS BY NUMBER

Same size grade.—Because of several advantages, some workers have used ratios by number in preference to percentages by number. Unless the similarity or differences between ratios can be determined by an objective measure, however, correlations and other conclusions based on ratios must be a matter of personal opinion.

To determine whether statistical measures based on laws of probability can be applied, ratios between selected minerals at Bosque were computed, and the differences between them were compared with the differences to be expected from random sampling. Because of time limitations, such comparisons were made only for samples A and G.

TABLE 15.—*Ratios and their errors for selected heavy minerals in samples A and G at Bosque*

Minerals	Size Grade					
	.175–.124 mm.		.124–.088 mm.		.088–.061 mm.	
	A	G	A	G	A	G
	Same specific gravity					
Zircon/Ilmenite	0.072 ± .040	0.025 ± .024	0.059 ± .013	0.045 ± .009	0.131 ± .013	0.116 ± .016
Blue-green hbl./Pyroxene	1.96 ± .25	1.04 ± .14	1.29 ± .15	1.32 ± .19	.82 ± .12	1.09 ± .19
Green hbl./Pyroxene	.68 ± .11	.53 ± .09	.95 ± .12	.98 ± .14	.73 ± .12	.94 ± .14
Brown hbl./Pyroxene	.26 ± .07	.26 ± .06	.25 ± .06	.21 ± .06	.29 ± .07	.43 ± .09
Green hbl./Bl. gr. hbl.	.35 ± .05	.82 ± .10	.74 ± .09	.74 ± .10	.90 ± .15	.86 ± .13
Brown hbl./Bl. gr. hbl.	.14 ± .03	.25 ± .06	.19 ± .04	.16 ± .04	.35 ± .09	.39 ± .07
Brown hbl./Green hbl.	.39 ± .09	.48 ± .12	.26 ± .06	.21 ± .06	.39 ± .09	.45 ± .09
	Different specific gravity					
Pyroxene/Ilmenite	1.26 ± .19	.18 ± .02	.211 ± .024	.103 ± .014	.075 ± .010	.114 ± .016
Bl. gr. hbl./Ilmenite	2.48 ± .35	.19 ± .02	.273 ± .027	.136 ± .016	.058 ± .009	.123 ± .016
Green hbl./Ilmenite	.86 ± .14	.10 ± .02	.202 ± .024	.101 ± .013	.055 ± .009	.106 ± .015
Brown hbl./Ilmenite	.33 ± .08	.05 ± .01	.052 ± .013	.021 ± .006	.022 ± .006	.048 ± .011
Garnet/Ilmenite	.12 ± .05	.10 ± .02	.084 ± .013	.031 ± .008	.039 ± .008	.058 ± .012
Zircon/Pyroxene	.057 ± .032	.137 ± .049	.28 ± .06	.43 ± .08	1.73 ± .23	1.02 ± .15
Garnet/Pyroxene	.09 ± .04	.55 ± .09	.40 ± .07	.30 ± .07	.51 ± .09	.51 ± .10
Zircon/Bl. gr. hbl.	.029 ± .016	.132 ± .047	.21 ± .05	.33 ± .06	2.12 ± .31	.94 ± .13
Garnet/Bl. gr. hbl.	.05 ± .02	.53 ± .09	.31 ± .06	.23 ± .05	.63 ± .12	.47 ± .09
Zircon/Green hbl.	.08 ± .05	.26 ± .09	.29 ± .06	.44 ± .08	2.36 ± .36	1.09 ± .16
Garnet/Green hbl.	.14 ± .06	1.04 ± .20	.42 ± .08	.31 ± .07	.70 ± .13	.55 ± .10
Zircon/Brown hbl.	.21 ± .12	.54 ± .19	1.13 ± .27	2.09 ± .58	6.1 ± 1.5	2.4 ± .5
Garnet/Brown hbl.	.36 ± .15	2.15 ± .54	1.63 ± .37	1.45 ± .41	1.79 ± .44	1.20 ± .26
Garnet/Zircon	1.7 ± .9	4.0 ± 1.3	1.44 ± .32	.70 ± .16	.29 ± .05	.50 ± .09

The sampling error (σ) of a ratio between two minerals is dependent on (1) the value of the ratio, and (2) the number of grains of the two mineral species that are counted (not on the total number of grains of all species counted). For the ratio A/B, σ may be calculated from the equation

$$\sigma_{A/B} = \pm \frac{\sqrt{\dfrac{A \cdot \sigma_B{}^2}{B} + \sigma_A{}^2}}{B}, \quad \text{(Mellor, 1926)}$$

where A and B are the number of grains of the two minerals that were counted, and σ_A and σ_B are their standard deviations as computed from the equation σ_A or $\sigma_B = \sqrt{npq} = \sqrt{A \cdot B/(A + B)}$.

The ratios and their standard deviations for seven selected minerals in three size grades of samples A and G are presented in Table 15. In this table, ratios between minerals of about the same specific gravity have been separated from ratios between minerals of different specific gravity. For minerals of about the same specific gravity, the difference between ratios exceeded the larger of the two standard deviations for 9 of the 21 pairs (43 per cent); for minerals of differing specific gravity, for 39 of the 42 pairs (93 per cent).

To evaluate more accurately the effect of density and to determine whether the differences for the same and different density heavies were greater than those to be expected from chance errors of sampling, further calculations were made. For each pair of ratios, t was determined from the equation

$$t = (A/B - A'/B')/\sqrt{\sigma_{A/B}^2 + \sigma_{A'/B'}^2},$$ [13]

where A and A' and B and B' are the numbers of the two minerals in the two samples. The probability that the difference could have occurred as a result of random sampling was found from table of t values.

Most of the differences between pairs of ratios could have occurred as a result of sampling, others could not. For each of the two density groups, however, t value of .82 should be exceeded 50 per cent of the time, and a t value of 2.9 should be exceeded 10 per cent of the time as a result of random sampling errors. For ratios between minerals of about the same density, $t = .82$ was exceeded 9 times in 21 (43 per cent) and $t = 2.9$ was exceeded 2 times in 21 (9.5 per cent). It may be concluded that differences in ratios between minerals of about the same density do not exceed those to be expected in random sampling. In contrast, for ratios between minerals of different density, $t = .82$ was exceeded 39 times in 42 (93 per cent) and $t = 2.9$ was exceeded 15 times in 42 (36 per cent). Differences in ratios between minerals of different density cannot be caused by errors of random sampling. An objective statistical measure of similarity or difference can be applied to some ratios but not to others, the critical factor being the difference in specific gravity of the minerals forming the ratio.

In a preceding section, the assumption was made, and later partially confirmed, that grains of hydraulic equivalent size are transported and deposited together. It also was shown that specific gravity and the hydraulic equivalent size of a mineral are closely related. If this is true, and if two minerals of about the same specific gravity, and therefore about the same hydraulic equivalent size, are being transported by a stream, deposition of one mineral will be accompanied by deposition of the other in the proportion in which they are being transported. The ratio between them will be constant, regardless of the absolute amount of both that is deposited at any place. In contrast, minerals of different specific gravity, and therefore differing hydraulic equivalent size, will not be deposited in the proportions in which they are transported, and, therefore, the ratios will differ as the hydraulic conditions at the site of deposit vary.

If this explanation is correct, ratios between minerals of about the same density should be the same, regardless of the average grain size and degree of sorting of the deposits in which they occur. Such ratios should be more widely applicable than any other number basis of comparing heavy minerals. Because ratios by number are

[13] Croxton and Cowden (1939), pp. 329–330, 875. Tables of t values may be found in most modern texts on statistics. For comparing two ratios, $n = 1$ and therefore the standard deviation and the standard error of each ratio are the same. Two degrees of freedom are used in determining the probabilities.

more easily obtained than hydraulic ratios, they should be used whenever possible in preference to hydraulic ratios. It should be clearly recognized, however, that such ratios by number do not give all available data for comparative purposes, since the weight of heavies as well as their proportion by number may vary. Clearly a correlation between deposits having similar ratios by number, but very different total weights of heavies, would be questionable. Also, as will be shown, ratios by number cannot be used to evaluate the relative importance of sediment sources.

For ratios between minerals of different density, it is very doubtful whether an objective measure of similarity or difference can be used, unless the average size and degree of sorting of the deposits is considerably less than that at Bosque. Such ratios are no more useful than number percentages and, like number percentages, may lead to invalid conclusions in some types of investigations. In considering the change in mineral composition with distance of transport, for example, a progressive change in ratio similar to that shown by the dashed line in Figure 9B might be obtained. This change would be due to change in hydraulic conditions, not to sorting, wear, or other factors. Similarly, if the ratio does not change, sorting or wear may have occurred but may have been compensated by changes caused by varying hydraulic conditions.

Same relative size grade.—For size grades having the same relative position with respect to the size distribution of the samples, data on ratios by number are not available from the Bosque investigation. It is known, however, that ratios by number between minerals of the same density measure the relative availability of the two minerals. If the relative availability changes with grain size, as it commonly does (Table 15), the ratios will reflect this difference in availability. When the difference in relative availability is large, ratios in closely related samples might appear to differ significantly. It seems reasonable to conclude that ratios between minerals in the same relative size grades will be more limited in applicability than ratios between minerals in the same size grades.

Entire sample.—Unfortunately, the extent to which ratios based on examination of entire samples may be used could not be determined by statistical tests of the Bosque data. Because the mineral composition in the entire samples was computed and not determined experimentally, the number of grains is not known, and the standard deviations cannot be estimated. Some comments are possible, however. As noted above, ratios by number between minerals of about the same density are measures of relative availability of the two minerals. If an entire sample is examined, the ratio will give an average relative availability for all sizes present. However, if the relative availability changes markedly with size (as for zircon at Bosque), a new factor is introduced. This factor becomes more important as the difference in average grain size and sorting of the samples increases. For some textural range, now unknown, an objective measure of similarity or difference cannot be used.

For ratios between heavies of different density, the examination of entire samples may reduce the variations due to hydraulic conditions and other factors for some textural ranges but may increase the variations for others. The usefulness of an objective measure for such ratios is questionable.

RATIOS BY WEIGHT

For minerals of about the same specific gravity, the ratios by weight in a size grade will be essentially the same as ratios by number. Therefore, for a size grade common to all samples number and weight ratios will have essentially the same advantages and disadvantages in comparing heavy-mineral compositions. Ratios by number, being more easily obtained, will be more widely used than ratios by weight.

For minerals of different specific gravity, the ratios by weight for a single size grade will not be the same as the ratios by number. Due to varying hydraulic conditions, inherent differences between the ratios by weight will be present. Consequently, an objective measure cannot be used to determine the statistical similarity or dissimilarity of samples.

In the same relative size grade, the ratios by weight will be the same as the ratios by number for minerals of the same density, but not for minerals of different densities. Ratios by weight will have the same advantages and disadvantages as ratios by number in the same relative size grades.

If entire samples are examined, there may be notable differences in the ratios by number and ratios by weight for minerals of the same density. Grains of large size may have almost no effect on the ratios by number but may be very important in determining the ratios by weight. Because the ratios by weight will measure the average relative availability of the two minerals in the entire sample, such ratios may be expected to have essentially the same advantages and disadvantages as ratios by number for entire samples.

WEIGHT PERCENTAGES

Same size grade.—In a single sample, weight percentages will be essentially the same as hydraulic ratios for those size grades that are on or just finer than the "straight section" of the light mineral cumulative curve. For two or more samples, the weight percentages will be essentially the same as the hydraulic ratios only in those size grades for which the light mineral cumulative curves of all samples are straight. In other size grades, the weight percentages will be too large or too small (Table 11), and the variations about these values will be large (Table 8). Therefore, in practice, weight percentages will be restricted to comparing one or two $\sqrt{2}$ size grades in suites of samples that have a narrow range in average grain size. If the use of weight percentages are not so restricted, misleading changes or differences in mineral composition of the type illustrated by Figure 9A may be obtained.

Where applicable, however, weight percentages have several advantages: (1) An objective measure of similarity or difference may be used, provided the standard deviation is increased by using the corrective factor C_x (Fig. 8); (2) samples having the same number frequencies but different absolute amounts of heavy minerals may be differentiated; (3) the relative importance of sediment sources may be evaluated quantitatively, an objective that cannot be attained with methods based on numbers.

Same relative size grade.—Like ratios by number or weight, and hydraulic ratios, weight percentages are measures of relative availability. But relative availability changes with grain size. Consequently, in comparing the same relative size grades of samples that differ greatly in texture, notably different weight percentages might be obtained, even if the samples actually are closely related. Because the change in relative availability with size cannot be predicted in advance for a stream or formation, suitable corrections that will permit the application of a measure of similarity or dissimilarity cannot be made.

Entire sample.—Weight percentages for the entire sample will measure the average relative availability in the sample as a whole. For samples from the same source but of differing texture, these average relative availabilities may differ considerably. The use of weight percentages for entire samples must be restricted to comparing samples of about the same texture.

HYDRAULIC RATIOS

Hydraulic ratios are determined for the same size grades or smaller units within different samples, but not for the same relative size grades, or for the entire samples. Provided the estimated standard deviation of hydraulic ratios are increased by using the corrective factor C_x, the similarity or dissimilarity of samples may be determined by objective statistical measures.[14] For reasons now unknown, C_x is larger for heavy minerals of high density than for heavies of low density.

Because any of the six objectives set forth under Methods in Common Use may be attained through use of hydraulic ratios, such ratios appear to be of wider applicability than any other method of representing mineral compositions. In comparison with the most widely applicable method based on numbers—namely, ratios between minerals of about the same density in a size grade common to all samples—hydraulic ratios have the following advantages: (1) Samples having the same ratios by number but different absolute amounts of heavy minerals may be differentiated. (2) Data on all minerals—not just those of the same density—may be obtained. Thus, for objectives 1, 2, 3, 5, and 6, more information is provided. (3) If the hydraulic ratios are known for all size grades in which heavy minerals occur, the size distribution by weight of any heavy mineral may be computed for any sample of known or postulated size distribution. This should be of value in the discovery and exploitation of economically useful minerals (objective 6). (4) Accurate quantitative data may be obtained on the relative importance of sediment sources.

The fourth advantage merits further discussion. Below the junction of a tributary with a main stream, the mineral composition depends not only on the relative frequencies by number of the various heavies in the tributary and main stream sands but also on the absolute amounts of heavy minerals in the two sands. This may be illustrated by a simple example. Assume that sand X carries 10 per cent by weight of heavies, of which half is apatite and half is hornblende. Because both of these heavies have essentially the same density and shape, the number frequencies would be essentially the same—namely, apatite 50 per cent; hornblende, 50 per cent. Assume also that sand Y carries only 2 per cent by weight of heavies, of which one-tenth is apatite and nine-tenths is hornblende. The number frequencies would be apatite, 10 per cent; hornblende, 90 per cent. A mixture of equal quantities

[14] To determine the similarity or dissimilarity of two samples: (1) Compute C_s from the equation $C_s = \sqrt{npq}/np$ (or a graph giving C_s for a range of n and p values); (2) determine C_{hr} from Figure 8; (3) multiply the hydraulic ratio (hr) by C_{hr} to give the estimated standard deviation (σ_{hr}) Compute t from the equation $t = (hr - hr')/\sqrt{\sigma_{hr}^2 + \sigma_{hr'}^2}$. From a table of t values (Croxton and Cowden, 1939, p. 875) determine the probability (when n or df is 2) that the existing difference could be due to chance errors of sampling ($P = .05$ indicates the difference is significant; $P = .01$, highly significant).

of sands *X* and *Y* would actually yield a sand containing 6 per cent by weight of heavies, of which apatite would be about 43.3 per cent and hornblende would be 56.7 per cent by weight. Yet, if the amounts of *X* and *Y* needed to produce this mixture were computed from the number percentages alone, the apparent contribution would be 5 parts of *X* to 1 part of *Y*.

If ratios are used, two answers are possible, neither of which is correct; the apatite/hornblende ratio indicating 2.8 parts of *X* to 1 part of *Y*; the hornblende/apatite ratio indicating 25 parts of *X* to 1 part of *Y*. Actual differences in percentage by weight between some of the tributary and main stream sands in the Middle Rio Grande Valley show that the weight percentages in foregoing example are not extreme (Rittenhouse, 1943b; 1943c). In Krynine's (1937) use of heavy minerals to evaluate sediment sources, it was tacitly assumed that the weight of heavies from all sources was the same. For most investigations this assumption is not valid.

SUMMARY

(1) The heavy-mineral composition in six samples from a transverse cross section of the Rio Grande was studied quantitatively. The number percentages in the nonmagnetic separates of seven size grades of each sample were determined by identifying a total of 27,630 grains. From these number percentages and the weights of heavies in each size grade, the size distributions by weight of 12 heavies in each of the 6 samples were computed.

(2) The size distributions by weight of the heavy and light minerals vary systematically among samples.

(3) These systematic variations are caused by (1) varying hydraulic conditions at the places and times of deposition, (2) the hydraulic equivalent size of each heavy mineral, (3) the relative availability for deposit of each mineral, and (4) some factor or factors now unknown.

(4) The hydraulic equivalent sizes of 16 minerals were determined and found to range from .2ϕ to 1.0ϕ. Hydraulic equivalent size of a mineral appears to be closely related to its density. Shape appears to be of secondary importance.

(5) For most of the 16 minerals, the relative availability changes with size.

(6) The magnitude of the variation due to unknown factors was determined. These variations are much greater for high- than for low-density heavy minerals but do not differ with size of grain. Different proportions of suspended-load and bed-load sediment in the six samples studied may cause part or all of the unexplained variation. For low-density minerals, all or part of the unexplained variation may be caused by random errors of sampling.

(7) The magnitude of variations due to hydraulic conditions was determined. These variations are greater for high- than for low-density minerals and, in contrast to the unexplained variations, are greater for some size grades than for others.

(8) The hydraulic ratio is recommended as a new and widely applicable method of representing heavy-mineral compositions. The hydraulic ratio is the same in the stream load and in all deposits from the stream load.

(9) Changes in the hydraulic ratios due to wear, sorting, and other factors are discussed.

(10) Methods of representing heavy-mineral compositions are discussed and their

limitations outlined. Diagnostic minerals, scales of abundance, percentages by number, ratios by number, ratios by weight, weight percentages, and hydraulic ratios were considered. Analysis was made for mineral compositions in a size grade common to all samples, in the same relative size grade, and in entire samples.

(11) No single method of representing mineral compositions is best for all heavy-mineral problems. For each problem, the method that will yield the needed information with a minimum of time and effort should be selected. If the limitatons of various methods are not observed, however, invalid conclusions may be reached.

(12) Of the methods based on numbers of grains, ratios between heavies of about the same specific gravity in a size grade common to all samples appear to be most widely applicable. Using such ratios, the statistical similarity or dissimilarity of samples may be established by objective statistical measures. Percentages by number are probably no better than scales of abundance, because objective statistical measures can be used only when the range in average grain size and sorting is slight.

(13) The hydraulic ratio is more widely applicable than any other method based on weights and can be used to solve some problems for which no number method can be used. Using hydraulic ratios, the statistical similarity or dissimilarity of samples may be established.

(14) It is believed that the principles of heavy-mineral transportation and deposition, and the limitations of various methods in representing heavy-mineral compositions discussed in this paper, will be applicable with some modification to beach and dune sands. Hydraulic equivalent sizes in such deposits will probably not be the same as those in fluvial sediments.

REFERENCES CITED

Cogen, William M. (1935) *Some suggestions for heavy mineral investigations of sediments*, Jour. Sedim. Petrol., vol. 5, p. 3–8.

Croxton, F. E., and Cowden, D. J. (1939) *Applied general statistics*, p. 941, New York.

Dryden, A. L. (1931) *Accuracy in percentage representation of heavy mineral frequencies*, Nat. Acad. Sci., vol. 17, p. 233–238.

——— (1935) *A statistical method for the comparison of heavy mineral suites*, Am. Jour. Sci., 5th ser., vol. 29, p. 393–408.

Einstein, Dr. sc. techn. H. A. (1937) *Der geschiebetrieb als wahrsheinlichkeits-problem*, Mitt. Versuchsanstalt Wasserbau Eidg. techn. Hochschule Zurich, p. 111, Zurich.

Eisenhart, C. (1935) *A test for the significance of lithological variations*, Jour. Sedim. Petrol., vol. 5, p. 137–145.

Ho, Dr. Eng. Pang-Yung (1939) *Abhangigkeif der Geschiebebewegung von der Kornform und der Temperatur*, Mitt. Preus. Versuchsanstalt Wasser-, Erd-, und Schiffbau, Heft 37, p. 43.

Krumbein, W. C. (1936) *Application of logarithmic moments to size frequency distributions of sediments*, Jour. Sedim. Petrol., vol. 6, p. 35–47.

———, and Pettijohn, F. J. (1938) *Manual of sedimentary petrography*, p. 549, New York and London.

———, and Rasmussen, W. C. (1941) *The probable error of sampling beach sand for heavy mineral analysis*, Jour. Sedim. Petrol., vol. 11, p. 10–20.

Krynine, Paul D. (1937) *Glacial sedimentology of the Quinnipiac-Pequabuck lowland in southern Connecticut*, Am. Jour. Sci., 5th ser., vol. 33, p. 111–139.

Mellor, J. W. (1926) *Higher mathematics for students of chemistry and physics*, 4th ed., p. 641, London.

Otto, G. H. (1933) *Comparative tests of several methods of sampling heavy mineral concentrates*, Jour. Sedim. Petrol., vol. 3, p. 30–39.

——— (1937) *The use of statistical methods in effecting improvements on a Jones sample splitter*, Jour. Sedim. Petrol., vol. 7, p. 101–133.

Pettijohn, F. J. (1931) *Petrography of the beach sands of southern Lake Michigan*, Jour. Geol., vol. 39, p. 432–455.

———, and Ridge, J. D. (1933) *A mineral variation series of beach sands from Cedar Point, Ohio*, Jour. Sedim. Petrol., vol. 3, p. 92–94.

———, and Lundahl, A. C. (1943) *Shape and roundness of Lake Erie beach sands*, Jour. Sedim. Petrol., vol. 13, p. 69–78.

Rittenhouse, Gordon (1942) *The size of sand as measured by sieving and microprojection* (Abstract), Geol. Soc. Am., Bull., vol. 53, p. 1930.

——— (1943a) *The relation of shape to passage of grains through sieves*, Indus. and Engin. Chem., Anal. Ed., vol. 15, p. 153–155.

——— (1943b) *The heavy mineral method for evaluation of sediment sources*, Submitted for publication as a U. S. Dept. Agric. Tech. Bull. (Publication deferred for the duration of the war.)

——— (1943c) *The source of modern sand in the Middle Rio Grande Valley, N. M.*, Jour. Geol., In press.

———, and Bertholf, W. E., Jr. (1942) *Centrifuge vs. gravity separation of heavy minerals*, Jour. Sedim. Petrol., vol. 12, p. 85–89.

Rubey, W. W. (1933a) *Settling velocities of gravel, sand, and silt particles*, Am. Jour. Sci., 5th ser., vol. 25, p. 325–338.

——— (1933b) *The size-distribution of heavy minerals within a water-laid sandstone*, Jour. Sedim. Petrol., vol. 3, p. 3–29.

Russell, R. Dana (1936) *The size-distribution of minerals in Mississippi River sands*, Jour. Sedim. Petrol., vol. 6, p. 125–142.

——— (1937) *Mineral composition of Mississippi River sands*, Geol. Soc. Am., Bull., vol. 48, p. 1307–1348.

———, and Taylor, Ralph E. (1937) *Roundness and shape of Mississippi River sands*, Jour. Geol., vol. 45, p. 225–267.

Smithson, Frank (1939) *Statistical methods in sedimentary petrology. I. Percentage composition of assemblages and their graphical study. II. Grain-size measurements and their graphical study. III. Cartographic methods*, Geol. Mag., vol. 76, p. 297–309, 348–360, 417–427.

Snedecor, George W. (1940) *Statistical methods*, 3d ed., p. 422, Ames.

Thiel, George A. (1940) *The relative resistance to abrasion of mineral grains of sand size.* Jour. Sedim. Petrol., vol. 10, p. 103–124.

[*Editor's Note:* Rittenhouse's 1943b reference is not listed in any subsequent bibliographic reference and is presumed to have never been published. The completed reference for Rittenhouse 1943c is *Jour. Geology* **52:**145–183.]

7

HEAVY MINERAL CONCENTRATING PROCESSES AND DENSITY/SHAPE/SIZE EQUILIBRIA IN THE MARINE AND COASTAL DUNE SANDS OF THE APALACHICOLA, FLORIDA, REGION[1]

FRANK W. STAPOR, JR.
Department of Geology, Florida State University, Tallahassee, Florida 32306

ABSTRACT: *Two* marine concentrating processes operate to produce the heavy mineral deposits found on beaches in the vicinity of Apalachicola, Florida. The process operating in the more energetic Gulf of Mexico tends to remove the coarser grains from initial populations. This results in a relatively fine-grained black sand concentrate being delivered to the beach for deposition. The reverse occurs in the less energetic bays and sounds; the process tends to remove the finer grains and the deposits are coarser. The quartz associated with both types of deposits is in equilibrium, but not settling equivalence, with ilmenite, staurolite, and kyanite. Settling equivalence is present among the ilmenite, staurolite, and kyanite populations of the sheltered beach deposits only. Zircon and rutile are too small in grain size to be in equilibrium with the other minerals. This 'too small' nature most probably reflects smaller variances of the initial populations.

The dune deposits were concentrated by 'gentler than normal' winds reworking beach sands containing areally limited heavy mineral deposits. Ilmenite, staurolite, kyanite, and quartz are in settling equivalence; the zircon and rutile are too small. Good evidence exists for a shape-based decrease in grain size for kyanite and rutile during wind transport.

DESCRIPTION OF DEPOSITS

Concentrations of heavy minerals can be found in three separate environments of the Apalachicola, Florida region: 1) open beaches facing the Gulf of Mexico, 2) sheltered beaches facing sounds and bays, and 3) coastal dunes adjacent to the Gulf of Mexico beaches.

The deposits of the open beaches occur as heavy mineral-rich layers concordant with the bedding of the surrounding quartz sand and having sharp, well defined contacts with the quartz sand. The black sand layers range in thickness from less than a centimeter to 20 cm, extend tens of meters parallel to the beach, and are 2 to 4 m wide. The thicker heavy mineral-rich layers contain pebble- to cobble-size wood fragments, glass bottles, and pebble- to cobble-size shells (predominantly *Venericardia*). The heavy mineral weight percent ranges from 10 to 80 and averages approximately 20%. Within the Apalachicola area these deposits are limited to those sections of the coast which are undergoing net erosion (see Fig. 1 for locations of the samples: two from St. George Island and 12 from St. Joseph Spit).

Deposits of heavy minerals on sheltered beaches occur as high berms of limited extent, less than 5 m parallel to the beach and 1 m wide. The thickness of the heavy mineral-rich sand ranges from 10 to 32 cm; the dip angles of these deposits range from 8° to 11°. Along the bay coast of Little St. George Island and the northern shore of St. Vincent Sound these deposits are localized in coves formed between tree stump and root protected headlands. The deposits sampled on the bay coast of St. Joseph Spit and at Yent's Bayou are similar in geometry to those of the open beaches. The heavy mineral weight percent ranges from 26 to 76 and averages 62%. As on the open beaches, these deposits are limited to those sections of the bay and sound coasts which are undergoing net erosion (see Fig. 1 for locations of the samples: four from St. Joseph Spit, four from Little St. George Island (bay side), one from St. Vincent Island (bay side), and one from the Yent's Bayou area).

The dune concentrations of heavy minerals occur as isolated, thin (2 to 5 mm thick) layers of black sand concordant with the slip-face bedding of the quartz sand; the boundaries were not as sharp and planar as those of the beach deposits. Seven samples were obtained from steep, slip face, beds from dunes along St. Joseph Spit.

[1] Manuscript received July 7, 1972; revised November 17, 1972.

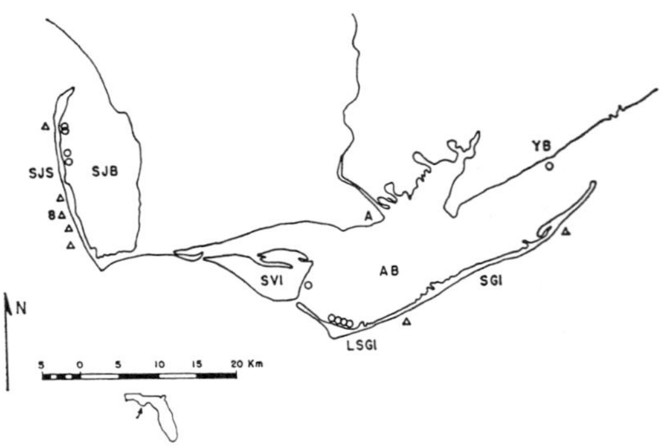

FIG. 1.—Location diagram of the heavy mineral deposits sampled for this study: SJS is St. Joseph Spit; SJB is St. Joseph Bay; SVI is St. Vincent Island; A is Apalachicola, Florida; LSGI is Little St. George Island; AB is Apalachicola Bay; SGI is St. George Island; and YB is the Yent's Bayou area. Triangles locate the open beach samples; circles locate the sheltered beach samples.

Beach sands in this region average 0.10% heavy minerals by weight; the offshore sands average 0.46% (Tanner, Mullins, and Bates, 1961). Therefore, these heavy mineral-rich layers are the result of an efficient concentration mechanism which deposits material during discrete periods of time. The place of operation, whether on the beach face or offshore, and the nature of the concentrating mechanism as indicated by the granulometry of the resultant deposits, are the subjects of the present paper.

SAMPLING PROCEDURE AND LABORATORY TECHNIQUE

Great care was taken to insure that material from just the heavy mineral-rich layer was included within each beach sample, in order to restrict each sample to one distinct sedimentation unit. As the boundaries of the heavy mineral-rich layers in the dunes are somewhat gradational, and as a volumetrically large sample is needed to provide the necessary heavy mineral concentration for effective mineral separation, a zone approximately one centimeter thick enclosing these heavy mineral-rich layers was sampled at each of the seven dune localities.

All samples were treated in the following manner to segregate the six dominant mineral constituents: (1) tetrabromoethane ($CHBr_2$-$CHBR_2$) flotation to remove the quartz, (2) magnetic separation of the ilmenite, (3) magnetic separation of the staurolite, (4) thallous formate-malonate[1] ($TICO_2H$ and $CH_2(COOTI)_2$) flotation to remove the kyanite, and (5) magnetic separation of the rutile and zircon. All samples had an initial minimum weight of about 150 grams. Magnetic separations were accomplished with a Frantz Isodynamic Separator. Each monomineralic segregate was seived with a nest of quarter phi Tyler screens for 30 minutes. The straight line character of the probability plots and the lack of extreme skewness and kurtosis moment measures require that these mineral size populations be considered Gaussian, or normal.

It should be emphasized that this technique of first mineralogically segregating the sample and then sieving each mineral separate to determine its statistical moment measures is quite different from techniques of many previous workers. Walger (1966) first sieved his samples on a set of 0.1 zeta screens (approximately one third phi) for 11 minutes, separated light minerals from heavies by tetrabromoethane flotation, and then microscopically differentiated grain mounts made of the respective 0.1 zeta heavy mineral factions. Ljunggren and Sundborg (1968) first sieved on a set of half-phi screens (time length unreported), and then mineralogically segregated the sieved fractions by tetrabromoethane and thallous formate-malonate flotation and magnetic properties. However, not all of these mineral segregations are monomineralic; Ljunggren and Sundborg (1968) report monomineralic grain size distributions only for garnet and magnetite. In addition, Walger (1966) and Ljunggren and Sundborg (1968) used frequency/cumulative frequency curves to obtain median sizes and not statistical moment measures as did the present writer.

McIntyre (1959) and Cordes (1966) microscopically differentiated and sized their samples. Hand (1967) and White and Williams (1968) obtained settling velocities of monomineralic segregations by means of a settling tube. These settling velocities were converted to grain sizes by using Rubey's (1933) fall velocity equation.

The study of polymineralic hydrodynamic equilibria is completely dependent on the precision with which the density and grain size parameters of the minerals comprising the sand are known. The precision of the former can be maximized by examining minerals of a constant or restricted density and by employing separatory techniques sensitive to density—such as flotation and magnetic properties. Inosilicates should be the last minerals considered for hydrodynamic equilibria studies because of their potential for extreme density variation. Optical differentiation of heavy mineral grain mounts most probably is the least effective separatory

[1] Great precautions should be taken when using this solution—fume hood and rubber gloves.

technique for maximizing density precision. To date the most common and most precise grain sizing technique is sieving on a set of quarter-phi screens for suitable time periods. (See Mizzutani (1963), for an analysis of the concept of 'suitable time periods.') Settling tube analysis should provide similarly precise grain size data.

The present writer has maximized the precision of both density and grain size parameters by using minerals of a constant or restricted density (no inosilicates or garnets), separating these minerals on the basis of density, and by sieving each respective mineral segregation on a set of quarter-phi screens for 30 minutes. The grain sizing technique of Hand (1967) and White and Williams (1967)—settling tube analysis—most probably comes the closest to matching the precision of the present writer's technique. As to the matter of monomineralic separations only the precision of Hand (1967) and White and Williams (1967) comes close to matching that of the present writer. However, Hand examined both garnet and hornblende and White and Williams examined black tourmaline—all minerals subject to potential density variation.

DETERMINING DENSITY/GRAIN SIZE HYDRODYNAMIC EQUIVALENCE

It should be emphasized that the omnipresent problem of grain size availability is fundamental to the discovery of hydrodynamic equivalence in polymineralic sands. The availability effect may be identified by: (1) comparing grain sizes of two or more similarly dense minerals, (2) using empirically derived equivalence formulae, and (3) attempting to relate the greatest number of mineral populations by the simplest mathematical expression. The second technique is somewhat independent of the sample and so the matter of availability can be rather directly attacked; the third technique maximizes the data employed and can show gross departures from any general trend.

The method of determining hydraulic equivalent sizes developed by Rittenhouse (1943), recently employed and verified by Walger (1966), is not recommended because the mineral populations are not treated as basically Gaussian distributions, but are analyzed rigidly divided into the intervals arbitrarily obtained through sieving. The described method as utilized by Rittenhouse does not consider, in effect, the problem of grain size availability. Furthermore, this technique does not maximize utilization of the data, but rather, as each mineral is determined separately, minimizes it.

IMPLICATIONS OF A CONCENTRATE DEPOSIT

Given the extremely low concentrations of heavy minerals relative to quartz for this region, 0.1% by weight for beaches and 0.46% by weight for offshore sands (Tanner, Mullins, and Bates, 1961), either the initial quartz population has been depleted or the initial heavy mineral population has been greatly increased during concentration, or some combination of both. The granulometry of the heavy minerals in the offshore and beach sands has not been undertaken because of the extremely low concentration of heavy minerals in these sands. This low concentration would necessitate volumetrically large samples, leading to problems in the correct sampling of discrete sedimentation units. However the granulometry of the offshore and beach quartz populations is known, and so the matter of concentration will be approached from the standpoint of changes in the *quartz populations.*

In order to concentrate heavy minerals at the expense of the quartz population, the mean grain size of the initial quartz population must be different from that quartz mean grain size in equilibrium with the resultant heavy mineral concentrate. The absolute value of the quartz mean grain size shift between that of the initial population and the resultant heavy mineral concentrate, as well as the weight percent of quartz within the concentrate, would indicate the efficiency/duration of the concentrating mechanism. The direction of the quartz mean grain size shift should help to identify the character of the concentrating mechanism.

PLACE OF OPERATION OF CONCENTRATING MECHANISM

Heavy minerals are concentrated primarily in one of two places: either on the beach face, or offshore with subsequent delivery to the beach face as a slug of heavy mineral-rich sand. The concordant bedding of the heavy mineral-rich and quartz-rich layers of the open beach deposits as well as the extremely low concentration of heavy minerals in the beach sand argue against the black sand layers having formed as lag deposits from the erosion of quartz-rich beach sand. In addition, the quartz grain size within the heavy mineral-rich layers is less than that of the quartz-rich beach sand (see Table 1), just the opposite of what an erosion/lag mechanism should produce. Furthermore, it should be noted that with the 0.1% heavy mineral concentration of the quartz-rich beach sand, 1.60 m of this sand would have to be winnowed *on the beach face* to produce an

TABLE 1.—*'X' mean and standard deviations for the open beach samples. The 't' tests evaluate the equality between the observed 'X' and the 'r' predicted by settling equivalence based on ilmenite*

	'X' mean (phi)	'X' Standard Deviation (phi units)	'r' mean on ilmenite (phi)	t-value	D.F.
Quartz	2.125	0.1365	1.934 (1.917)	4.34*	25
Kyanite	2.423 (2.449)	0.1362 (0.0971)	2.374 (2.357)	1.17 (2.88*)	26 (24)
Staurolite	2.476 (2.478)	0.1268 (0.0700)	2.424 (2.407)	1.24 (2.63*)	26 (23)
Rutile	2.763	0.0907	2.584 (2.567)	5.26*	26
Zircon	3.018	0.0859	2.704 (2.687)	9.52*	26
Ilmenite	2.714 (2.697)	0.0883 (0.0646)			
			'r' mean on staurolite (phi)		
Kyanite	2.423 (2.449)	0.1362 (0.0971)	2.426 (2.428)	0.06 (0.64)	26 (23)
Offshore, St. Joseph Spit	1.997	0.1518			
Open Beach heavy mineral-associated quartz	2.125	0.1365		2.17*	22
St. Joseph Spit Beach Sand	1.898	0.1577			
Open Beach heavy mineral-associated quartz	2.125	0.1365		4.20*	30

* Significant at the 0.05 level, two-tailed test.
(. . .) refers to modified open beach data.

observed black sand layer 5 cm thick having a 28% heavy mineral concentration. Likewise, the low concentration of heavy minerals in the offshore sands (0.46%) together with the occurrence of many interbedded heavy mineral-rich and quartz-rich layers in these open beach heavy mineral deposits probably precludes swash zone concentration of incoming sands from being important to the origin of these deposits. The berm deposits of the sheltered beaches are constructional features located high on the beach face and are not the result of in situ erosion/concentration. These data strongly argue that the observed black sands, on both open and sheltered beaches, were *not* concentrated on the beach face, but rather, were delivered to the beach face as a slug of heavy mineral-rich sand the concentration of which had occurred offshore.

THE OPEN BEACH HEAVY MINERAL-RICH LAYERS

As ilmenite is the most dense, most abundant heavy mineral present, it was chosen to be the basis for settling equivalence comparisons. It can be argued, however weakly, that as deposition is an inverse analogue of entrainment, all grains equivalent to the most difficult ones to entrain should also be entrained if they are available. These settling equivalence comparisons were made using McIntyre's (1959) formula which is based on Rubey's (1933) fall-velocity equation. For each of the six analysed minerals an overall mean grain size 'X' was computed using the 14 mesaured mean grain sizes; the ilmenite 'X' was used in McIntyre's formula to generate settling equivalent mean sizes 'r' for each analysed mineral. Using all of the data, only kyanite and staurolite are in settling equivalence with ilmenite in these deposits (Table 1). However, when extreme values which tend to greatly inflate the variance of kyanite/staurolite are removed (underlined points in Fig. 2), these populations also are not in settling equivalence with ilmenite (modified values of Table 1). The 'X'–'r' differences for zircon, rutile, kyanite, staurolite, and quartz indicate that these minerals are too small to be in settling equivalence with ilmenite, or conversely that the ilmenite is too coarse. Now, as being 'too coarse' is hydrodynamically impossible, either grain size availability and/or hydrodynamic processes other than settling are responsible for these populations being 'too small.'

If a hydrodynamic process is largely responsible, then the difference between 'X' and 'r' should be directly related to the density contrast between each mineral and the mineral chosen to be the basis of comparison. Quartz, theoretically, should have the largest 'X'–'r' difference and zircon the smallest. These differences are 0.314 phi units for zircon, 0.196 phi units for rutile, 0.071 phi units for staurolite, 0.092 phi units for kyanite, and 0.191 phi units for quartz.

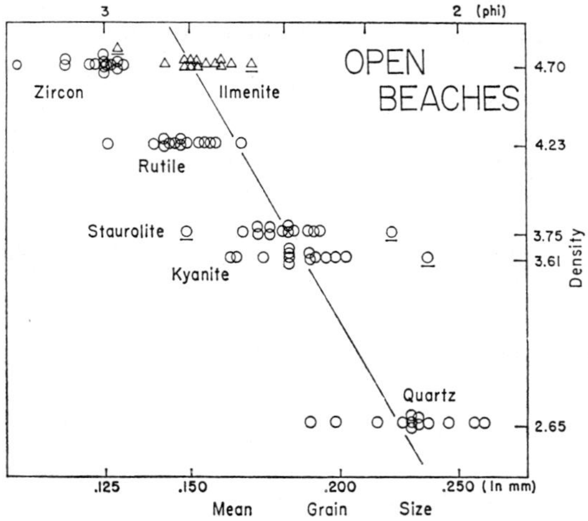

FIG. 2.—Plot of density versus mean grain size (natural log of grain size) for the open beach samples. The slant line is the regression line constructed on all of the ilmenite, staurolite, kyanite, and quartz data.

The zircon and rutile differences do not follow the theoretical trend and so, regardless of what hydrodynamic process effects may be reflected in their respective populations, the problem of availability is paramount. In addition, zircon and ilmenite have the same density and thus should have the same grain size. Hence, zircon and rutile are 'too small' relative to ilmenite because of grain size availability and the magnitude of this availability effect masks any hydrodynamic process effects present.

The quartz 'r' size demanded by ilmenite is 1.934 phi which cannot be statistically differentiated from the mean grain size of the off-short quartz sands. Availability does not appear to be responsible for the 'too small' character of the heavy mineral-associated quartz and so, the effects of a non-settling hydrodynamic process are most probably reflected in this 'X'–'r' difference.

The staurolite and kyanite populations cannot be so tested as to the importance of availability. However, as the 'X'–'r' differences are rather small, consonant with the theoretical effect of a non-settling hydrodynamic process, and the simple exponential relationship with ilmenite and quartz (Fig. 2), availability is most probably a minor factor. It is interesting to note that kyanite is in settling equivalence *relative to staurolite* (Table 1). This implies that a density contrast of 0.14 g/cm^3 and shape differences of rectangles and plates are insufficient to cause settling differences on a beach face and, by analogy, entrainment from offshore sands.

A semi-log plot of density versus measured mean grain size demonstrates a simple exponential relationship among ilmenite, staurolite, kyanite, and quartz (Fig. 2). The slant line is a regression line determined on all of the data of these minerals. This line accounts for 80.7% of the variance and is the best fit for the data:

$$g = 0.038e^{-0.196p} \quad p = \text{density}$$
$$g = \text{grain size in centimeters}$$

In summary, the ilmenite, staurolite, kyanite, and quartz populations are considered to be in hydrodynamic equilibrium (*non-settling*). Zircon and rutile are 'too small' because of grain size availability. These heavy mineral-rich layers of the open beach deposits most probably resulted from a concentrating process which tended to remove the coarser particles from the parent quartz-rich population. The lighter the mineral, the more efficient the removal.

THE SHELTERED BEACH HEAVY MINERAL-RICH LAYERS

Relative to ilmenite—the most dense, most abundant mineral present—staurolite and kyanite are in settling equivalence; quartz, zircon, and rutile are too small (Table 2). Using arguments similar to those employed in the open beach case, zircon and rutile are 'too small' for reasons of grain size availability and are not in equilibrium with the other minerals.

The quartz 'r' value demanded by ilmenite, 1.676 phi, is far coarser than the off-shore sand mean grain size in either Apalachicola Bay, 2.876 phi (Kofoed, 1961), or St. Joseph Bay, 2.258 phi (Stewart, 1962). A grain size availability problem clearly precludes the establishment of settling equivalence between quartz and ilmenite. The heavy mineral-associated quartz is significantly coarser than the mean grain size of the parent offshore quartz-rich sands (Table 2).

A semi-log plot of density versus measured mean grain size demonstrates an exponential relationship among ilmenite, staurolite, kyanite, and quartz (Fig. 3). The slant line is a regression line determined on all of the data of these minerals. This line accounts for only 55.8% of the variance and is *not* the best fit for the data, indicating that the equilibrium existing between these minerals cannot be described adequately by a simple exponential function.

In summary, the ilmenite, staurolite, and kyanite populations are in settling equivalence; zircon, rutile, and quartz are 'too small' for reasons of grain size availability. Quartz is in

TABLE 2.—*'X' mean and standard deviations for the sheltered beach samples. The 't' tests evaluate the equality between the observed 'X' and the 'r' predicted by settling equivalence based on ilmenite*

	'X' mean (phi)	'X' Standard Deviation (phi units)	'r' mean on ilmenite (phi)	t-value	D.F.
Quartz	2.024	0.2094	1.676	4.48*	17
Kyanite	2.175	0.1108	2.116	1.23	18
Staurolite	2.234	0.1231	2.166	1.34	18
Rutile	2.497	0.0898	2.326	3.98*	18
Zircon	2.741	0.1034	2.446	6.41*	18
Ilmenite	2.456	0.1013			
			'r' mean on staurolite (phi)		
Kyanite	2.175	0.1108	2.184	0.17	18
Offshore, St. Joseph Bay	2.258	0.3709			
Sheltered Beach, heavy mineral-associated quartz	2.052	0.2184		0.93	71
Offshore, Apalachicola Bay	2.873	0.6796			
Sheltered Beach heavy mineral-associated quartz	2.005	0.2184		4.38*	21

* Significant at the 0.05 level, two-tailed test.

non-settling hydrodynamic equilibrium with ilmenite/staurolite/kyanite. These heavy mineral-rich layers of the sheltered beach deposits most probably resulted from a concentrating process which tended to remove the finer particles from the parent quartz-rich population. This process of fine grain removal may have operated on the heavy minerals as well.

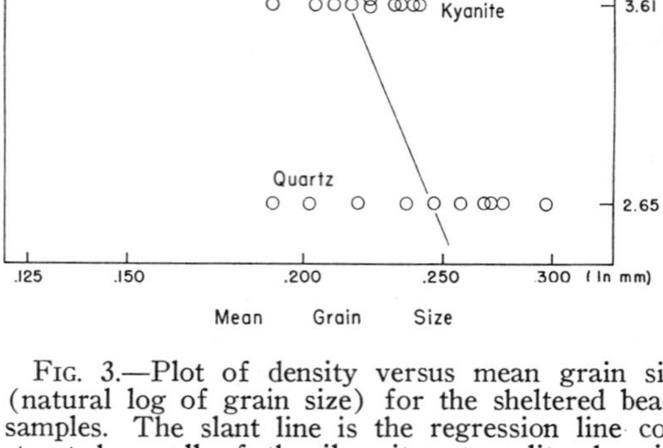

FIG. 3.—Plot of density versus mean grain size (natural log of grain size) for the sheltered beach samples. The slant line is the regression line constructed on all of the ilmenite, staurolite, kyanite, and quartz data.

COMPARISON OF THE OPEN AND SHELTERED BEACH HEAVY MINERAL-RICH LAYERS

These two types of heavy mineral deposit can be contrasted on three important bases: (1) the weight percent of heavy minerals, (2) the direction and magnitude of the quartz mean grain size shift—initial grain size minus that associated with the heavy minerals, and (3) the mean grain sizes of the respective mineral constituents. A relatively minor difference is that the deposits on the open beaches are more planar in geometry and much larger in areal extent than those of the bays and sounds.

Basic to any analysis concerning the genesis of these two types of heavy mineral deposits is the assumption that energy conditions capable of entraining heavy minerals are consistently more vigorous in the open Gulf of Mexico than in the shallow bays and sounds of limited fetch. The weight percentage of heavy minerals in the sheltered beach deposits is approximately three times that of the open beach deposits—60% versus 20%. In addition, the magnitude of the quartz mean grain size shift is larger for the sheltered beach samples than for those of the open beach: 0.87 phi units for Apalachicola Bay, 0.20 phi units for St. Joseph Bay, and 0.12 phi units for the open beach samples.

Frequent visits to St. Joseph Spit have established that these open beach deposits are found after either one storm or a week to 10 days of higher than normal wave energy. Visits to portions of Apalachicola Bay indicate that al-

TABLE 3A.—*Comparison between the open beach and the sheltered beach mineral populations*

	OPEN BEACH		SHELTERED BEACH			
	'X' mean (phi)	Standard Deviation (phi units)	'X' mean (phi)	Standard Deviation (phi units)	t-value	D.F.
Quartz	2.125	0.1365	2.024	0.2094	1.43	21
Kyanite	2.423	0.1362	2.175	0.1108	4.86*	22
Staurolite	2.476	0.1268	2.234	0.1231	4.66*	22
Rutile	2.763	0.0907	2.497	0.0898	7.00*	22
Zircon	3.018	0.0859	2.741	0.1034	7.10*	22
Ilmenite	2.714	0.0883	2.456	0.1013	20.74*	22

TABLE 3B.—*Comparison between the sheltered beach deposits of Apalachicola Bay and St. Joseph Bay*

	APALACHICOLA BAY		ST. JOSEPH BAY			
	'X' mean (phi)	Standard Deviation (phi units)	'X' mean (phi)	Standard Deviation (phi units)	t-value	D.F.
Quartz	2.005	0.2184	2.052	0.2243	0.33	8
Kyanite	2.159	0.0942	2.198	0.1441	0.52	8
Staurolite	2.219	0.0774	2.258	0.1852	0.47	8
Rutile	2.477	0.0842	2.526	0.1024	0.83	8
Zircon	2.750	0.0995	2.728	0.1233	0.31	8
Ilmenite	2.452	0.1025	2.463	0.1147	0.16	8

* Significant at the 0.05 level, two-tailed test.

though heavy mineral deposits are found after storms or periods of high wave activity, deposit-producing storms are very infrequent and not of exceptional violence relative to the storms which don't produce deposits.

The weight percentage data and the magnitude of quartz mean grain size shift suggest that the concentrating process is either more efficient or of longer duration in the bays and sounds than it is in the open Gulf of Mexico. The assumption that both types of deposits are concentrated and deposited by one higher energy event of limited time duration may not be as valid for the sheltered beach deposits as it is for the open beach deposits. If both are truly the result of one event of limited time duration, then it is difficult to understand why efficiency should be greater in the shallow bays and sounds of limited fetch than it is in the more energetic Gulf of Mexico. We may argue that the normal wave and hence, energy regimes of the bays and sounds may be sufficiently low that no or limited reworking of a partially concentrated heavy mineral-rich sand mass occurs. This would enable sands within the bays and sounds to be progressively concentrated before they are driven ashore for deposition. The significantly better sorting in the sheltered beach heavy mineral populations supports this argument. Conversely, the higher normal wave regime of the open beaches would rework any concentrated sand thus prohibiting a progressive enrichment of heavy minerals relative to the quartz. Furthermore, the more energetic storms of the open beaches would drive shoreward and hence deposit any heavy mineral concentrate much more rapidly than would the less energetic storms of the bays and sounds.

The direction of the quartz mean grain size shift can be related to the constituent mean grain sizes of these two types of heavy mineral deposit. The shift in the case of the sheltered beach samples is to make the heavy mineral-associated quartz coarser than the initial quartz, indicating that the concentrating process removes the finer grain sizes. The opposite is true for the open beaches: the shift is to make the heavy mineral-associated quartz finer than the initial quartz, indicating that the concentrating process removes the coarser grain sizes. The existence of two types of concentrating processes, fine removal and coarse removal, is not surprising. However the location of the fine-removing process in the lower energy bays and sounds and the coarse-removing process in the higher energy open Gulf of Mexico leads to a most unexpected *inverse* association of the grain size of heavy mineral deposits and energy level. The heavy minerals are significantly coarser in the sheltered beach than in the open beach deposits (Table 3A). This inverse association is not a function of the bays' and sounds' geographic proximity to a river mouth; the heavy mineral mean grain sizes are statistically identical for St. Joseph and Apalachicola Bays (Table 3B) and *no river flows* into St. Joseph Bay.

The nature of the concentrating process operating in the bays and sounds can be explained by appealing to a differential in the entrainment capability between the landward and seaward strokes generated by shoaling waves (May, in press). In this case the landward stroke is sufficient to move all the sediment whereas the seaward stroke is sufficient to move only the finer grains. This results in the coarser grains having a more rapid net landward movement than the finer grains.

The concentrating process operating in the open Gulf of Mexico cannot be explained in such a simple manner. Here, one must add to the landward-versus-seaward stroke differential, some other mechanism allowing the coarser grains to be more easily moved than the finer ones. Perhaps a dispersive stress analogous to the one described by Bagnold (1954) would allow the coarser grains to be preferentially elevated and moved more readily by the seaward stroke. This might result in the finer grains having a more rapid net landward movement than the coarser grains.

The observation that zircon and rutile grain sizes, sufficiently coarse to be in equilibrium with the open beach ilmenite, are present in the sheltered beach samples, strongly suggests that the initial variances of these two minerals are significantly smaller than those of the other heavy minerals. Hence, regardless of the energy level, the concentrating process never operates long enough for zircon and rutile to reach equilibrium with the other minerals.

In summary, the differences in granulometry between the sheltered and open beach heavy mineral deposits are caused by the difference in concentrating process, fine grain removal in the case of the former and coarse grain removal in the latter. The fine grain removing process results in a coarse-grained deposit and the coarse grain removing process in a fine grained deposit. This leads to an unexpected inverse association of concentrate grain size and energy level. The greater weight percentage content of heavy minerals in the sheltered beach deposits may be a result of their being progressively concentrated by many storms before being driven ashore. The open beach deposits most probably were driven ashore much more quickly; normal wave energy of the offshore region should rework sands and prevent progressive concentration.

THE COASTAL DUNE HEAVY MINERAL-RICH LAYERS

Relative to ilmenite—the most dense, most abundant mineral present—staurolite, kyanite, and quartz are in settling equivalence; zircon and rutile are too small (Table 4). As in the open and sheltered beach cases this 'too small' nature of zircon and rutile is a function of grain size availability; these two populations are not in equilibrium with the other minerals.

A semi-log plot of density versus mean grain size demonstrates the simple exponential relationship among ilmenite, staurolite, kyanite, and quartz (Fig. 4). The slant line is the regression line computed on all of the data of these minerals; it accounts for 86.2% of the variance and is the best fit for the data:

$$g = 0.036e^{-0.194p} \qquad p = \text{density}$$
$$g = \text{grain size in centimeters}$$

TABLE 4.—*'X' mean and standard deviations for the St. Joseph Spit coastal dune samples. The 't' tests evaluate the equality between the observed 'X' and the 'r' predicted by settling equivalence based on ilmenite*

	'X' mean (phi)	'X' Standard Deviation (phi units)	'r' mean on ilmenite (phi)	t-value	D.F.
Quartz	2.176 (2.234)	0.1333 (0.0956)	2.202 (2.2223)	0.50 (0.27)	12 (8)
Kyanite	2.497 (2.540)	0.0766 (0.0202)	2.489 (2.510)	0.24 (1.71)	12 (8)
Staurolite	2.539 (2.551)	0.0396 (0.0300)	2.529 (2.550)	0.44 (0.05)	12 (8)
Rutile	2.812 (2.840)	0.0585 (0.0411)	2.642 (2.663)	6.09*(7.34*)	12 (8)
Zircon	3.022 (3.047)	0.0583 (0.0458)	2.749 (2.770)	9.78*(10.82*)	12 (8)
Ilmenite	2.749 (2.770)	0.0460 (0.0352)			
			'r' mean on staurolite (phi)		
Kyanite	2.497 (2.540)	0.0766 (0.0202)	2.499 (2.511)	0.06 (2.50*)	12 (8)
St. Joseph Spit beaches	1.898	.1577			
St. Joseph Spit dunes	2.041	.0922		3.33*	27

(...) Refers to modified dune date.
* Significant at the 0.05 level, two-tailed test.

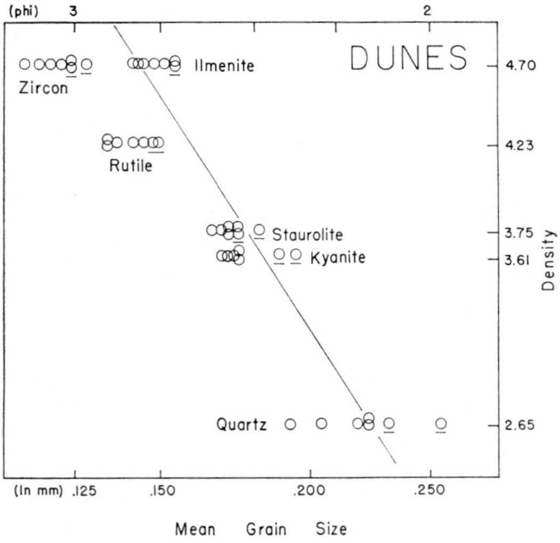

FIG. 4.—Plot of density versus mean grain size (natural log of grain size) for the St. Joseph Spit coastal dunes. The slant line is the regression line constructed on all of the ilmenite, staurolite, kyanite, and quartz data.

The grain size reduction suffered by an original marine population during eolian transport is largely a function of the horizontal and vertical distance traversed. Grain shape plays a definite though minor role. Along St. Joseph Spit the mean grain sizes of the beach and dune sands can be statistically differentiated (Table 4). Using the modified open beach data—mean unchanged, variance reduced—ilmenite and staurolite are significantly finer grained in the coastal dunes (Table 5). This presents a paradox inasmuch as kyanite and staurolite are in settling equivalence in the dunes. If one is significantly finer grained then the other should be also. An examination of the kyanite population in Figure 4 shows two extreme values which unduly influence both the mean and variance of this population. These two samples are from topographically low dunes (less than 1.5 meters); the other samples are from topographically much higher dunes (greater than 5 m). Furthermore, the coarsest values of the other minerals are likewise from these two samples (the underline points in Fig. 4). Rejecting the two samples from the topographically low dunes permits the resolution of the kyanite paradox. The ilmenite, rutile, staurolate, and kyanite populations are significantly finer grained in the higher dunes than in the modified open beach deposits (Table 5, values in parentheses).

It is of some interest that the kyanite and staurolite of the higher dune samples are not in settling equivalence. This cannot be a function of non-availability, because the predicted kyanite 'r' size of 2.511 phi is well within one standard deviation of the modified open beach mean of 2.449 phi. This difference may represent the effect of shape on the transportability of kyanite relative to staurolite. The former has two good cleavages which produce rectangular plates, the latter has only one much poorer cleavage. Of even more interest is the shift in rutile mean grain size; as the 2.763 phi rutile mean grain size of the modified open beaches is much finer than the 2.663 phi settling equivalence prediction for the higher dunes, it could be argued that, based on size/density, all of the rutile should be transported by the wind. Rutile, however, has a good prismatic cleavage which may account for this observed shift.

There are no heavy minerals available for settling equivalence with the 2.041 phi quartz of the St. Joseph Spit coastal dunes. All of the open beach heavy minerals, the only source, are much too small. There is quartz available, however, within the dune sand population to be in settling equivalence with the open beach heavy minerals. Why then the reduction in heavy mineral grain size in the dunes? Concentration of heavy minerals demands a reduction in the volume of associated quartz. The quartz size demanded for settling equivalence with the

TABLE 5.—*Comparison between the coastal dune and the modified open beach heavy mineral populations*

	COASTAL DUNES		MODIFIED OPEN BEACHES			
	'X' mean (phi)	'X' Standard Deviation (phi units)	'X' mean (phi)	'X' Standard Deviation (phi units)	t-value	D.F.
Kyanite	2.497 (2.540)	0.0766 (0.0202)	2.449	0.097	1.12 (2.04*)	19 (15)
Staurolite	2.539 (2.551)	0.0396 (0.0300)	2.478	0.0700	2.09* (2.22*)	19 (15)
Rutile	2.812 (2.840)	0.0585 (0.0417)	2.763	0.0907	1.29 (1.81*)	19 (17)
Zircon	3.022 (3.047)	0.0583 (0.0458)	3.018	0.0859	0.11 (0.71)	19 (17)
Ilmenite	2.749 (2.770)	0.0460 (0.0352)	2.697	0.0646	1.88* (2.36*)	19 (16)

(...) Refers to modified dune data.
* Significant at the 0.05 level, one-tailed test.

open beach heavy minerals is 2.150 phi and represents a reduction in dune sand volume of 88.5%; the quartz associated with the dune heavy minerals represents a reduction of 98.2%. It is hypothesized that relatively gentle winds, capable of moving only the finer sizes of dune sand, blowing across beaches spotted with areally limited black sand deposits, effect the concentration of the eolian heavy mineral-rich layers.

COMPARISON WITH PREVIOUS STUDIES

McIntyre (1959) analyzed five mineral species from three heavy mineral-bearing beach layers located along the northern shore of Lake Ontario. The size analyses were accomplished by microscopic examination. Using garnet as the basis for comparison, the heavy minerals (hornblende, orthopyroxene, and clinopyroxene) were in Rubey settling equivalence for all three layers. However, the quartz in two of the layers was too small. McIntyre attributed this result to differential removal of the coarser quartz grains during the reworking process which he believed formed the deposits. As no data concerning the relative weight percents of the minerals present or any information concerning the mineralogic character of either the parent beach or offshore population were presented, no analysis can be made as to the concentrating process responsible for these layers.

Hand (1967) compared settling velocities of ilmenite, garnet, hornblende, and quartz from beaches and dunes along the New Jersey coast. The beach heavy minerals were not in settling equilibrium with the quartz: the heavies were too small. Hornblende and garnet settled with the same velocity, ilmenite was slower, and quartz had the most rapid settling velocity. No data were presented on mineral percentages or initial population characteristics and so no analysis can be made of the concentrating process responsible for these layers. The ilmenite size necessary for equilibrium with garnet-hornblende might not have been available. The equality of garnet/hornblende suggests that the shape difference between these minerals—prismatic cleavage for hornblende and no cleavage for garnet—was of little significance in settling and, by analogy, entrainment. This is consonant with the observed equality of kyanite/staurolite in the marine deposits of the Apalachicola region.

Hand (1967) argues that as the quartz is much larger in grain size it should tend to shield the smaller heavy mineral grains during entrainment. Hence, the heavy minerals entrained equally with the quartz should be somewhat smaller than equilibrium settling requires. This slight reduction in grain size could serve to to increase the ratio of fluid drag to grain mass, greatly aiding the entrainment ability of the grain. However, the garnet/hornblende data did not support this; knowledge of the nature of the offshore sands or the initial beach population might aid in the testing of this hypothesis. This shelter process would probably be most effective in quartz-rich sands and/or during the initial functioning of a heavy mineral concentrating process.

The dune data presented by Hand (1967) show the garnet, ilmenite, and hornblende coarser in the dunes than in the beaches; the quartz remained the same. Furthermore, the quartz, garnet, and hornblende were very nearly in settling equilibrium. The heavy mineral mean grain size shift to coarser sizes in the dunes is rather puzzling if one assumes that the dunes have been derived from the analyzed beach heavy mineral populations. The lack of a shape-based departure of settling equilibrium between garnet and hornblende is not consonant with the staurolite/kyanite and rutile data from the St. Joseph Spit dunes.

Hood, Hood, and Oleson (1971) stated that the heavy mineral deposits of Sand Key, Florida, are associated with small quartz grain sizes. However, no size data were presentd for individual heavy minerals; their result was based on a composite (14 species) heavy mineral mean grain size. These writers attributed this quartz grain size deficiency to selective removal of the coarser quartz by eroding wave action, implying that these deposits were the result of in situ erosion/lag concentration. No information was provided concerning the geometry of these deposits and so this hypothesis cannot be tested adequately.

Cordes (1966) concluded that the heavy mineral deposits (placers) at Skagen, Denmark, result from surf and wave reworking/concentrating of dune sand; both the reworking/concentrating as well as the deposition taking place on the beach face. These deposits are geometrically very similar to the ones of the Apalachicola region open beaches and occur in an area of net coastal erosion located downdrift from the jetties protecting the Skagen harbor. One significant difference between the Apalachicola and Skagen localities is the heavy mineral content of the eroding coastal material: 12 to 20 weight percent heavies for the dunes and 1 to 2 weight percent heavies for the beach sands of the Skagen area. This is a far greater initial abundance than that present in the Apalachicola area.

An examination of reported grain size data—see Cordes (1966), Table 1, page 20—indicates that the grain sizes of the beach placers are almost identical to those of the normal beach sand and quite dissimilar to the grain sizes of the dune heavy minerals. Furthermore, the garnet, amphibole, and quartz of these beach placers are very nearly in settling equivalence; the opaque population (dominantly ilmenite) is too fine-grained, probably a function of grain size availability. The quartz of the beach placers is approximately 0.20 phi units finer than that of the normal beach sand and 0.10 units coarser than that of the dune sand.

A process of fine grain removal has most likely altered the dune parent mineral populations to those found in the beach placers during the course of the net coastal erosion. It should be expected then, that the greatest concentration would result along with the greatest degree of fine grain removal which would occur during the change from normal dune quartz sand (1.69 phi) to normal beach quartz sand (1.39 phi). However, the normal beach sands are not the ones most concentrated and so, this process cannot be directly responsible for the origin of the beach placers.

The present writer suggests that reworking of normal beach sand during periods of high wave energy, either offshore or in the surf zone, by a process of coarse grain removal concentrates these placers which are then driven ashore and deposited on the beach face. The process of fine grain removal which has altered the dune parent populations may be considered to be the normal energy event, the coarse grain removal process which may have formed the beach placers the abnormal event.

CONCLUSIONS

Granulometric analysis of ilmenite, zircon, rutile, staurolite, kyanite, and quartz of heavy mineral deposits from open beaches, sheltered beaches, and coastal dunes of the Apalachicola, Florida, region has made possible the identification of mineral equilibria and two types of marine concentrating processes. The process operating in the open Gulf of Mexico tends to remove the coarser grains; the process operating in the bays and sounds tends to remove the finer grains. This results in an inverse association of heavy mineral grain size and energy level of the incident waves: the higher energy regime (open Gulf of Mexico) produces the finer heavy mineral concentrates.

Ilmenite, staurolite, kyanite, and quartz are in equilibrium; zircon and rutile are too small to be in equilibrium with the other minerals. This 'too small' nature is a result of non-availability, probably reflecting smaller variances of the parent populations. The equilibrium between the heavy minerals and quartz is not settling equivalence—the quartz is too small. Ilmenite, staurolite, and kyanite are in settling equivalence only in the sheltered beach deposits. Kyanite and staurolite are in settling equivalance in both the open and sheltered beach deposits, reflecting a minimum influence of shape on the settling and entrainment of these similarly dense grains.

Ilmenite, staurolite, kyanite, and quartz are in settling equivalence in the dune deposits. All of the heavy minerals except zircon have a smaller grain size in the dunes than their respective initial marine populations. Kyanite in the higher dunes is too fine to be in settling equivalence with staurolite, probably reflecting a shape-based additional decrease in grain size during wind transport. The grain size reduction of rutile likewise probably reflects a shape-based decrease.

ACKNOWLEDGMENTS

The present writer would like to thank Barbara Keating and Thomas Kiernan for their help in the granulometric analysis of these samples. Dr. R. S. McEwen (Chemistry Dept., F.S.U.) provided the hood and laboratory space for the thallous formate-malonate separations. Dr. Duane Meeter (Statistics Dept., F.S.U.) gave advice on the statistical evaluation of the granulometric data.

Special thanks go to Dr. W. F. Tanner (Geology Department, F.S.U.) for his critical reading of the manuscript and general comments during the course of this investigation.

REFERENCES

Bagnold, R. A., 1954, Experiments on a gravity-free dispersion of large solid spheres in a Newtonian fluid under shear: Proc. Royal Soc. London, Proc., A, v. 225: 49–63.

Cordes, E., 1966, Aufbau und Bildungsbedingungen der Schwermineralseifen bei Skagen (Danemark): Meyniana, v. 16, p. 1–35.

Hand, B. M., 1967, Differentiation of beach and dune sands, using settling velocities of light and heavy minerals: Jour. Sed. Petrology, v. 37, p. 514–521.

Hood, W. C., Hood, S. D., and Oleson, S. M., 1971, Heavy minerals of northern Sand Key, Pinellas County, Florida: Southeastern Geology, v. 13, p. 187–198.

Kofoed, J. W., 1961, Sedimentary environments in Apalachicola Bay and vicinity, Florida: Unpub. Master's thesis, Florida State Univ., Tallahassee, Fla., 78 p.

LJUNGGREN, P. AND SUNDBORG, A., 1968, Some aspects on fluvial sediments and fluvial morphology. II. A study of some heavy mineral deposits in the valley of the river Lule Alv: Geog. Annaler, v. 50, p. 121–135.

MAY, J. P., in press, Selective transport of heavy minerals by shoaling waves: Sedimentology.

MCINTYRE, D. D., 1959, The hydraulic equivalence and size distribution of some mineral grains from a beach: Jour. Geology, v. 67, p. 278–301.

MIZUTANI, S., 1963, A theoretical and experimental consideration on the accuracy of sieving analysis: Jour. Earth Sciences, Nagoya Univ., v. 11, no. 1, p. 1–27.

RITTENHOUSE, G. W., 1943, Transportation and deposition of heavy minerals: Bull. Geol. Soc. America, v. 54, p. 1725–1780.

RUBEY, W. W., 1933, Settling velocities of gravel, sand, and silt particles: Am. Jour. Sci., v. 25, p. 325–338.

STAUBLE, J., 1971, The bathymetry and sedimentation of Cape San Blas shoal and shelf off St. Joseph Spit, Florida: Unpub. Master's thesis, Florida State University, Tallahassee, Fla., 87 p.

STEWART, R. A., 1962, Recent sedimentary history of St. Joseph Bay, Florida: Unpub. Master's thesis, Florida State University, Tallahassee, Fla., 82 p.

TANNER, W. F., MULLINS, A., AND BATES, J. D., 1961, Possible masked heavy mineral deposit, Florida panhandle: Econ. Geology, v. 56, p. 1079–1087.

WALGER, E., 1966, Untersuchungen zum Vorgang der Transportsonderung von Mineralen am Beispiel von Strandsanden der westlichen Ostsee: Meyniana, v. 16, p. 55–106.

WHITE, J. H., AND WILLIAM, E. G., 1967, The nature of a fluvial process as defined by settling velocities of heavy and light minerals: Jour. Sed. Petrology, v. 37, p. 530–540.

[*Editor's Note:* On page 407 of the article the completed reference for May, 1973 is *Sedimentology* **20**(2):203-211.]

8

Reprinted from *Jour. Sed. Petrology* **47**:753-770 (1977), by permission of the publisher

THE EFFECTS OF ENTRAINMENT ON THE HYDRAULIC EQUIVALENCE RELATIONSHIPS OF LIGHT AND HEAVY MINERALS IN SANDS[1]

RUDY L. SLINGERLAND
Department of Geosciences, Pennsylvania State University,
University Park, Pa. 16802

ABSTRACT: A semi-empirical critical velocity equation and critical suspension velocity equation have been developed to explain the settling velocity relationships and concentrations of light and heavy minerals in sands. Based on solutions to these equations and supporting experimental data, a fourfold classification of constant terminal settling velocity (*CTSV*) ratios of light and heavy minerals and heavy mineral concentrations has been constructed using ranges of $R*$, the boundary Reynolds number, and d/BKS, the ratio of grain size in question to the bottom roughness grain size. If $R*$ is less than five and d_h/BKS is about one, a heavy enriched, well size-sorted, fine to medium size sand deposit should result, with entrainment equivalent size ratios of heavy to light grains up to four times smaller than those predicted by settling laws. If $R*$ is greater than five but less than 70 and d_h/BKS is about one, a fine to medium deposit should be formed of very heavy enriched, very well size-sorted sand, whereas if $R*$ is greater than 70 and d_h/BKS is less than one, a coarse more poorly sorted light mineral deposit should result. Garnet and quartz in heavy and light layers from a Presque Isle, Lake Erie beach have equal sizes in the heavy layers and equal settling velocities in the light layers as predicted. A beach heavy and light mineral size budget is proposed which incorporates these mechanisms.

INTRODUCTION

Hydraulic equivalence was first stated by Rubey (1933) as the assumption that "whatever the conditions may have been which permitted the deposition of quartz grains of a certain size, these conditions would also permit the deposition of magnetite grains (of a size) that had the same settling velocity." Using known settling velocity laws he deduced that the grain diameter ratios of hydraulically equivalent minerals in a deposit would be inversely proportional to a power of the ratio of their densities. Subsequent studies, however, have revealed the inapplicability of the hypothesis as exactly stated by Rubey; light and heavy minerals in most sands are not in hydraulic equilibrium in his sense of the terms (Lowright et al., 1972).

Explanations for deviations from hydraulic equivalence have followed two lines of reasoning as summarized by Lowright et al. (1972): 1) hydraulic inequivalence is the result of differentially inherited size restrictions from the source rocks, or 2) hydraulic inequivalence is the result of differential transport of the various minerals in a sand. Rittenhouse (1943), van Andel (1950), Shanov (1964), and Briggs (1965) have presented data supporting the first line of reasoning. These data may be summarized by the work of Rittenhouse and Briggs.

Rittenhouse defined the hydraulic ratio of a mineral as "100 times the weight of a mineral in a known range of sizes, divided by the weight of light minerals of equivalent hydraulic size" (Rittenhouse, 1943, p. 1743). He measured hydraulic ratios for different sand sizes of the Rio Grande, showing that for the sands studied, the hydraulic ratio was lowest for the coarser sizes (0.25–0.175 mm). He interpreted this to mean that there existed a deficiency of larger sized heavies in the source rocks relative to larger sized lights.

Briggs (1965) documented that the modal separation between heavy and light settling velocities increased with increasing grain size for sample diameters larger than 0.125 mm.

[1] Manuscript received February 4, 1976; revised August 1, 1976.

He concluded that this reflected unrestricted size ranges available for the lights, as opposed to a restricted upper size limit for the heavies.

However, Feniak (1944) computed average grain areas from thin sections of 200 granitoid rocks in a study of the source size ranges of different minerals. The order of average area (estimated grain size) of minerals pertinent to this study is (largest to smallest): feldspars, pyroxene, olivine, hornblende, quartz, magnetite, sphene, and zircon. Thus it is not obvious that heavies always have smaller size distributions in the source, especially relative to quartz, as Briggs and others have assumed.

Subsequent work by Lowright et al. (1972) has shown that differential transport and not source restrictions can explain these observed deviations from hydraulic equivalence. They determined the settling velocity distributions of lights and heavies from Pleistocene tills (source), rivers, offshore sediments, and beaches and dunes in a complete sedimentary dispersal system near Presque Isle, Lake Erie. Their data show an equivalent range of sizes in the tills for the lights and heavies implying no source availability problem for the coarser heavies. However, in the Presque Isle beach and dune deposits after 20 miles of river and longshore transport, the largest and smallest heavies were no longer present. This was interpreted by them to indicate differential entrainment of the heavies versus the lights.

Differential entrainment of grains has been studied by von Engelhardt (1940), McIntyre (1959), Hand (1964), and Grigg and Rathbun (1969). Von Engelhardt determined from 14 modern dune and beach sands that the size distributions were explainable by a "process model" and that the influence of the source grain size distribution was slight. He demonstrated that the hydraulically equivalent sizes of quartz versus garnet and magnetite in suspension would be different from hydraulically equivalent sizes which were rolling.

Similarly, McIntyre hypothesized that smaller than predicted separations between quartz and garnet sizes from swash sands on Presque Isle were the result of selective removal of the coarser quartz by rolling.

Hand extended this idea to explain why heavy minerals in light layers on three New Jersey beaches had lower settling velocities than predicted by settling theory. He reasoned that because the heavies were smaller than the lights they would be shielded by the larger grains and also not protrude into zones of higher velocities. Thus the heavy that moves with a given light would settle slower than its larger hydraulically equivalent size, by an amount proportional to the degree of difficulty of entrainment. It is not apparent though, without invoking source size restrictions, why heavy grains should initially be smaller than the lights determining the bottom roughness.

Finally, Grigg and Rathbun (1969) showed again, that grains settled together will not necessarily be entrained together. Their analysis, however, used the Shield's criterion for the critical shear stress just needed to entrain sediment, which is not strictly valid for different sized grains in a sediment distribution.

Thus, deviations from hydraulic equivalence of light and heavy minerals in sands have been well observed, and differential entrainment has been investigated as a cause. However, no entirely adequate physical mechanisms have been proposed and no equations derived which can quantitatively predict the expected settling velocity distributions. This report presents quantitative theoretical and experimental support for deviations from hydraulic equivalence as a result of differential mineral entrainment. A critical velocity equation is developed which puts boundaries on the hydraulic equivalent sizes of light and heavy minerals in sands. The critical velocity concepts are then applied to the hydrodynamics of beach and foreshore light and heavy mineral deposition and tested with experimental data. As a result, a classification of heavy and light mineral settling velocity relationships and amount of heavy enrichment is proposed. Finally, in application to a natural system, the light and heavy minerals in beach foreshore deposits on Presque Isle, Lake Erie, are analysed in light of the theoretical and experimental data and a model is developed for the concentration of heavy mineral deposits in the beach foreshore zone.

THEORETICAL ANALYSIS

Velocity Profiles near Walls

The boundary velocity distributions used here are the Karman-Prandtl equations for the smooth and rough regimes of flow as defined by the boundary Reynolds number,

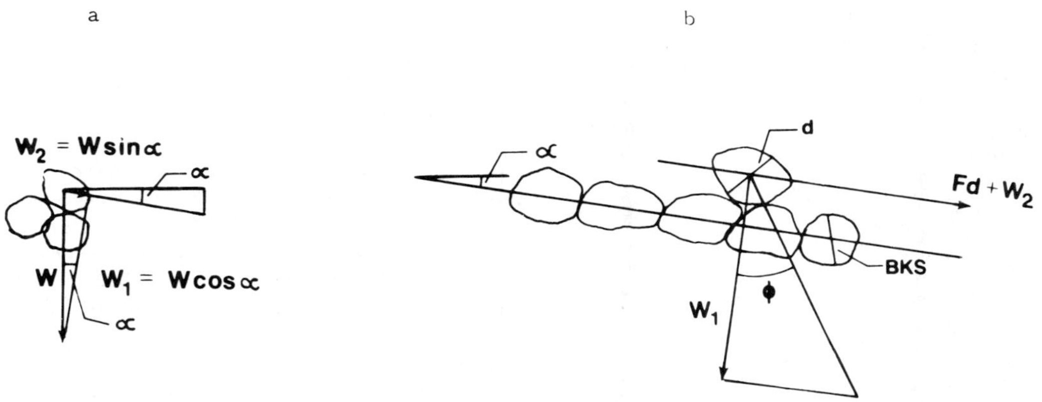

FIG. 1.—Force diagram for cohesionless particles of diameter d, on a bed of roughness BKS, slope $\alpha°$. ϕ is the reactive force point angle.

$R*$, which equals $U*BKS/\nu$ where $U*$ is the friction velocity, BKS is a parameter describing the size of the roughness elements on the bottom, and ν is the kinematic viscosity. With the modifications of Einstein (1950) to include hydraulically smooth boundaries ($R* < 5$), transitional ($5 < R* < 70$), and rough boundaries ($R* > 70$), the velocity distribution becomes:

$$V_y = 5.75 \log_{10}\left(\frac{30.2xy}{BKS}\right)U* \tag{1}$$

where V_y are mean point velocities at a high y from the bed datum, and x is a correction factor from Einstein (1950, Fig. 4).

Force Balance on a Grain on a Bed

Now consider a bed of loose cohesionless sand over which flows a fluid of uniform physical characteristics. For prediction of sediment transport, the forces to be considered are 1) inertial and viscous fluid drag forces, 2) fluid lift forces, and 3) reactive forces, the most important of which is the submerged grain weight.

Inertial force or form drag and viscous or surface drag are combined in the drag equation representing the force imparted to a stationary spherical particle within a moving fluid:

$$F_d = C_d A_p \frac{\rho_f V^2}{2} \tag{2}$$

where: C_d = the coefficient of drag
A_p = the projected cross-sectional area of the particle
ρ_f = the density of the fluid
V = the fluid velocity on the grain.

The influence of lift is not considered in this study.

The reactive forces oppose the resultant of all active forces through the point of contact between a grain and the grains upon which it rests. The primary reactive force is the grain submerged weight which is assumed to act through the grain center of gravity, whereas the point of application of drag forces will vary with the grain shape and relative magnitudes of viscous to inertial forces. From Fig. 1a, the resolved submerged weight of a spherical grain on a slope of angle α is equal to:

$$W_1 = \frac{\pi d^3}{6}(\rho_p - \rho_f)g \cos \alpha. \tag{3}$$

Parallel and down the slope is the resolved weight component of the grain:

$$W_2 = \frac{\pi d^3}{6}(\rho_p - \rho_f)g \sin \alpha. \tag{4}$$

From Fig. 1b, the resolution of the forces about the reactive force point yields the condition of incipient motion:

$$\tan \phi \propto \frac{F_p}{F_n} \propto \frac{W_2 + F_d}{W_1} \tag{5}$$

where: F_p = force parallel to slope
F_d = down slope drag force
F_n = force normal to a slope inclined α degrees.

This is a two dimensional balance of moments at the grain centroid and at the reactive force point and does not consider frictional forces between adjacent particles.

Substituting equations 2, 3, and 4 in 5, one obtains for incipient motion of a spherical particle of diameter d on an equidimensional bed of diameter BKS:

$$\tan \phi \propto \frac{3C_d V^2 \rho_f}{4dg \cos \alpha(\rho_p - \rho_f)} + \tan \alpha. \quad (6)$$

The critical velocity may be defined as the fluid velocity at which sediment particle motion commences, measured at 'y' units from a datum on which the surface sediment rests. Solving equation 6 for Vc, yields:

$$\mathrm{Vc} \propto \sqrt{\frac{4dg \cos \alpha(\rho_p - \rho_f)(\tan \phi - \tan \alpha)}{3C_d \, \rho_f}} \quad (7)$$

Since Vc is the critical velocity of a grain of diameter d, on a bed of roughness BKS, Vc, as a first approximation can be made equal to $U*$, the boundary friction or shear velocity. In other words, the needed transport energy represented by a velocity, Vc, must equal the amount of energy lost to transport, represented by $U*$. This is not strictly true since according to Chezy for open channel flow (Leliavsky, 1966):

$$\tau_o = C^2 V^2$$

or:

$$V = \sqrt{\tau_o/C^2} = \sqrt{U*^2 \rho_f/C^2} = \frac{\sqrt{\rho_f}}{C} U*$$

where: V = mean fluid velocity
τ_o = intensity of tractive force of moving water per unit area of bed,
C = Chezy C.

But since $C = 1.49 \, R_h^{1/6}/n$, and the variation of these parameters is of small order of magnitude compared to the range of the variables of interest, the approximation is assumed to be valid.

Thus, combining equations 1 and 7 yields:

$$Vc \propto 5.75 \log_{10}\left(\frac{30.2xy}{BKS}\right) \sqrt{\frac{4dg \cos \alpha(\rho_p - \rho_f)(\tan \phi - \tan)}{3C_d \rho_f}} \quad (8)$$

which is the critical erosion velocity measured at a height y from the bed datum.

Some Additional Assumptions

In addition to the already stated and generally accepted assumptions, these implicit assumptions have been made. 1) The relationship $\tan \phi \propto F_d/F_n$, is an assumption that $F_d = f(\tau_0 A_p)$ since the force about the reactive force point is actually due to a torque from an applied shear stress. 2) Theoretically, the velocity term in equation 8 is the fluid velocity impacting on the cross-sectional area of a sphere free from wall effects, and may be functionally related to Vc.

Equation 8 includes the C_d which many investigators take to be constant. The increase in drag on smaller sized particles will cause a reduction in Vc for which the C_d will account. Although the general Cd vs. Re curve was developed for a sphere falling freely through a fluid, it has been shown that the dynamics of the drag force caused by water on a sphere resting on a surface is similar (Coleman, 1967).

Quantification of Sediment Parameters

Equation 8 represents the readily quantifiable parameters relevant to sediment transport. Before application to a natural system, the following more complex variables must be accounted for: random turbulent velocity fluctuations, grain shape, the point of application of the drag forces, packing, cohesiveness, and quantification of the reactive force angle.

Kalinske (1943, p. 274) found that, "fluctuations equal to twice the mean velocity at a point in the turbulent zone near a boundary can be readily expected." Similar measurements by Grass (1967) confirm that as the flow regime passes from the transition zone into the hydraulically smooth regime with a viscous sublayer, it is reasonable that the turbulent velocity fluctuations at the level $y = BKS$ (cf. Yalin, 1972, p. 98) will decrease and become negligible. There is no theoretical reason why this decrease should not be continuous, and as a first approximation, a coefficient β_1 will be introduced to account for the variation in critical velocity due to turbulence. β_1 will equal 0.5 at $R* > 70$, increasing to 1 at $R* < 5$. Thus from the equation for a straight line:

$$\beta_1 = -0.0077R* + 1.038 \quad (9)$$

for $5 < R* < 70$.

For mathematical simplicity, grain shape has been assumed to be spherical. In this study quartz and garnet were used because they were reasoned to be subequant and nearly identical in shape characteristics.

The point of application of the fluid forces can be reasoned to be some function of the ratio BKS/δ', the size of the roughness elements to the thickness of the viscous sublayer, since increased immersion in the viscous sublayer will increase viscous forces which tend to raise the point on the grain through which the force acts. Experimental data from White (1940) indicate that for small BKS, $\alpha\eta = 0.37$ where α is a coefficient reflecting the point of application of fluid forces and η is a packing coefficient. Since small BKS implies close packing or $\eta \approx 1$, α must be on the order of 0.3 to 0.4. Since BKS/δ' is proportional to $R*$, a second coefficient β_2 is defined as:

$$\begin{aligned} \beta_2 &= k_1 \qquad R* < 70 \\ \beta_2 &= k_2 \qquad R* > 70. \end{aligned} \tag{10}$$

In this study the packing coefficient is assumed to be constant and equal to 1.0.

Below $d = 600\ \mu$ sediment may exhibit both frictional and cohesive properties (Sundborg, 1956). However, for sediment in the size range composed mainly of sand, Grass (1970) has shown that for d down to $50\ \mu$ the critical velocity curve can be expected to flatten but still decrease for decreasing grain size where grain cohesion is not an important property.

For noncohesive sediments, the wide range of grain sizes and possible combinations of grains in contact with each other will yield a wide range of reactive force angles. This angle, ϕ, according to Miller and Byrne (1966), or Ippen and Eagleson (1955), can be defined by:

$$\tan\phi = \frac{0.866}{\left(\left(\frac{d}{BKS}\right)^2 + 2\left(\frac{d}{BKS}\right) - \frac{1}{3}\right)^{1/2}} \tag{11}$$

for the most likely path of rolling of a grain, where d, BKS, and ϕ are defined in Fig. 1. It has been found that allowing $BKS = d_{65\%}$[2] in equation 11 yields for measurements made on natural sands a spread of values around 28 degrees, the exact value of which depends on the sorting of the size distribution and the grain size in question.

Substituting the preceding coefficients into equation 8 gives the final approximation of the critical erosion velocity needed for a particle of size d on a bed of roughness BKS:

$$Vc = 5.75 \log_{10}\left(\frac{30.2xy}{BKS}\right)\beta_1\beta_2 \sqrt{\frac{4dg\cos\alpha(\rho_p - \rho_f)(\tan\phi - \tan\alpha)}{3C_d\rho_f}} \tag{12}$$

where: $\tan\phi$ = reactive force angle (equation 11)
C_d = coefficient of drag
x = Einstein correction factor (Einstein, 1950, Fig. 4)
BKS = roughness size = $d_{65\%}$
α = bed slope in degrees
d = grain diameter
g = gravitational acceleration
ρ_f = density of fluid
ρ_p = density of particle
β_1 = turbulent velocity fluctuation coefficient
β_2 = coefficient for point of application of fluid force.

Solution of the Equation

Equation 12 as presented is not readily solvable since the Vc is a function of the coefficient of drag, which is a function of the fluid flow around the grain. The solution to the equation is obtained by using the Schiller equation (1933, *in* Graf, 1972), relating the coefficient of drag to the Reynolds number for a falling sphere:

$$C_d = \frac{24}{Re}(1 + 0.15Re^{0.687}). \tag{13}$$

We may define a critical Reynolds number as:

$$Re_c = \frac{V_c\, d}{\nu} \tag{14}$$

Substitution of equation 14 in 13 and 13 in 12 gives a complex polynomial in Vc which may be solved by the Newton-Raphson Method (James et al., 1967, p. 139).

Comparison of Theoretical with Experimental Results for the Critical Velocity Equation

We must now evaluate the assumptions in the previous section and the correctness of the

[2] $d_{65\%}$ is defined as that grain size for which 65% of the distribution is finer.

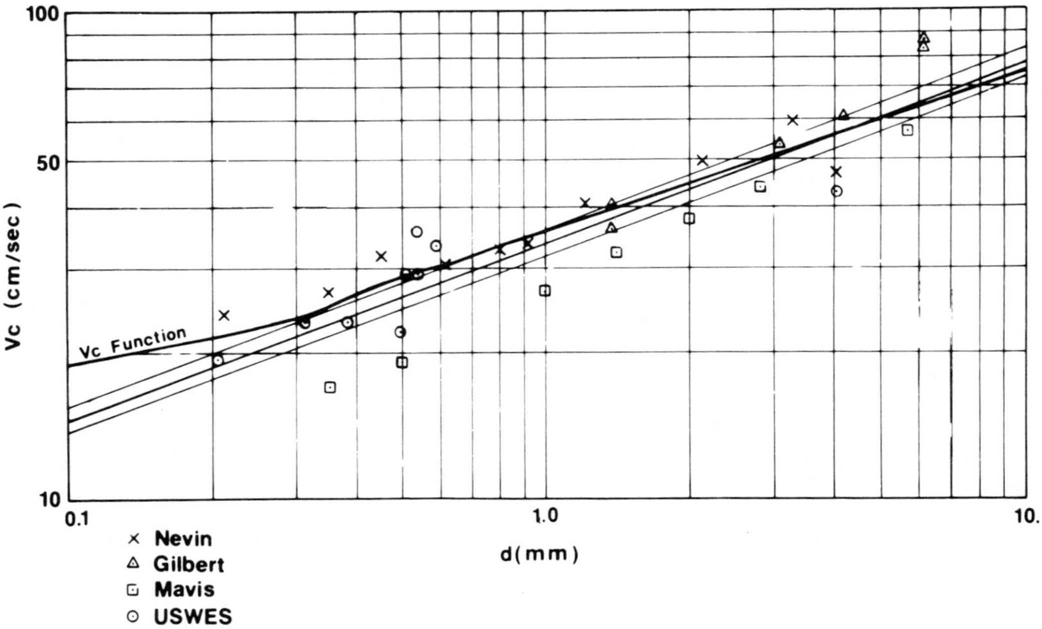

FIG. 2.—Plot of the Vc function from this report for 1-cm depth, with experimental values from Nevin (1946), Gilberg (1914), Mavis (1935), and United States Waterways Experiment Station (1935). Included is the simple linear regression equation for the data: $Vc = \log_{10}Vc = 1.53 + 0.36 \log_{10}d$, with the 95% confidence limits on the Vc intercept.

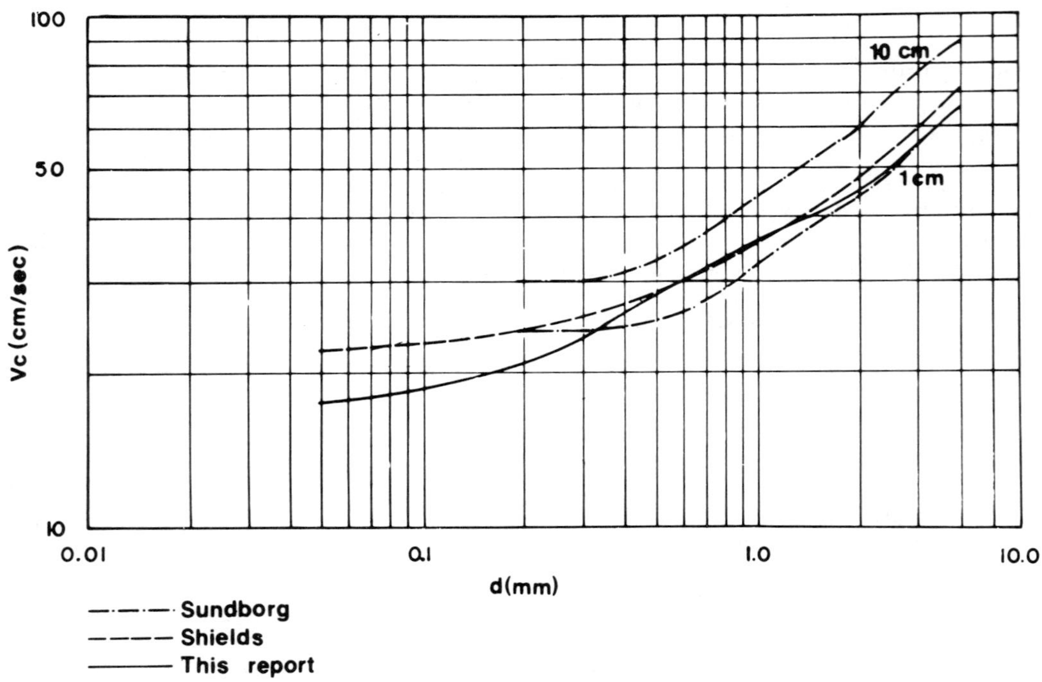

FIG. 3.—Plot of the Vc function from this report with those of Sundborg (1956) and Shields (1935 in Grigg and Rathbun, 1969). The Sundborg plots are for a depth of 1 cm and 10 cm as labeled.

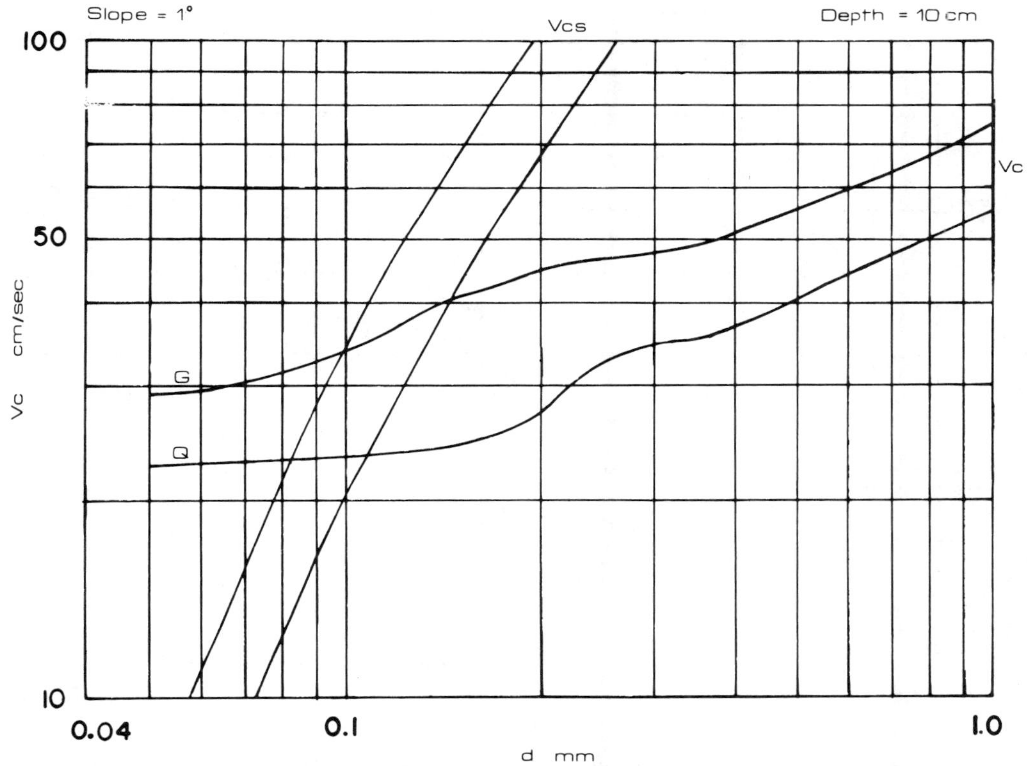

FIG. 4.—Plot of the critical velocity (Vc) and critical suspension velocity (Vcs) equations vs. grain diameter for quartz (Q) and garnet (G), for depth = 10 cm, slope = 1°.

β_1 function as well as determine the constants of β_2. Figure 2 shows the Vc function with experimental data from Nevin (1946), Mavis (1935), Gilbert (1914), and USWES (1935), recalculated to standard conditions of 1-cm depth, $\rho_p = 2.65$, $\rho_f = 0.9979$, slope $= 0.0004$, and $BKS = d_{65\%}$ of the bottom size distribution. This fit is with $\beta_2 = 0.3$ for $R* < 70$ and $\beta_2 = 0.33$ for $R* > 70$. These values are commensurate with expected values of 0.3–0.4 as discussed in the section on quantification of sediment parameters. Also, for uniform sediments, β_1 was best estimated as $(f(Re))^{1/2}$ thus limiting its variation from 0.71 to 1.0 over the range $R* = 5$ to 70.

Figure 3 is a plot of the Vc function (equation 12) with the entrainment functions of Sundborg (1956) and Shields (1935, *in* Graf, 1972). Equation 12 predicts lower values for the finer sizes (<0.4 mm) than either of the other two equations as a result of the influence of the increasing value of C_d at lower Re_c. Figure 2 indicates that the linear regression line of the experimental data actually predicts lower Vc values for the smaller sizes down to 0.2 mm than the entrainment function. The entrainment function ($V_c = f(d)$) is at least as accurate in predicting Vc values for this range as the Shields and Sundborg functions. Above 0.6 mm the entrainment function maps within the 95% confidence limits of the regression line for the experimental data.

Critical Suspension Velocity

The critical suspension velocity is herein defined as the mean fluid velocity measured at a distance y from the bed, which just maintains suspension of a particle of constant terminal settling velocity ($CTSV$) Vs, where Vs is the settling velocity of the particle in still water. Experimental evidence from Lane and Kalinske (1939, p. 640), indicates that $U*$ must be greater than or equal to Vs to initiate suspension of a particle of $CTSV$, Vs. As previously mentioned, $U*$ can be seen to be a function of the intensity of the turbulent

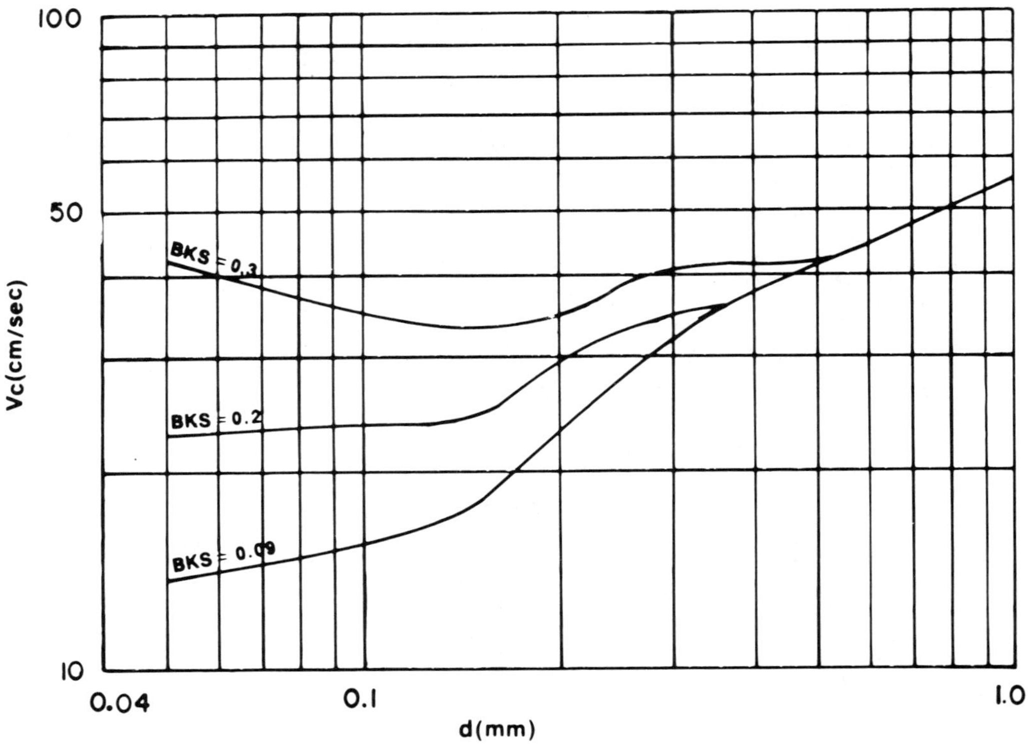

FIG. 5.—Plot of Vc function vs. grain diameter for three different bed roughness sizes (BKS) in mm, with density = 2.65, slope = 7°, and depth = 10 cm.

velocity fluctuations. It is then reasonable to assume that $U*$, as a measure of the vertical velocity components of the turbulent flow, would have to equal Vs, for suspension. Substituting Vs for $U*$ in equation 1 yields:

$$Vcs = 5.75 \log_{10}\left(\frac{30.2xy}{BKS}\right)Vs \qquad (15)$$

where Vcs is the critical suspension velocity which is dependent on the point of measurement, the boundary roughness, and the $CTSV$ of the particle in question. This allows comparison of horizontal flow velocities needed to maintain suspension of a particular sized particle with that velocity needed for entrainment.

REPRESENTATIVE SOLUTIONS OF THE CRITICAL VELOCITY EQUATIONS

Figure 4 is a solution of the Vc entrainment and suspension functions for a depth of 10 cm (the point of measurement), grain densities of quartz and garnet, $BKS = d$, and bed slope = 1°. The form of the entrainment function varies for different densities, in contrast to the usual representation of the Vc equation as displaced parallel curves. This form difference is a result of the higher $U*$ needed for denser grains. The relationship between Cd and $R*$ changes, thus changing the form of the function.

Figure 5 is a plot of Vc vs. d (mm) for three different ϕ roughness sizes equal to 0.09, 0.2, and 0.3 mm with $\rho_p = 2.65$ and water depth = 10 cm. The increase in size of the bed upon which a grain rests causes an increase in Vc for all sizes up to 0.25 mm larger than the given roughness size. With a ϕ roughness size of 0.3 mm the increase in Vc is greater for the smallest grains decreasing to 0.15 mm after which it increases again becoming equal to Vc for a ϕ roughness size of 0.2 mm at $d = 0.55$ mm. For a ϕ roughness size of 0.09 mm, the decrease in Vc is especially great for the smaller sizes since the grains no longer are hidden amongst larger grains which cause larger ϕ angles.

Figure 6 is a plot of the Vc function vs. grain size for $\alpha = 7°$, depth = 1 cm, $BKS = 0.2$

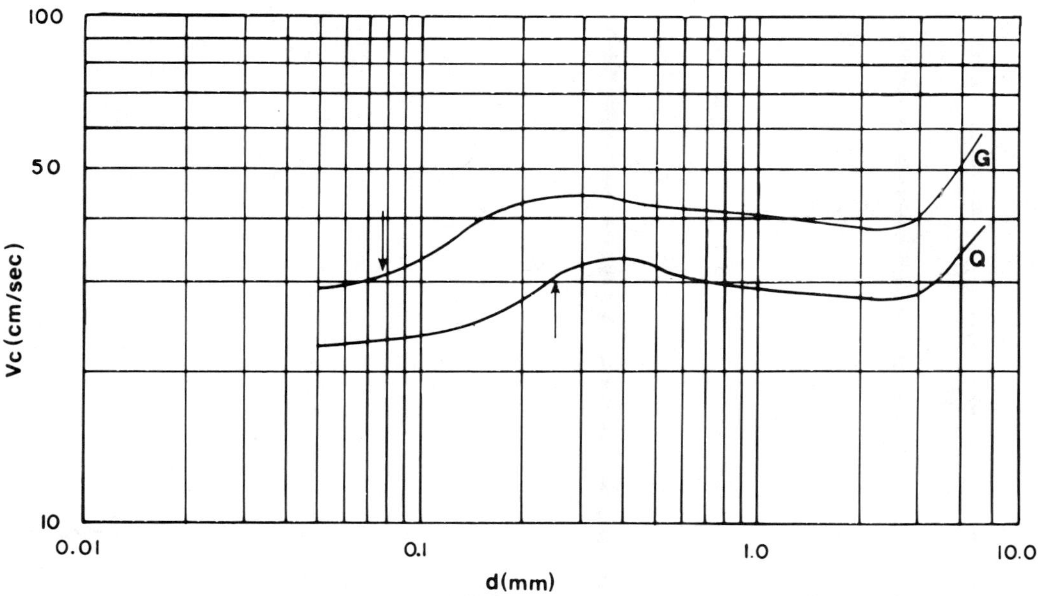

FIG. 6.—Plot of the *Vc* function vs. grain diameter for quartz (Q) and garnet (G) with the slope = 7°, depth = 1 cm, where ϕ is unrestricted to 3°.

mm, and where ϕ is unrestrained to 3°. This is assumed to reflect the situation where grains of a wide range of sizes are resting on a bed of fairly uniform smaller grains on a steep slope, for example, an upper beach swash face. The form of the function indicates that in this situation some grains larger than about 0.35 mm are easier to entrain than smaller grains. The function is of this form because between d = 0.15 to 3.5 mm, the decrease in Cd for increasing Re offsets the variation of ϕ and β_1, causing higher critical velocities. As Cd approaches a constant value, the decrease in ϕ and β_1 balances the increase in force needed to move grains of larger size and the critical velocity decreases slowly.

EXPERIMENTAL DATA

Water Table Studies

A recirculating water table usually used for aerodynamic shock-wave studies was modified for observing the relative transportation rates of sand-sized heavy and light minerals. The table consisted of a 1.5 meter wide plate glass surface of slope = 0.001, at the upper end of which was a metal plate with a uniformly beveled lower edge. By rotation of the plate about its upper edge, the velocity and type of flow over the table bottom could be regulated. Because of the table construction, the flow should approach the theoretical case where a viscous sublayer exists at the bed, becoming turbulent above.

The grains used were monomineralic natural populations size to ¼ϕ intervals. At the beginning of a test, 1.5 g of each sieve fraction were placed along a line parallel to the flow regulation gate. Flow of the water was increased to 0.66 m/sec (depth = 2–4 mm at 0.3 m downcurrent from the gate), and held at this velocity for approximately one minute, whereupon the flow was stopped. The distances of movement downcurrent of the various size fractions were measured.

Figure 7 is a plot of the *Vc* function and the observed distance of movement of the various fractions for quartz and garnet (some experimental data from James Warg, personal communication). Although the distance moved is taken to be proportional to the *Vc* for a given size sediment, the position of the curves on the graph is arbitrary, thus only the relative forms of the curves are meaningful. The close agreement between the predicted and observed critical velocities supports the theoretical conclusions of this report at least for the ideal case, concerning the variation in Cd, β_1, β_2 and ϕ. The viscous flow over the smaller

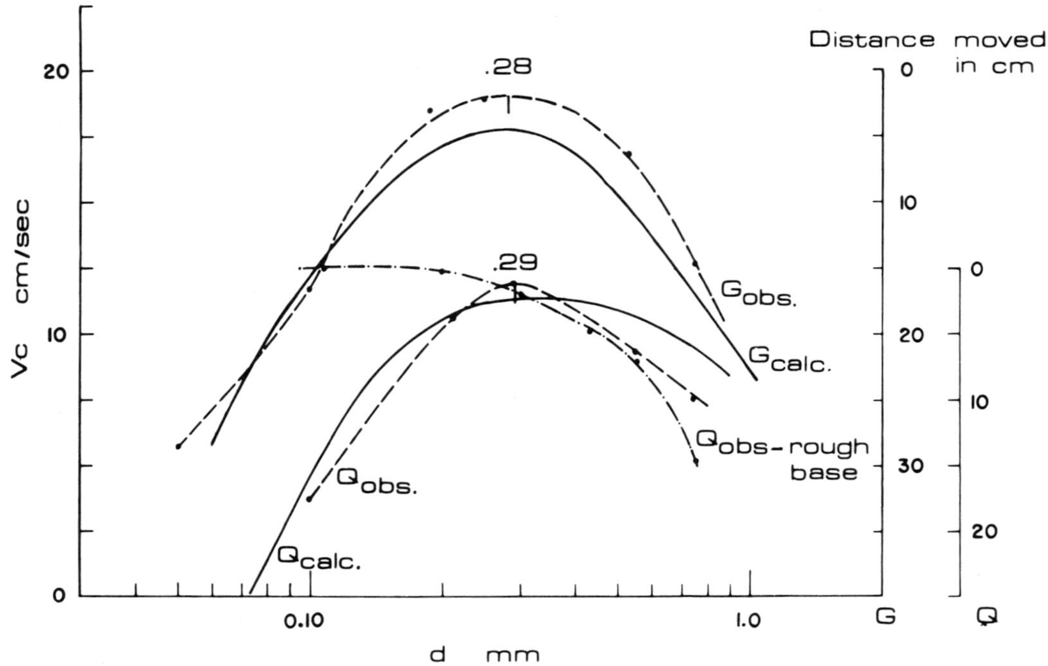

FIG. 7.—Plots of the calculated Vc function (calc) and observed distance of movement (obs) vs. grain diameter for quartz (Q) and garnet (G). Also plotted is the observed distance of movement of quartz grains on a rough base, with BKS = 0.102 mm (Q_{obs} – rough base).

grains will create more drag and hence a larger Cd than for larger grains. Larger grains, however, will experience higher turbulent velocity fluctuations, a higher point of application of fluid force, and larger ϕ angles. Thus there will be some relationship between Cd, β_1, β_2, and ϕ angle where the force imparted to a grain is at a minimum. This is the size for which the Vc must be highest in Fig. 7. The size least easily moved is not the largest or smallest but some intermediate size, depending upon the sediment and fluid conditions.

This conclusion is contrary to the generally accepted Vc size relationship as presented in Hjulstrom (1935) and Inman (1949) who showed 0.18 mm to be the size most easily entrained. But the Hjulstrom data are a mixture of Vcs for cohesive clays as well as fine sands. In that case the Vc *will* increase below 0.18 mm. Also, the analysis of Inman that the redistribution of drag at $R* < 5$ (0.18 mm size at standard conditions) will cause an increase in the Vc is contrary to the data of Grass (1970) and this report. Finally, neither Inman or Hjulstrom analyzed explicitly the differences in Vc owing to variations in ϕ as a result of specific differences in size distribution.

With a Rough Base

To determine the degree of deviation from ideal conditions, the glass plate bottom of the water table was replaced with a plate upon which sand of uniform size (0.104 mm) had been glued. The quartz curve for a typical run under these conditions is superimposed on Fig. 7. As predicted from Fig. 5, grains with diameters up to 0.25 mm greater than the base roughness size had ϕ angles increased and velocity gradients reduced so that their Vc exceeded the fluid velocity. Therefore no grain movement occurred with grain diameters less than 0.20 mm.

Wave Tank Studies

These experiments were conducted in a wave tank 0.32 m deep, 1.2 m wide, and 4 m long. One end contained a dislocation-type wave generator; on the opposite end was located a gradually sloping beach of quartz sand.

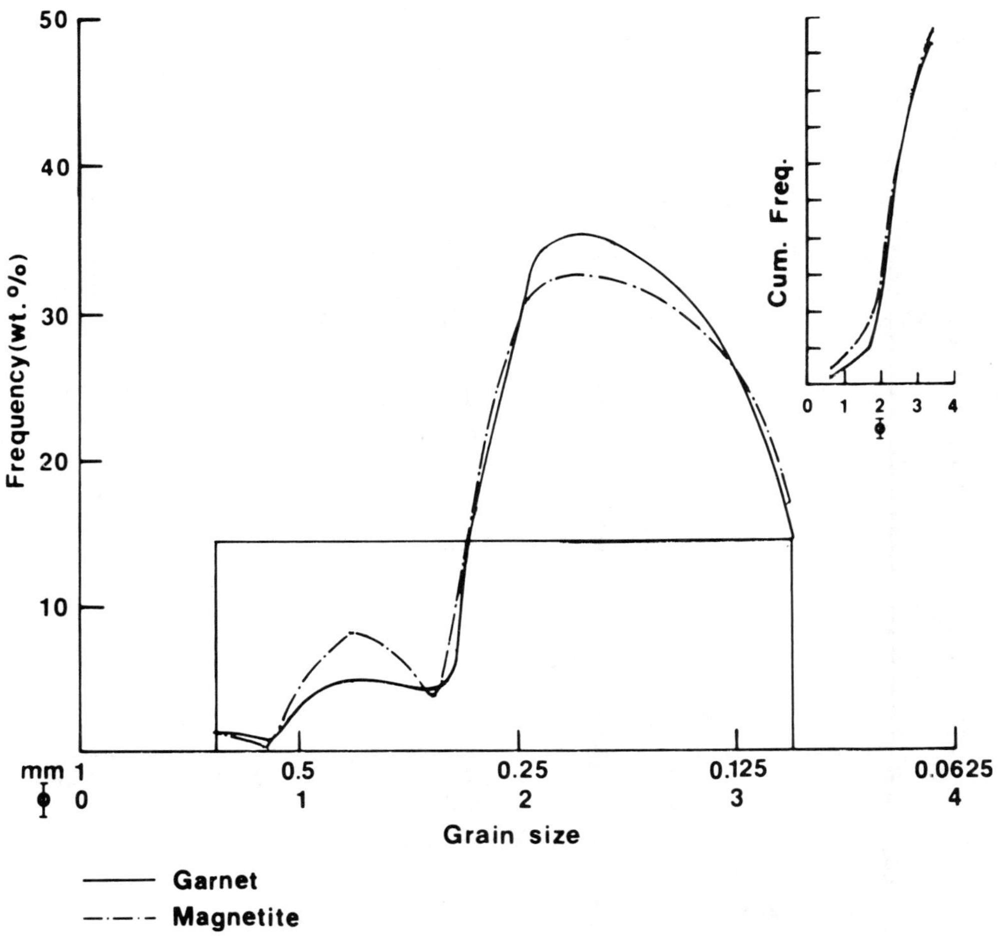

FIG. 8.—Frequency–grain size and cumulative frequency–grain size plots of garnet and magnetite for Run 1. Initial distribution is rectangular; swash modified distributions are bimodal.

In the first set of experiments, rectangular distributions (sizes 0.888 to 0.707 mm) of garnet and magnetite (run 1), and quartz, garnet, and magnetite (run 2), were introduced on the beach of bed size 0.22 mm. Samples of run 1 were located about ⅔ of the way up the swash face under destructive wave conditions, and the samples from run 2 were located immediately under the plunge point of the waves under constructive conditions. After 27 waves had passed, the remains of the initial samples were resampled and dried. The garnet and magnetite were separated from the quartz by heavy liquids, and the magnetite from the garnet by magnetic separation. All fractions were sieved and weighed. Figures 8 and 9 give the frequency distributions and cumulative weight percent curves for the various samples and runs.

In all the runs the larger sizes (0.324 to 0.707 mm) were selectively removed and always the minerals were removed by an amount inversely proportional to their densities. Figure 6 predicts for quartz that sizes from 0.3 to 0.5 mm will be harder to entrain than from 0.5 to 0.7 mm which is verified by the minor mode in Figs. 8 and 9. Combining a *BKS* between 0.2 and 0.3 mm from Fig. 5 with Fig. 6 will also produce the situation experimentally observed in Fig. 7 (quartz observed, rough base) which is additionally verified in Figs. 8 and 9. Also, from run 2 for samples at the plunge point, less of the larger sizes and more of the finer sizes were removed, possibly indicating that be-

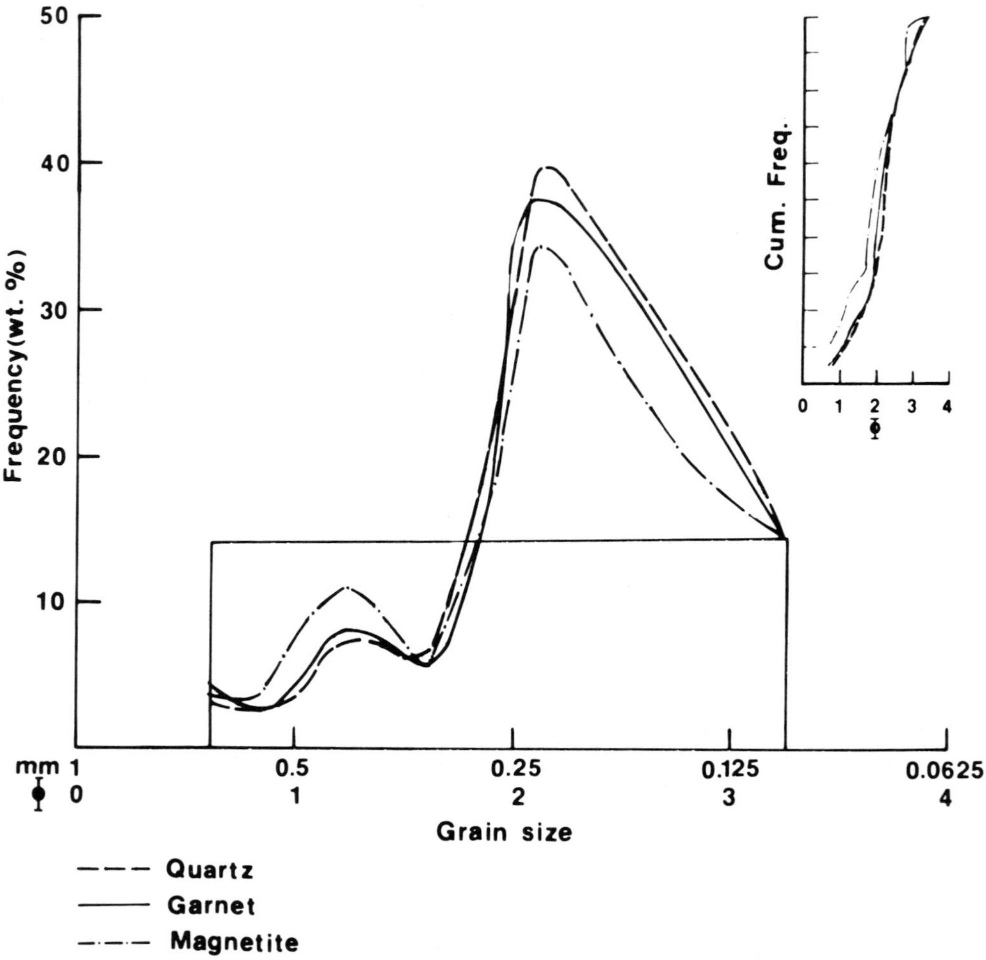

FIG. 9.—Frequency–grain size and cumulative frequency–grain size plots of quartz, garnet, and magnetite for Run 2. Initial distributions are rectangular; swash modified distributions are bimodal.

cause of the greater $R*$ and the increased turbulence, the smaller grains were plucked out from between the larger grains by a vertical velocity component.

SHE Distributions of Quartz and Garnet

To test the influence of the roughness size in the swash zone on transportation of SHE sized grains, equal amounts by weight of standard hydraulically equivalent (SHE) garnet and quartz (sieve sizes 0.148 and 0.210 mm, respectively) were introduced on an artificial base of sand cemented to plexiglass. Eight swashes passed over the samples, which were then resampled, dried and weighed. Three different base roughness sizes were used; all were well sorted. The data are presented in Table 1. The increase in the relative proportions of quartz to garnet is an indication of the change from where the boundary roughness was such that the larger quartz could roll away to the situation where the finer heavies were entrained by fluid turbulence.

DISCUSSION

Based upon this theoretical analysis and experimental data, a model of heavy mineral selective sorting may now be proposed. As grains move over a given area of the bed, they

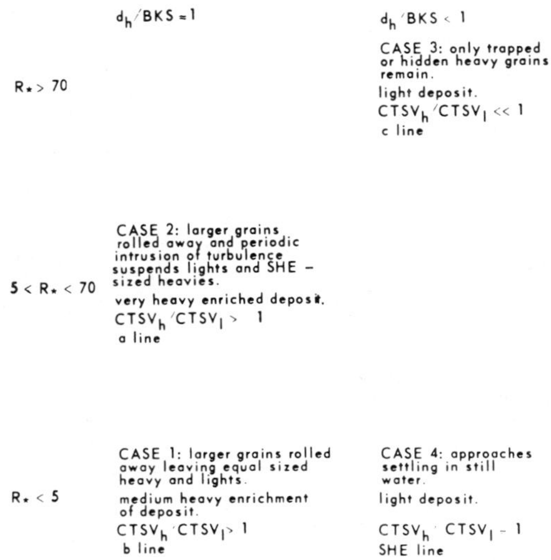

FIG. 10.—A classification of sediment-fluid conditions controlling the relationships of light and heavy minerals in sands.

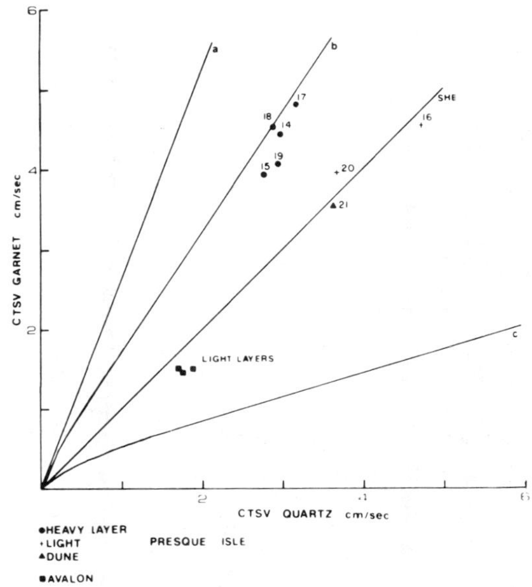

FIG. 11.—*CTSV* garnet/*CTSV* quartz ratio lines as given in Fig. 10 plotted with samples from Presque Isle, Pennsylvania and Avalon, New Jersey.

test the flow and interparticle geometry for stable positions. Given a certain bottom roughness and current velocity, Figs. 5, 6, and 7 show that the size already established on the bed will select like-size particles in preference to larger and smaller sizes. Thus the tendency here is to minimize the effect of hydraulic equivalence by settling and select only by size, since β_1, β_2, and ϕ are determined predominantly by size.

However, from Fig. 4, the frequency with which grains of a certain density will test conditions on the bed is a function of the *Vc*s. Heavier grains of a given size will test the bed more than lighter ones of the same size, and thus have a better chance of filling a position. Also, from Fig. 4, if a heavy grain and a light grain happen to have the same ϕ angle, their hydraulically equivalent entrainment sizes will be quite far apart. These processes then emphasize density. Thus two different controls—one depending on size, and one on density—are attempting to differentiate the sediment.

TABLE 1.—*The relationship between base roughness and quartz/garnet ratios*

Wt. Qtz. / Wt. Gnt.	Base Roughness
0.656	0.149
0.823	0.190
1.13	0.210

The variables exerting this control are functions either of the *d/BKS* ratio or a measure of the fluid flow at the bed. It is reasonable then, to use *d/BKS* and a $R*$ to classify the *CTSV* relationship of light and heavy minerals in sands. Figure 10 shows the combinations of these two parameters thought to be important.

Case 1.—$R* < 5$ and the roughness size of the bed is nearly the same as the median size of the immediately available heavies ($d/BKS \sim 1$). As argued above, selection of heavy and lights is more by size than density. Thus the population will tend towards equal-sized heavy and lights because the hydraulically equivalent larger lights have lower critical velocities due to lower ϕ angles and greater turbulent influences from outside the viscous sublayer. Heavy enrichment has now occurred by rolling away the larger sized lights even though they are hydraulically equivalent in Rubey's sense of the term to the remaining heavies. This case is represented by the b line of equal *sized* heavy and lights in Fig. 11.

Case 2.—$5 < R* < 70$, but at the lower end, and $d/BKS \sim 1$. Case 1 entrainment

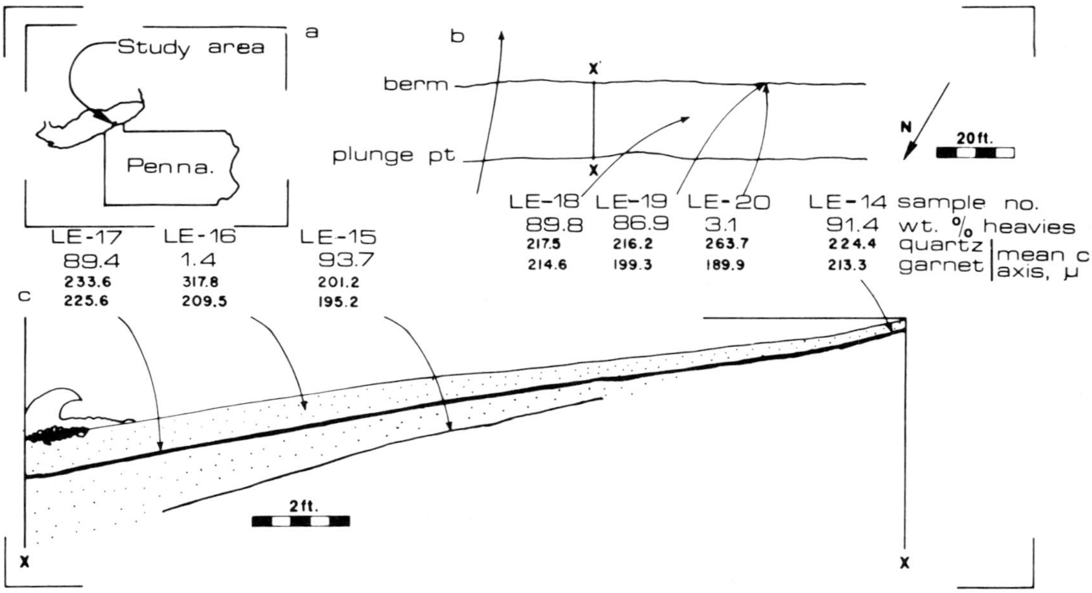

FIG. 12.—Location of study area (a); plan view of Presque Isle beach #10 in September, 1972 showing sample sites (b); cross section X–X′ from b (c).

mechanisms will attempt to produce a uniformly sized heavy and light deposit, but in addition, periodic intrusion of turbulence from above into the viscous sublayer will suspend the lights and smaller (suspension equivalent) heavies. The deposit will then consist of a well size-sorted, even more heavy enriched population. The heavy grains now are equal to or larger in size than the lights because the light grains, if present at all, are trapped in the heavies. Although this case would produce $CTSV_h/CTSV_l$ ratios anywhere above the b line in Fig. 11, line a is chosen as representative of this situation.

Case 3.—$R* > 70$, and median size of heavies is smaller than the roughness determining population ($d/BKS < 1$). Only entrainment equivalent heavy sizes or heavies that are small enough to lodge within the larger population will remain, since fluid turbulence will act to remove all suspension equivalents. If the bed conditions were those of Fig. 6 for example, with $BKS = 0.2$ mm and median size of garnet about 0.07 mm, the entrainment equivalent sizes of quartz would be 0.25, 0.7, and 4.8 mm. Taking the first case gives a $CTSV_g/CTSV_q$ ratio of 0.8/3.33 which is much less than 1. The other two entrainment equivalent sizes and the case of trapped heavy grains produce ratios smaller than this. Therefore line c is drawn through the point (0.8/3.33) to illustrate Case 3 ratios.

Case 4.—$R* < 5$, $d/BKS < 1$, or the fluid velocity is sufficiently small for fluid conditions to approach still water. This is represented in Fig. 4 by all diameters to the left of the Vcs line. This case is represented in Fig. 11 by the standard hydraulically equivalent (SHE) line, which may also represent the set of entrainment conditions such that the Vc of the heavy and light grains produces SHE sizes.

APPLICATION TO NATURAL SYSTEMS

The preceding theoretical and experimental development will now be tested on the nearshore-beach-dune complex of Presque Isle, Lake Erie. Samples of heavy and light mineral layers were collected in September 1972 from existing beach and dune deposits. Figure 12 gives the sampling locations, percent heavies, and mean c axial lengths.

The sampling procedure was designed to obtain single sedimentation unit samples by minimizing the thickness to a few mm. One half of each sample was dried, weighed and the heavy and light fractions separated by heavy liquids. The weight percent of heavies and lights was then calculated. To measure grain size, the other half of each sample was split with an Otto microsplit until a quantity small enough to mount on a gelatinous-coated slide was obtained. Slide points were chosen by determining stage coordinates selected from a random number table. The apparent a and c axes of the five quartz and garnet grains closest to a point were then measured using a microscope with a 10× objective for a total of 50 grains per slide.

A multivariate discriminate function on both the a and c axes was used for comparison of the mean sizes of the lights and heavies. It finds the greatest distance between the two clusters (garnet and quartz) of multivariate points (a and c axial lengths) such that each cluster also has the least inflation (multivariate variance). Limited departure from normalcy does not seriously affect the function but the means and variances of the axes must be independent. This was shown to be fulfilled for these samples (Slingerland, 1973).

A measure of the separation between the two multivariate means, Mahalanobis D^2, may be tested for significance by an F test. The null hypothesis is that the two means are equal, or that the distance between them is not significantly different from 0. The F value at the 95% level of significance is 3.2 with $\nu_1 = 2$ and $\nu_2 = 92$. Figure 13 is a plot of the F values for D^2 for the samples collected.

There is no statistical difference between the multivariate means of the a and c axes between quartz and garnet of the samples from heavy layers (numbers 14–15 and 17–19). When the converted *CTSV*s calculated from the mean c axial length are plotted in Fig. 11 they fall on or near the b line, as is consistent with the postulated conditions of Case 1 (Fig. 10). The light layers (samples 16 and 20) plot along the SHE line, indicating that the heavies were not trapped but apparently were deposited under fluid conditions allowing suspension equivalent sizes to come to rest together. In addition, after settling, the *BKS* and fluid flow conditions must have been such that $R* < 70$ since entrainment of the smaller sizes did not occur. Neither was $R*$ small enough to cause removal of the larger lights by rolling. Also plotted in Fig. 11 are three samples from light layers at Avalon Beach, N.J. (data from Hand, 1964) illustrating that even at smaller sizes, the $CTSV_h/CTSV_l$ ratios fall as predicted.

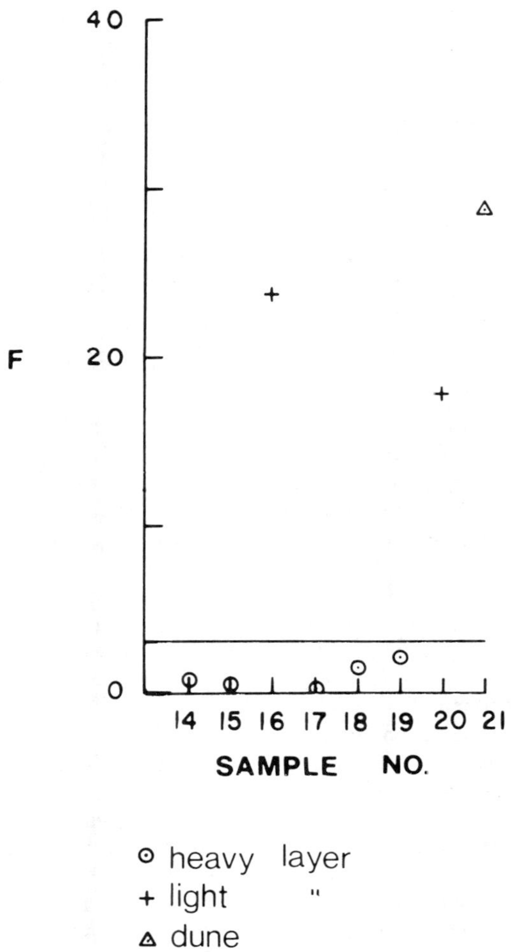

FIG 13.—F test values on Mahalanobis D^2 for Presque Isle samples; F significance value at 95% confidence level = 3.2.

Heavy and Light Mineral Budget for a Presque Isle, Lake Erie Beach

Results of this paper plus those of Lowright et al. (1972) may now be used to construct a heavy and light mineral size budget for the Presque Isle beach. Modification of the ultimately available sizes has already occurred by the time the sands have reached the lake bottom offshore from the beach, producing the settling velocity distributions in Fig. 14 at point y.

Deposition in the swash zone of sizes from these distributions will proceed in the following manner. SHE sizes in suspension (aa) and

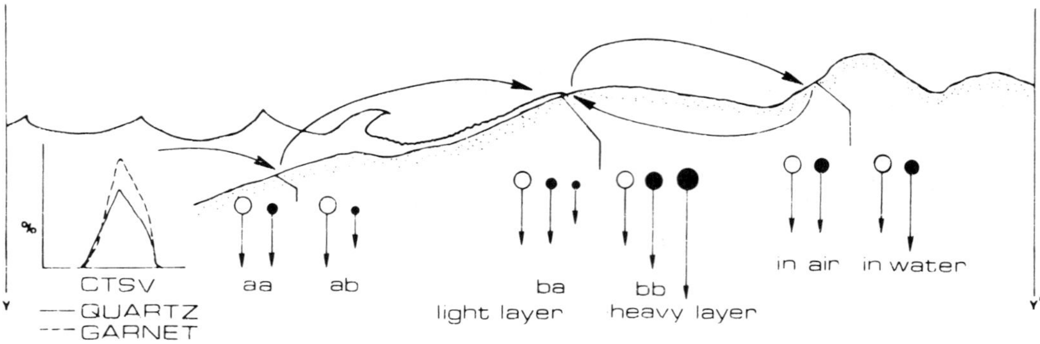

FIG. 14.—Schematic heavy and light mineral size budget for Presque Isle beach.

entrainment sizes in saltation (ab) with *CTSV*s between SHE and the c line will settle out up the beach face in proportion to the decrease in velocity with the coarser sizes settling first. Because the swash must decrease to 0 velocity, and given the average order of magnitude of swash velocities on sandy beaches, the top of the swash is the most likely place to have a fluid velocity and grain size such that $R* < 70$, which may even approach $R* < 5$. (For $R* = 70$ and with a grain size of $BKS = 0.7$ mm, the $\bar{V}$ at 1 cm depth = 161 cm/sec.). Also because the distribution of sizes will settle out according to equation 15, the upper portion of the swash will contain the finest sizes of the population and thus the *BKS* grain size may be of the same order of magnitude as the size of the heavies. As deduced in a previous section this represents optimum conditions for the enrichment of heavy minerals, since all larger-sized grains may be rolled away. Consider now, the fact that some swashes will be higher than others, and will thus produce a higher than usual velocity for a point near the top of the swash. Since the heavy and light grains are now more nearly of the same size, the lights may be easily suspended and removed, as evidenced by the *CTSV* relationship (ab), thus enriching the deposit even further. If these deposits are now dried out and incorporated by winds into the dunes as would occur after a storm, the dunes also would become enriched (ca).

As has been argued by Hand (1964) and others, the difference in density between air and water causes the heavy/light size ratios for dunes to be larger than those in light layers of waterlaid deposits. Re-erosion of the heavy-enriched dunes by the upper swash will result in the formation of relatively thick heavy mineral layers. In addition, the coarser quartz may be rolled away because the $R*$ values are below 70, the *BKS* at the top of the swash is more nearly the size of the heavies, and because suspension will select out the lights which are smaller than those which would be the SHE size of the heavies.

This analysis offers an alternative explanation to that of Stapor (1973) who studied the marine, coastal, and dune sands of Apalachicola, Florida. He argued that the "black sands" were deposited as slugs of heavy mineral rich sand from offshore. The enrichment of these slugs was presumably due to more rapid shoreward movement of the finer (more heavy rich fractions) than the coarser. However, the fact that the heavy deposits were found as *high berms,* and the quartz grain size within the heavy layer was smaller than within the light layers indicates precisely those conditions expected for concentration of heavies in the upper swash by the processes indicated in this paper.

SUMMARY

1. A theoretical analysis of fluid flow and sediment properties has led to the development of a critical entrainment and critical suspension velocity equation which allows the prediction of the sizes of heavy and light minerals which will be deposited together in sands. The entrainment critical velocity equation:

$$Vc = 5.75 \log_{10}\left(\frac{30.2xy}{BKS}\right)\beta_1\beta_2 \sqrt{\frac{4dg \cos\alpha(\rho_p - \rho_f)(\tan\phi - \tan\alpha)}{3C_d\rho_f}}$$

predicts a decrease in critical velocity with decreasing size for single sized sediments from 10 to 0.05 mm. For mixed sizes, the function predicts that depending upon the roughness size of the bed and the size of the particle in question, larger sizes may have lower critical velocities than smaller, and that entrainment equivalent size ratios of heavies to lights may be up to four times smaller than those predicted by the settling laws.

The critical suspension equation:

$$Vcs = 5.75 \log_{10}\left(\frac{30.2xy}{BKS}\right)Vs$$

where Vs is the constant terminal settling velocity in still water, gives the horizontal flow velocity needed to maintain suspension of a particular sized particle and thus allows a direct comparison with entrainment critical velocities.

2. Experimental data show that for sizes 0.05 to 0.8 mm when $d = BKS$ and surface fluid velocities equal about 66 cm/sec, the intermediate sizes (around 0.3 mm) are least easily moved. Thus depending upon the BKS and $R*$ values, the 0.18 mm size is not necessarily the most easily entrainable size. In fact under certain $R*$ conditions and when $BKS < 0.18$, 0.18 mm is the most difficult size to move. Thus the use of generalized Vc size functions such as the Hjulstrom or Inman curves will give erroneous results for sands. For samples of size 0.088 to 0.707 mm on a bed of natural quartz of size 0.22 mm, more of the larger sizes (0.324 to 0.707 mm) are removed than finer for low $R*$ but as $R*$ increases higher amounts of fines are removed. The same occurs with constant flow velocity but increasing BKS.

3. The CTSV relationships of light and heavy minerals in sands may be classified by four combinations of the ratio of median heavy size to bottom roughness size (d/BKS) and the boundary Reynolds number $R*$, as presented in Fig. 10. These may produce $CTSV_h/CTSV_l$ ratios which range from about 0.25 to 2.7.

4. Garnet and quartz mineral sizes for a beach at Presque Isle, Lake Erie, show no statistical size difference for samples from heavy layers as predicted by Fig. 10. The light layers have size ratios predictable by settling theory.

5. A beach size budget based on these results and applied to the Presque Isle beach shows that SHE sizes brought in suspension and entrainment equivalent sizes brought in by bedload will be deposited up the swash with size relationships such that the heavies are concentrated at the top of the swash by Fig. 10 case 1 mechanisms. Incorporation into the dunes of this heavy-rich fine deposit will cause increased heavy concentration in the dunes. The erosion of dunes by water has a high probability of producing large heavy mineral concentrates because 1) the heavy/light $CTSV$ ratios are greater than 1, 2) the dunes have an increased concentration of heavies, and 3) erosion of the dunes by the upper swash provides a favorable BKS-$R*$ combination.

APPLICATIONS

1. These results impose serious restrictions on the use of heavy and light mineral $CTSV$ ratios to indicate the medium of deposition of a sediment. The delta values of Hand (1964) used to differentiate dune (air-settled) from beach (water-settled) sands may reflect entrainment equivalent sizes rather than the different densities of the depositing medium.

2. The location of heavy mineral placer deposits might now be predicted by prospecting for those $R*$-BKS conditions most favorable for heavy mineral enrichment as outlined in Fig. 10. The use of $CTSV$ ratios as illustrated in Lowright et al. (1972) and applied by Pretorius (personal communication), however, is further restricted by the findings of this report. Deviations from SHE are indications not only of distance from source but primarily of hydraulic and boundary roughness conditions.

ACKNOWLEDGMENTS

I express my gratitude to Dr. E. G. Williams, Department of Geosciences, and Dr. A. Miller, Department of Civil Engineering, Pennsylvania State University, for guidance and healthy criticism of this research. Mr. James Warg made available much of the water table data. M. D. Menzie assisted in the field work, and J. D. and S. Breen provided computer support. This research was supported by NSF Grant GA-13012, and computer funds were supplied by the College of Earth and Mineral Sciences, PSU.

REFERENCES

BRIGGS, L. I., 1965, Heavy mineral correlations and provenances: Jour. Sed. Petrology, v. 35, no. 4, p. 939–955.

COLEMAN, N. L., 1967, A theoretical and experimental study of drag and lift forces: Int. Assoc. Hydro. Res., 12th Cong., Ft. Collins, Colo.

EINSTEIN, H. A., 1950, The bedload function for sediment transportation in open channel flows: U.S. Dept. of Agriculture Tech. Bull. No. 1026, 71 p.

FENIAK, M. W., 1944, Grain sizes and shapes of various minerals in igneous rocks: Am. Mineralogist, v. 29, p. 415–421.

GRAF, W. H., 1972, Hydraulics of sediment transport: McGraw-Hill, New York, 513 p.

GRASS, A. J., 1970, Initial instability of fine bed sand: ASCE Hydraulics Division, v. 96, p. 619.

———, 1967, The measure of turbulent velocity fluctuations close to a boundary in open channel flow: Proc. 12th IAHR, v. 2, p. 201.

GILBERT, G. K., 1914, The transportation of debris by running water: U.S. Geol. Survey, Prof. Paper 86.

GRIGG, N. S., AND R. E. RATHBUN, 1969, Hydraulic equivalence of minerals with a consideration of the reentrainment process: U.S. Geol. Survey, Prof. Paper 650B, p. B77–B80.

HAND, B. M., 1964, Hydrodynamics of beach and dune sedimentation: unpublished Ph.D. Thesis, Pennsylvania State University, 159 p.

HJULSTROM, F., 1935, Studies of the morphological activity of rivers as illustrated by River Fyris: Bull. Geol. Inst. Upsala, v. 25, Chapt. III.

INMAN, D. L., 1949, Sorting of sediments in the light of fluid mechanics: Jour. Sed. Petrology, v. 19, p. 51–70.

IPPEN, A. T., AND A. S. EAGLESON, 1955, A study of sediment sorting by waves shoaling on a plane beach: MIT Technical Report No 18.

JAMES, M. L., G. M. SMITH, AND J. C. WOLFORD, 1967, Applied numerical methods for digital computation with Fortran: International Textbook Company, Scranton, Pa., 514 p.

KALINSKE, A. A., 1943, The role of turbulence in river hydraulics: U. of Iowa Studies in Eng. Bull. 27.

LANE, E. W., AND A. A. KALINSKE, 1939, The relation of suspended to bedload material in rivers: Trans. Am. Geophys. Union, v. 20, p. 637–641.

LELIAVSKY, S., 1966, An introduction to fluvial hydraulics: Dover Pub., New York, 257 p.

LOWRIGHT, R., E. G. WILLIAMS, AND F. Dachille, 1972, An analysis of factors controlling deviations in hydraulic equivalence in some modern sands: Jour. Sed. Petrology, v. 42, no. 3, p. 635–645.

MAVIS, F., T. C. HO, AND Y. TU, 1935, The transportation of detritus by flowing water: U. Iowa Studies in Engineering, Bull. 5.

MCINTYRE, D. D., 1959, The hydraulic equivalence and size distributions of some mineral grains from a beach: Jour. Geology, v. 67, no. 3, p. 278–301.

MILLER, R. L., AND R. J. BYRNE, 1966, The angle of repose for a single grain on a fixed rough bed: Sedimentology, v. 6, p. 303–314.

NEVIN, C., 1946, Competancy of moving water to transport debris: Geol. Soc. America Bull., v. 57, p. 651–674.

RITTENHOUSE, G., 1943, The transportation and deposition of heavy minerals: Geol. Soc. America Bull., v. 54, p. 1725–1780.

RUBEY, W. W., 1933, The size distribution of heavy minerals within a water-laid sandstone: Jour. Sed. Petrology, v. 3, p. 3–29.

SHANOV, V. N., 1964, The distribution of minerals in the size fractions of sand deposited by water and wind: Universitat, Vestnik Leningradskii 6, p. 155–163.

SLINGERLAND, R. L., 1973, Transportation and hydraulic equivalence relationships of light & heavy minerals in sands: unpublished M.S. Thesis, Pennsylvania State University, 120 p.

STAPOR, F. W., 1973, Heavy mineral concentrating processes and density/shape/size equilibria in the marine and coastal dune sands of the Apalachicola, Fla. region: Jour. Sed. Petrology, v. 43, no. 2, p. 396–407.

SUNDBORG, A., 1956, The River Klaralven; a study in fluvial processes: Geografiska Annaler, Ser. A., no. 115, p. 125–316.

UNITED STATES WATERWAYS EXPERIMENT STATION, Paper # 17 USWES, 1935.

VAN ANDEL, T., 1950, Provenance, transport, and deposition of Rhine River sediments: Doctoral Thesis, University of Groningen.

VON ENGELHARDT, W., 1940, The distinction of water and wind-sorted sands on the basis of the grain-size distribution of their light and heavy components: Chemie der Erde, v. 12, p. 445–465.

WHITE, C., 1940, The equilibrium of grains on the bed of a stream: Proc. Roy. Soc. London, 174-A.

YALIN, M., 1972, Mechanics of sediment transport: Pergamon Press, New York, 290 p.

9

INVERSE GRADING AND HYDRAULIC EQUIVALENCE IN GRAIN-FLOW DEPOSITS[1]

ASBURY H. SALLENGER, JR.
U.S. Geological Survey
Menlo Park, California 94025

ABSTRACT: Inversely graded grain-flow deposits are characterized by a hydraulic equivalence that differs from that based on settling velocities or entrainment. Dispersive equivalence, derived from the dispersive pressure hypothesis on how inverse grading develops, was found to agree reasonably well with observed relationships between grain sizes and densities in grain-flow deposits. Furthermore, observed relationships in deposits formed in subaerial and subaqueous environments were found to be independent of fluid density as is required by dispersive equivalence. The results suggest that dispersive pressure controls the development of the inverse grading common to beach foreshore laminations, slip-face foreset strata, the basal parts of some coarse-grained turbidites, and other diverse deposits.

INTRODUCTION

A highly concentrated flow of sand-size and larger particles commonly produces inverse grading, a gradational decrease in grain size from top to bottom in the layer deposited by the flow. Such grading has been observed in numerous kinds of sedimentary deposits including beach foreshore and berm overwash laminations (Sanders, 1965; Clifton, 1969; Schwartz, 1975; Leatherman et al., 1977), foresets of eolian and subaqueous dunes (Bagnold, 1941; Inman et al., 1966; Hunter, 1976), basal parts of some coarse grained turbidites and other mass flow deposits (Fisher, 1971; Walker, 1975) and even some meteorite ejecta deposits on the moon (Lindsay, 1974). The mechanism that produces inverse grading has been the subject of some discussion.

For a grain flow in which inertial stresses dominate over viscous stresses, Bagnold (1954) found that grains are supported by dispersive pressure created by grain collisions. The grain flow may be either subaqueous or subaerial and can be driven either by gravity or by fluid flow. In an inertial grain flow, dispersive pressure is proportional to the square of the grain size and the square of the grain shear stress. At any horizon in the flow, Bagnold hypothesized that dispersive pressure would be higher on larger grains than on smaller grains since shear stress is constant. Thus, larger grains would migrate towards the surface of the flow into zones of less shear stress. Middleton (1970) disagreed and hypothesized that small grains simply fall downward between large grains during flow, forcing large grains upward. Dyer (1929) suggested a mechanism similar to that of Middleton (1970).

Rubey's (1933) hypothesis of hydraulic equivalence states that grains of different sizes and densities, but of the same fall velocity will be deposited together. However, subsequent studies have shown that this hypothesis is not universally valid. For example, McIntyre (1959) and Slingerland (1977) found that some beach-foreshore deposits were in settling equivalence as defined by Rubey (1933), but that many were not. These discrepancies have commonly been attributed to differences in entrainment potential between grains of the same fall velocity, but of different size and density (von Engelhardt, 1940; McIntyre, 1959; Hand, 1967; Lowright et al., 1972; Slingerland, 1977). This interpretation, however, may not be valid for inversely graded laminations. Determining the equivalence of grains in inversely graded deposits may resolve some of the observed discrepancies. This may also be valuable in understanding the concentration of heavy minerals in grain-flow environ-

[1]Manuscript received June 7, 1978; revised November 17, 1978.

ments and in interpreting ancient sand deposits.

This study derives the hydraulic equivalence that should result from each of the two mechanisms proposed to explain inverse grading. The two mechanisms predict different relationships that can be tested against observed relationships. The character of observed equivalence within natural and experimental grain-flow deposits clearly indicates which of the two proposed mechanisms best explains the development of inverse grading.

THEORY

Bagnold's (1954) hypothesis suggests that at a single horizon within a grain flow, grains of equal dispersive pressure will reside together. For a grain flow composed of grains of uniform size and density, the magnitude of dispersive pressure is given by:

$$P = a\lambda f(\lambda)\rho D^2(du/dy)^2 \cos\alpha \qquad (1)$$

where 'a' is a coefficient of proportionality, λ is linear grain concentration, ρ is grain density, D is grain diameter, du/dy is the rate of change of grain velocity with depth and α is the angle of internal friction. Even though no theoretical justification was presented, Bagnold assumed that equation 1 is also valid for grains of mixed size. The following derivation assumes that equation 1 is fundamentally correct for grains of mixed size and density. Along any one horizon of a grain flow composed of light and heavy grains of different sizes, du/dy, λ, α, and 'a' are assumed to be approximately constant. The diameter of a heavy grain (D_H) of density ρ_H that would reside at any given horizon with a light grain of diameter D_L and density ρ_L would then be given by:

$$D_H \approx D_L(\rho_L/\rho_H)^{1/2} \qquad (2)$$

Therefore, a heavy grain would be smaller than an associated light grain.

Middleton (1970) does not consider the effects of grain density in his hypothesis on the origin of inverse grading. It seems clear, however, that there is no basis for expecting a heavy grain to rise relative to a light grain of equal size at any horizon in an inversely graded flow. Indeed, one might expect a heavy grain to sink relative to its light counterparts. Thus, the heavy grain would reside with light grains of equal or, more probably, smaller size, a relation opposite from that predicted by equation 2.

Dyer (1929) conducted an experiment that appears to support the equivalence derived from Middleton's arguments. A mixture of wooden and steel spheres of various sizes was agitated. There was nearly a complete density separation, with the wooden spheres on top, and both types of sphere were inversely graded. Thus heavy grains resided lower in the deposit than light grains of the same size. However, the applicability of these results to mixtures of grains of less density contrast is unclear. After segregation, the two layers may have behaved independently and the inverse grading in each may have been controlled by dispersive pressure.

EXPERIMENTS

Subaerial Grain Flow

A subaerial grain flow was generated in the laboratory using an avalanche flume and natural beach sand. The channel of the flume was 8 cm wide and 46 cm long, and was artificially roughened. The channel was set at approximately the angle of repose and sand was introduced from a repository at the head of the channel. A sluice gate regulated the flow thickness. When the flow thickness exceeded approximately 0.7 cm, the sediment did not appear to shear throughout the flow, and a surficial layer of sediment flowed as a plug. Care was taken, therefore, to limit the flow thickness to less than 0.5 cm. The channel was hinged at its upper end, so that once a steady grain flow was established the channel could be rapidly rotated to horizontal, freezing the flow. This deposit was sampled by sequentially applying pieces of adhesive tape, 6 cm wide and 20 cm long, to its surface, thereby obtaining samples approximately one grain diameter thick through the thickness of the flow. Approximately 1 cm was trimmed from the borders of each piece of tape and the sediment was collected from the remaining tape.

Portions of each sample in the density ranges of < 2.9 and 3.1–3.3 gm/cm^3 were separated using heavy liquids. Using a binocular microscope with a micrometer eyepiece, long axes of approximately 400 grains in the 3.1–3.3 gm/cm^3 fraction and 400 grains of

TABLE 1.—*Data for the subaerial grain flow deposit*

Relative Depth[1]	Measured Mean Long Axis[2] (mm)		Dispersive Equivalent Sizes[3] (mm)	Percent Difference Between Calculated & Measured	Percent of Heavy Minerals	
	$\rho = 2.65$	$\rho = 3.1$–3.3			$\rho > 2.65$	$\rho = 3.1$–3.3
1	.682	—	—	—	2.0	0.3
2	.607	—	—	—	4.2	1.0
3	.545	.474	.496	+4.6	7.2	4.6
4	.519	.452	.472	+4.4	13.4	5.6
5	.513	.456	.467	+2.4	12.8	5.1
6	.503	.422	.458	+8.5	18.4	8.7
7	.473	.427	.430	+0.7	25.7	7.1
8	.467	.393	.425	+8.1	35.5	9.2
9	.434	.374	.395	+5.6	47.5	13.1
10	.399	.371	.363	−2.2	82.6	38.7
				Mean = +4.0		

[1]Number 1 is at the surface of the deposit, number 2 is the second sample taken, etc.
[2]Based on approximately 400 grains of each fraction.
[3]Calculations based on a density of 3.2 gm/cm^3.

visually identified quartz ($\rho = 2.65$) in the < 2.9 gm/cm^3 fraction were measured for each sample (Table 1). A few of the heavy subsamples in the upper part of the deposit contained significantly less than 400 grains and accordingly were omitted from subsequent analysis. Also, the percent by weight of all heavy minerals ($\rho > 2.9$ gm/cm^3) and of the 3.1–3.3 gm/cm^3 fraction were determined for each sample (Table 1).

The mean size for each fraction is plotted in Figure 1A against its relative vertical position in the deposit. The data clearly show inverse grading in both the heavy and light fractions. Moreover, at any one horizon in the flow, the heavy grains are considerably smaller than associated light grains. Such a relation is predicted by dispersive equivalence, but not by the relation derived from Middleton's arguments. The percent by

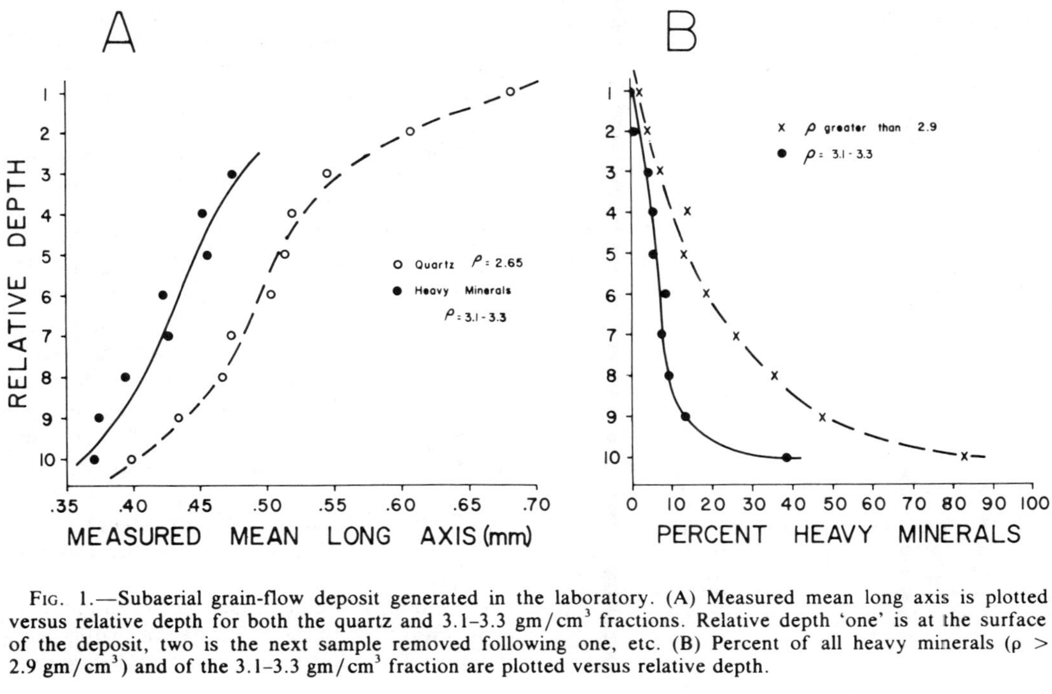

FIG. 1.—Subaerial grain-flow deposit generated in the laboratory. (A) Measured mean long axis is plotted versus relative depth for both the quartz and 3.1–3.3 gm/cm^3 fractions. Relative depth 'one' is at the surface of the deposit, two is the next sample removed following one, etc. (B) Percent of all heavy minerals ($\rho >$ 2.9 gm/cm^3) and of the 3.1–3.3 gm/cm^3 fraction are plotted versus relative depth.

weight of all heavy minerals ($\rho > 2.9$ gm/cm^3) and of the 3.1–3.3 gm/cm^3 fraction are plotted versus relative depth in Figure 1B. Heavy minerals generally increase in concentration with depth in the lamination. The trends shown in Figures 1A and 1B are similar qualitatively to those shown by Clifton (1969, Fig. 2) for a beach foreshore lamination.

Using dispersive equivalence (equation 2), the size of a heavy grain that should be associated with a measured light grain was computed for each sample (Table 1). These data are plotted versus the measured heavy grain sizes as solid circles in Figure 2. A reasonable correlation is evident, although the measured grain sizes are somewhat smaller than the computed sizes. The percent differences between computed and measured grain sizes are below +9% for all samples, whereas the mean difference for all samples is +4.0% (Table 1). This discrepancy may result from shape differences, which almost certainly would influence grain behavior.

Beach Foreshore Deposits

The hydraulic equivalence expressed in equation 2 is independent of fluid density, unlike those equivalences based on settling velocities or entrainment. Consequently, it would be of interest to also test the relationship on a subaqueous deposit as well as on a subaerial one. Presumably, if the grain shapes resemble those used in the subaerial experiment, then the results should be similar to the subaerial data plotted in Figure 2.

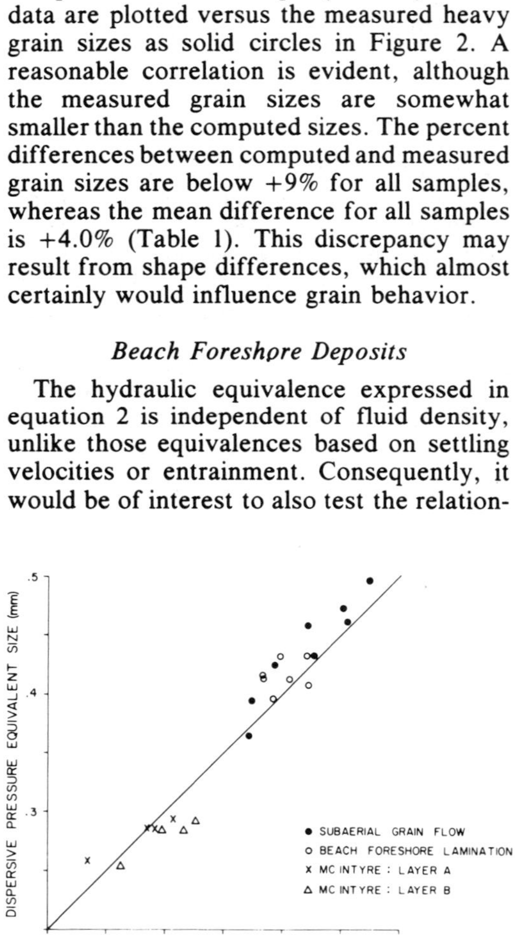

FIG. 2.—Measured mean long axis is plotted versus computed dispersive equivalent size (equation 2). Heavy minerals for the solid and open circles have a density range of 3.1–3.3 gm/cm^3. Calculations for these data are based on a density of 3.2 gm/cm^3. McIntyre (1959) measured garnet, hornblende, clino-pyroxene, and hypersthene.

A beach foreshore lamination found on the beach that supplied the sand for the subaerial experiment, was sampled using the same procedure as before. The lamination was visibly inversely graded, with predominantly coarse light minerals at the surface grading downwards to predominantly fine heavy minerals at the bottom. Seventeen samples were taken from the upper ~75% of the ~0.5 cm thick lamination. Sampling was not continued through the entire lamination because the sampled layers became visibly oblique to the lamination itself. Beginning with the first sample (surface), every other sample was processed as before, yielding nine data sets (Table 2).

The data are plotted in Figure 3A and B in the same manner as in Figure 1A and B. The light fraction is clearly inversely graded in the upper part of the lamination, but is more uniform below (Fig. 3A). If we had been able to sample the lower part of the lamination, where the fine tail of the overall sediment distribution should reside, then the trend probably would have curved again towards the finer grain sizes like that observed in Figure 1A, but more radically. As in Figure 1A, the data show that the heavy grains are somewhat smaller than the light grains at any one horizon. The percent of heavy minerals increases with depth (Fig. 3B) as it did in Clifton's (1969) observations and Figure 1B.

Measured values of D_H are plotted against values computed from equation 2 (Fig. 2, open circles). The grain sizes do not have a sufficient range to define a correlation line, but for any given sample the computed values fall reasonably close to the line of perfect correlation. As for the subaerial experiment, the computed sizes are somewhat larger than the measured. The mean percent difference between computed and measured sizes is +4.1%, which compares closely to the mean difference of +4.0% obtained for the subaerial experiment (Tables 1, 2).

The data indicate that the equivalence associated with these subaerial and subaqueous deposits is independent of fluid density.

TABLE 2.—*Data for the beach foreshore lamination*

Relative Depth[1]	Measured Mean Long Axis[2] (mm)		Dispersive Equivalent Sizes[3] (mm)	Percent Difference Between Calculated & Measured	Percent of Heavy Minerals	
	ρ = 2.65	ρ = 3.1–3.3			ρ > 2.9	ρ = 3.1–3.3
1	.573	—	—	—	1.3	0.3
3	.494	—	—	—	1.8	0.5
5	.475	.421	.432	2.6	3.4	0.8
7	.456	.383	.415	8.4	6.2	2.7
9	.459	.383	.418	9.1	6.7	3.3
11	.446	.412	.406	−1.5	7.8	2.3
13	.473	.398	.430	8.0	8.1	3.0
15	.431	.396	.392	−1.0	10.3	3.6
17	.463	.407	.421	3.4	11.0	4.8
				Mean = +4.1		

[1]Number 1 is at the surface of the deposit, number 3 is the third sample taken, etc.
[2]Based on approximately 400 grains of each fraction.
[3]Calculations based on a density of 3.2 gm/cm^3.

This and the reasonable quantitative agreements with equation 2 suggest that dispersive pressure controls both the development of inverse grading and hydraulic equivalence within these deposits.

OTHER INVESTIGATORS' DATA

McIntyre's Data

McIntyre (1959) sampled portions of three laminations from a beach foreshore. He

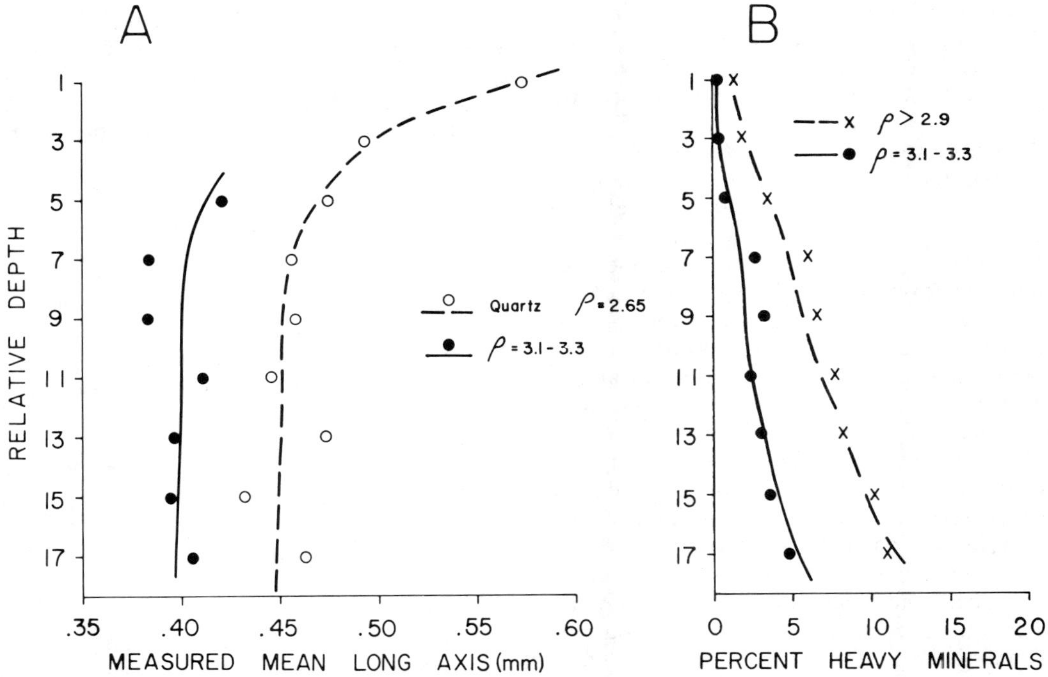

FIG. 3.—Beach foreshore lamination. (A) Measured mean long axis is plotted versus relative depth for both the quartz and the 3.1–3.3 gm/cm^3 fractions. Relative depth 'one' is at the surface of the deposit, three is the third sample removed, etc. (B) Percent of all heavy minerals (ρ > 2.9 gm/cm^3) and of the 3.1–3.3 gm/cm^3 fraction are plotted versus relative depth.

measured long axes of quartz, garnet, hornblende, clino-pyroxene and hypersthene for each sample. For one of the layers sampled the minerals were found to be in reasonable agreement with settling equivalence as defined by Rubey (1933). However, the minerals in the other two layers were found not to be in settling equivalence. McIntyre (1959) assumed that the two anomalous layers were deposited from bedload. He hypothesized that differences in entrainment among minerals of different sizes and densities could explain why these two layers were not in settling equivalence.

Using McIntyre's data for the two anomalous layers and his assumed densities for the minerals, dispersive equivalent sizes (equation 2) were computed for each heavy mineral. These computed sizes are plotted versus measured sizes in Figure 2. For layer A (x's), a good correlation is evident, whereas measured sizes for layer B (triangles) are somewhat greater than computed sizes. Overall, the data suggest that the relationship between sizes of heavy and light minerals was controlled by dispersive pressure.

The good agreement between McIntyre's data and dispersive equivalence may, however, be misleading due to his sampling methods. McIntyre did not report the thickness of the layers sampled. The thickness was, however, considerably greater than that of the present author's samples, which were restricted to approximately one grain layer (see Fig. 2, McIntyre, 1959). McIntyre's samples could have included multiple horizons of different grain sizes and densities, whereas equation 2 applies only to one horizon. Furthermore, McIntyre sampled a given layer at fifty different locations. These were analyzed separately but were arithmetically combined for use in his equivalence computations. Thus, the resulting data is subject to lateral variations in sediment characteristics. Data from each of the 50 samples for each layer were provided, but these were based on only 10 measurements of quartz, and 5 each of the heavy minerals. Thus, the precision of individual subsamples for a given layer was too low to warrant analysis.

Slingerland's Data

Slingerland (1977) sampled some beach foreshore and eolian deposits. The thickness of his samples was a few millimeters and thus approached the requirements of equation 2. Slingerland measured the apparent 'a' and 'c' axial lengths of 50 grains of quartz and garnet for each sample (Slingerland, 1973). He also determined the percent of heavy minerals in each sample. Slingerland found, as did McIntyre (1959), that some samples were in settling equivalence and some were not. The discrepancies were again attributed to differences in entrainment among minerals of different size and density.

Dispersive equivalent sizes (equation 2) of garnet were computed for all of Slingerland's long-axis data using 4.2 gm/cm^3 as the density of the garnet. These are plotted versus measured sizes of garnet in Figure 4. Also plotted are computations using all of McIntyre's garnet data. Correlation lines for dispersive and settling equivalence are shown. The latter is based on an approximate expression for settling equivalence which is similar in form to equation 2:

$$D_H = D_L(\rho_L - \rho_F/\rho_H - \rho_F)^{1.5} \qquad (3)$$

where ρ_F is the density of the fluid. Equation 3 was derived from McIntyre's (1959) fall velocity equation which is approximately valid for the grain sizes used here. Much of the data falls reasonably close to one of the two lines, but a significant portion of

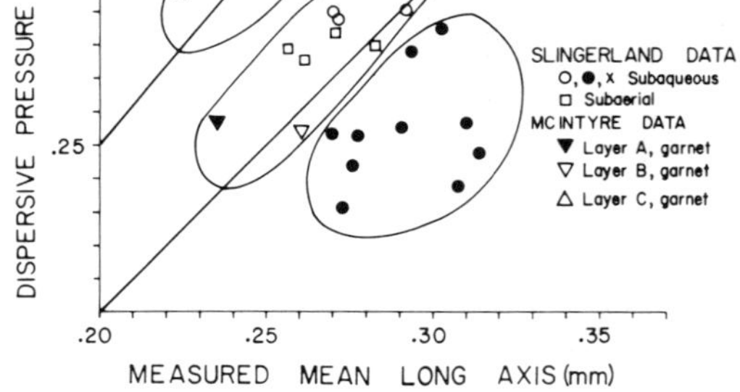

FIG. 4.—Measured mean long axis of garnet is plotted versus computed dispersive equivalent sizes for all of Slingerland's (1973) data and McIntyre's (1959) garnet data. Correlation lines are shown for both dispersive and settling equivalence (see text).

Slingerland's data falls well below the lines (solid circles).

Percent differences between dispersive equivalent sizes and measured sizes were calculated for all of Slingerland's data. These are plotted versus percent of heavy minerals in Figure 5. Three data points (x's) have percent differences greater than 15%. These data points fall reasonably close to the settling equivalence line in Figure 4 (same symbols). All samples that have a relatively low concentration of heavy minerals and have percent differences within −2 to +10% on Figure 5 (open circles and squares) fall reasonably close to the dispersive equivalence line on Figure 4. The mean difference for these data, which include samples gathered from both subaerial and subaqueous environments, is +4.5% (Fig. 5). This compares closely to the mean difference of +4.1% for the present author's subaqueous data and +4.0% for his subaerial data. This relatively close agreement may be because of similarities in grain shape. The subaerial deposits sampled by Slingerland need not have been formed by avalanches at the angle of repose. These deposits may have been formed by grain-flows generated under plane-bed conditions.

For most samples that have a high concentration of heavy minerals (Fig. 5, solid circles), dispersive equivalent sizes are smaller than measured sizes (Fig. 4). That is, the sizes of heavy grains are more nearly equal to the sizes of light grains than is predicted by dispersive equivalence. Slingerland argued that under certain conditions grain size controls equivalence more than grain density. Large light grains which project into the flow are differentially rolled away leaving a deposit of light and heavy grains of nearly equal size. By this process the deposit is enriched in heavy minerals. This may be the case for these data.

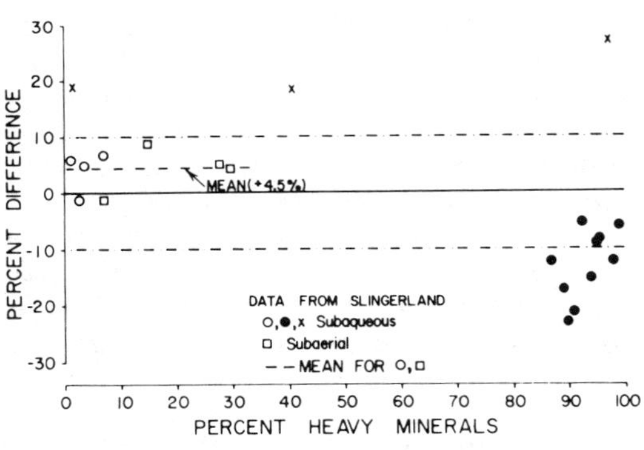

FIG. 5.—Percent differences between computed dispersive equivalent sizes and measured heavy mineral sizes are plotted versus percent heavy minerals for all of Slingerland's (1973) data.

DISCUSSION

Inverse Grading

The results suggest that the inverse grading observed was a product of grain dispersive pressure. The results may apply to other inversely graded deposits, provided certain conditions are met. First, Bagnold (1954) showed that dispersive pressure is not a function of grain diameter for viscous grain flow (that is, when the ratio of inertial to viscous stresses is small). Thus, inverse grading produced by dispersive pressure should not result from viscous grain flow. Second, the internal shear stress of the flow must exceed the yield strength of the flowing material (Hampton, 1975). Otherwise, a non-shearing plug of sediment will develop at the surface or at depth in the flow. If the plug extends to the surface, inverse grading can develop since the competence to transport sediment of a given size is greater in the plug than in the shearing sediment below (Hampton, 1975). If the plug is at depth in the flow, inverse grading in the lower part of the flow and normal grading in the upper part may develop. Third, it appears intuitively that the gradation in grain sizes must be reasonably restricted. If the grain size range is too large (for example, a flow composed of 10 cm diameter spheres and 1 mm diameter spheres), it is reasonably certain that the smaller spheres will simply fall between the larger as has been envisioned by Middleton (1970).

Everts (1973) proposed another hypothesis for the development of inverse grading. His experiments showed that large grains can be transported over a stable bed of smaller grains when the shear stress is *less* than that necessary to initiate movement of the smaller grains. Thus, for some circumstances, coarser grains may be more easily transported than finer grains. Applying this concept to transport on beach foreshores, Everts (1973) hypothesized that inverse grading develops because finer more difficult to transport grains will be deposited first followed by progressively coarser grains. Since heavy minerals are the most difficult to entrain (because of their fine size and high density), they will be deposited first at the base of the deposit. However, it is not clear how this hypothesis would apply to a grain flow many grain diameters thick where grain collisions must be important. Grain collisions are presumably important in the near-bed flow of grains driven by the backwash on beach foreshores as well as the flow of grains driven by gravity down an inclined plane at the angle of repose.

Inverse grading may develop in some environments by processes unrelated to those discussed above. Batemen (1953) demonstrated that small fragments of pumice will saturate and sink in water much more rapidly than will large fragments. He hypothesized that pyroclastics falling into water may explain some cases of inverse grading. Duffield et al. (1978), however, have observed inversely graded pyroclastic beds on steep slopes which may have been formed by mechanisms described herein.

Hydraulic Equivalence

The results also suggest that grain dispersive pressure controls the hydraulic equivalence of grain-flow deposits. Thus, there appears to be *at least* three kinds of equivalence associated with sedimentary deposits (i.e., settling, entrainment and dispersive equivalences). More work is needed to determine the relative importance of each kind of equivalence in various sedimentary environments.

The relevance of dispersive equivalence to various aspects of hydraulic equivalence studies is discussed below.

Entrainment.—The application of entrainment to explain discrepancies between observed size-denstiy relationships and settling equivalence appears inappropriate for deposits formed by highly concentrated flows of grains many diameters thick. That many of the deposits sampled by McIntyre (1959) and Slingerland (1977) were formed by grain flow is supported by the analyses reported herein. That the author's beach foreshore samples were formed by grain flow is supported both by the agreement between his data and the dispersive pressure hypothesis and that the lamination was inversely graded. Additional evidence would be the similarities between observed size-density relationships of these natural deposits and those of deposits formed in the avalanche flume, where it is difficult to perceive how entrainment could apply.

Availability.—Several investigators have suggested that some discrepancies between observed relationships and settling equivalence can be explained by the absence of minerals of a given size from the overall sediment population (Rittenhouse, 1943; van Andel, 1950; Briggs, 1965). Availability would apply to dispersive equivalence only in certain cases. Should there be a deficiency of a given mineral of a given size, those particles which would have been in dispersive equivalence with the given particles will occupy a horizon alone. Availability would apply only to the case where grains of a given dispersive pressure were too few to occupy a horizon alone. These grains may then cause discrepancies from dispersive equivalence by mixing with other horizons.

Sampling.—It is clear that the relevant sampling thickness to document dispersive equivalence is approximately one grain diameter. Sampling multiple horizons may result in discrepancies from dispersive equivalence. This may also result in the conclusion that minerals of a given size were not available. For example, analyses of a sample including the top four horizons shown in Figure 1A would result in the conclusion that the sediment is not in dispersive equivalence. This is because there is a deficit of heavy minerals in dispersive equivalence with the top two horizons (Fig. 1B). The mean size of quartz in the combined four horizons would be too large to be in dispersive equivalence with the mean size of heavy minerals in horizons 3 and 4. Sampling multiple horizons may not, however, always result in errors. For example, combining horizons 7–9 in Figure 3A would cause no errors since grain sizes are nearly uniform in each horizon. It is also clear that the above arguments for sampling a given horizon would apply to grains in settling equivalence as well, since normal grading would create the same kinds of problems as inverse grading.

CONCLUSIONS

1. Dispersive equivalence, derived from the dispersive pressure hypothesis on how inverse grading develops, is relevant to inversely graded grain-flow deposits, is independent of fluid density and differs from equivalence relationships based on settling velocity or entrainment.

2. Observed relationships within an inversely graded beach foreshore lamination and a subaerial grain-flow deposit generated in the laboratory were found to agree reasonably well with dispersive equivalence. Furthermore, observed relationships in these subaqueous and subaerial deposits were found to be independent of fluid density, as required by dispersive equivalence.

3. Observed relationships within most beach and eolian deposits sampled by other investigators were found to agree reasonably well with either settling equivalence or dispersive equivalence. Observed relationships for those subaqueous and eolian deposits in dispersive equivalence were also found to be independent of fluid density.

4. The results suggest that dispersive pressure controls the development of inverse grading common to beach foreshore laminations, slip-face foreset strata and other diverse deposits.

ACKNOWLEDGMENTS

I thank H. E. Clifton and R. E. Hunter for numerous valuable discussions, and David M. Drucker and Harold Gibson for many hours of measuring grain sizes. G. V. Middleton and M. A. Hampton reviewed the manuscript and provided valuable suggestions. This work was conducted in part while the author held a N.R.C.-U.S.G.S. Postdoctoral Research Associateship.

REFERENCES

BAGNOLD, R. A., 1941, The Physics of Blown Sand and Desert Dunes: London, Methuen, 265 p.

———, 1954, Experiments on a gravity free dispersion of large solid spheres in a Newtonian fluid under shear: Royal Soc. Proc. (A), v. 225, p. 49–63.

BATEMAN, P. C., 1953, Up-side-down graded bedding in right-side-up lacustrine pumice: Geol. Soc. America Bull., v. 64, p. 1499–1500.

BRIGGS, L. I., 1965, Heavy mineral correlations and provenances: Jour. Sed. Petrology, v. 35, p. 939–955.

CLIFTON, H. E., 1969, Beach lamination: nature and origin: Marine Geology, v. 7, p. 553–559.

DUFFIELD, W. A., BACON, C. R., AND ROQUEMORE, G. R., 1978, Origin of reverse-graded bedding in air-fall pumice, Coso Range, California: Jour. Vulcanology and Geothermal Research (in press).

DYER, F. C., 1929, Reverse classification by crowded settling in ore-dressing practice: Eng. Mining Jour., v. 127, p. 1030–1033.

EVERTS, C. H., 1973, Particle overpassing on flat granular boundaries: Am. Soc. Civil Eng. Proc., v. 99, p. 425–438.

FISHER, R. V., 1971, Features of coarse-grained, high concentration fluids and their deposits: Jour. Sed. Petrology, v. 41, p. 916–927.

HAMPTON, M. A., 1975, Competence of fine-grained debris flows: Jour. Sed. Petrology, v. 45, p. 834–844.

HAND, B. M., 1967, Differentiation of beach and dune sands using settling velocities of light and heavy minerals: Jour. Sed. Petrology, v. 37, p. 514–521.

HUNTER, R. E., 1976, Comparisons of eolian and subaqueous sand-flow cross-strata: Am. Assoc. Petroleum Geologists Bull., v. 60, p. 683–684.

INMAN, D. L., EWING, G. C., AND CORLISS, J. B., 1966, Coastal sand dunes of Guerrero Negro, Baja California, Mexico: Geol. Soc. America Bull., v. 77, p. 787–802.

LEATHERMAN, S. P., WILLIAMS, A. T., AND FISHER, J. S., 1977, Overwash sedimentation associated with a large-scale northeaster: Marine Geology, v. 24, p. 107–121.

LINDSAY, J. F., 1974, Transportation of detrital materials on the lunar surface: evidence from Apollo 15: Sedimentology, v. 21, p. 323–328.

LOWRIGHT, R., WILLIAMS, E. G., AND DACHILLE, F., 1972, An analysis of factors controlling deviations in hydraulic equivalence in some modern sands: Jour. Sed. Petrology, v. 42, p. 635–645.

MCINTYRE, D. D., 1959, The hydraulic equivalence and size distributions of some mineral grains from a beach: Jour. Geology, v. 67, p. 278–301.

MIDDLETON, G. V., 1970, Experimental studies related to problems of flysch sedimentation, *in* Lajoie, J., ed., Flysch Sedimentology in North America: Geol. Assoc. Canada Spec. Pub. 7, p. 253–272.

ROUSE, HUNTER, 1937, Nomogram for the settling velocity of spheres: Report of the Committee on Sedimentation, National Research Council, p. 57–64.

RITTENHOUSE, G., 1943, The transportation and deposition of heavy minerals: Geol. Soc. America Bull., v. 54, p. 1725–1780.

RUBEY, W. W., 1933, The size distribution of heavy minerals within a water-laid sandstone: Jour. Sed. Petrology, v. 3, p. 3–29.

SANDERS, J. E., 1965, Primary sedimentary structures formed by turbidity currents and related resedimentation mechanisms, *in* Middleton, G. V., ed., Primary Sedimentary Structures and their Hydrodynamic Interpretation: Soc. Econ. Paleontologists Mineralogists Spec. Pub. 12, p. 192–219.

SCHWARTZ, R. K., 1975, Nature and genesis of some washover deposits: U.S. Army Corps of Engineers, Coastal Eng. Research Center, Tech. Mem. 61, 69 p.

SLINGERLAND, R. L., 1973, Transportation and hydraulic equivalence relationships of light and heavy minerals in sands [unpub. M.S. thesis]: Pennsylvania State University, 120 p.

———, 1977, The effects of entrainment on the hydraulic equivalence relationships of light and heavy minerals in sands: Jour. Sed. Petrology, v. 47, p. 753–770.

van Andel, T., 1950, Provenance, transport, and deposition of Rhine River sediments [unpub. Ph.D. thesis]: University of Gronigen.

von Engelhardt, W., 1940, The distinction of water and wind-sorted sands on the basis of the grain-size distribution of their light and heavy components: Chemie der Erde, v. 12, p. 445–465.

Walker, R., 1975, Generalized facies models for resedimented conglomerates of turbidite association: Geol. Soc. America Bull., v. 86, p. 737–748.

[*Editor's Note:* On page 561 of the article the completed reference for Duffield et al. should be *Jour. Volcanology and Geothermal Research* **5:**35–48 (1979).]

Part III

STABILITY OF HEAVY MINERALS

Editor's Comments on Papers 10 Through 21

10 **EDELMAN and DOEGLAS**
Relikstrukturen detritischer Pyroxene und Amphibole

11 **BOND**
Solution Etching of Detrital Staurolite

12 **CARROLL**
Weatherability of Zircon

13 **DAVIDSON and COSGROVE**
On the Impersistence of Uraninite as a Detrital Mineral

14 **DRYDEN and DRYDEN**
Comparative Rates of Weathering of Some Common Heavy Minerals

15 **RAESIDE**
Stability of Index Minerals in Soils with Particular Reference to Quartz, Zircon, and Garnet

16 **FREISE**
Untersuchung von Mineralen auf Abnutzbarkeit bei Verfrachtung im Wasser

17 **BRAMLETTE**
The Stability of Minerals in Sandstone

18 **PETTIJOHN**
Persistence of Heavy Minerals and Geologic Age

19 **KRYNINE**
Provenance Versus Mineral Stability as a Controlling Factor in the Composition of Sediments

20 **TURNAU-MORAWSKA**
Excerpts from *Importance of Heavy Mineral Analysis in Solving Geological Problems*

21 **BLATT and SUTHERLAND**
Intrastratal Solution and Non-Opaque Heavy Minerals in Shales

Papers 10 to 15 deal with studies or weathering of particular species of heavy minerals. These papers show how the effects of physical and chemical weathering are often difficult to separate precisely. Papers 16 to 21 deal with more general problems in the study of stability of heavy minerals, including intrastratal solution and relative resistance to wear during transport.

Cornelius Hendrik Edelman (1901-1964) was a Dutch geologist and professor at the State Agricultural University at Wageningen, the Netherlands, whose studies on the soils and Quaternary geology of the Netherlands are known to sedimentologists throughout the world. In 1931 he first defined the concept of the sedimentary petrological province and founded the so-called Dutch school of sedimentary petrology, which educated many outstanding students. He was chairman of the Fourth International Soil Science Congress in Amsterdam (1950) and either chairman or member of various national and international committees—for example, the United Nations Food and Agricultural Organization.

Dirk Jakobus Doeglas (1905-1974), a colleague of Edelman, was for many years the head of the Department of Soil Science at the State Agricultural University at Wageningen. He was the first president of the International Association of Sedimentologists, founded in 1952. He was long associated with this society as a council member and later as general secretary, retiring in 1967.

Together, Edelman and Doeglas (Paper 10) conducted one of the earliest and most often cited studies of relict structures of detrital pyroxenes and amphiboles; the obviously corroded grains were found in unweathered rocks. They attribute the sawtooth etching so commonly seen in these minerals to the effects of intrastratal solution. The photomicrographs in this paper are classics.

Most papers describing the etching of detrital mineral grains concern grains that have been acted upon by water. Geoffrey Bond, who was vice-principal in the Department of Geology at the University of Zimbabwe at Salisbury, Zimbabwe, gave a short but noteworthy description of etched staurolite grains in an eolian formation directly overlain by a contemporaneous basalt flow (Paper 11). The etching was seen only in a sample taken immediately beneath the lava. Bond concluded that, because of marked difference in mineral assemblages from nearby formations, the lava supplied the agent that attacked the staurolite. Bond died in June 1983 in Harare, Zimbabwe.

Dorothy Carroll (1907-1970) was a native of Western Australia who later became a U.S. citizen. She was the first person to receive a doctoral degree from the University of Western Australia. She headed the first sedimentary petrology laboratory to be set up by the U.S. Geological Survey in Washington, D.C. Most of her early work dealt with optical analyses of heavy minerals; her later studies focused on geochemical relationships of minerals in sediments and soils. Carroll's long bibliography of studies in both Australia and the United States is testament to her illustrious career as a mineralogist.

One study (Paper 12) proved conclusively that zircon can indeed be subject to weathering, although it has long been considered among the stablest of all minerals. Of most significance was her findings that all zircon is not equally susceptible to corrosion and that the most strongly corroded grains occur in laterized soil profiles.

Charles Findlay Davidson (1911–1967) was a Scottish geologist most renowned for his expertise in uranium minerals and for his controversial advocacy of an epigenetic origin for the uraninite in the banket deposits of the Witwatersrand. He worked for the Geological Survey of Great Britain from 1934 to 1955, after which he accepted the chair of geology at St. Andrews University in Fife, Scotland (where he had earned his B.Sc. and D.Sc.). He was awarded the Lyell Medal (1965) and the Neill Prize (1968, posthumously).

He and Michael Edward Cosgrove, a professor of geology at the University of Southampton, England, used a somewhat different approach to document the impersistence of uraninite (Paper 13). Their study area was an old pitchblende-rich mine dump adjacent to a beach. The gravels around the mine dump were highly radioactive; the sands along the beach were not. A detailed mineralogical analysis of the beach sands, including sand intermixed with pitchblende-bearing pebble beds, showed that what radioactivity was present in the sands was due to the presence of either monazite or xenotime, not uraninite.

Of all the earlier studies of heavy-mineral weathering, perhaps the most often-cited paper is the classic by Lincoln and Clarissa Dryden (Paper 14). Abraham Lincoln Dryden, Jr. (1903–1977) was an outstanding teacher of geology who spent his entire teaching career at Bryn Mawr College in Pennsylvania; his pioneering work in statistical analysis of heavy minerals is referred to widely. During World War II he worked with the Military Geology Unit of the U.S. Geological Survey and later with the U.S. Army Corps of Engineers Beach Erosion Board. He was advisor to General MacArthur's headquarters staff on geologic matters; for this work he was awarded a War Department Citation. He also won the Christian R. and Mary F. Linback Award for Distinguished Teaching in 1970, upon his retirement.

Clarissa Compton Dryden, widow of Lincoln Dryden, received her B.A. and M.A. degrees in archaeology at Bryn Mawr College. After her marriage, she collaborated on all the heavy-mineral papers authored by her husband, even if she was not listed as a coauthor. She is now retired.

In Paper 14, the Drydens discussed the changes of relative abundances of severals minerals from fresh to weathered Wissahickon Schist from Pennsylvania and Maryland. From this study they derived comparative resistance-to-weathering ratios, using garnet as the common denominator. They explore the possibility that a mineral species may be destroyed in the weathering of the crystalline rock and therefore never made available for transport and deposition.

Problems of heavy-mineral stability in soils have been studied more commonly by pedologists than by geologists. J. B. Raeside, a soil scientist formerly with the Soil Bureau of the Department of Scientific and Industrial Research in New Zealand, examines three index minerals—quartz, zircon, and garnet—that have been regarded as stable enough to be used to evaluate changes in soil profiles (Paper 15). These minerals have different susceptibilities to weathering that depend upon climate, crystalline inclusions, or crystal strain; Raeside concludes that they are not equally useful everywhere.

One of the earliest papers to examine resistance to wear during water transport was published in 1931 by Friedrich (Frederico) W. Freise (Paper 16). Freise was a mining engineer who came to Brazil from Germany in 1908 and spent the remainder of his career in Rio de Janeiro. His last published paper appeared in 1938. The short paper included here details the results of experimental abrasion tests on important placer minerals, with a discussion of field observations. Keeping in mind that all the resistances are relative, the least resistant mineral was found to be monazite, the most resistant, tourmaline.

Milton Nunn Bramlette (1896–1977) was a leading sedimentary petrographer with the U.S. Geological Survey from 1921 to 1940; he made significant advances in the study of the effects of intrastratal solution. Intrastratal solution definitely affects heavy-mineral suites, but its relative geologic importance been debated ever since it was discovered. Bramlette (Paper 17), in his study of the Monterey Shale of California, points out that apatite, a mineral readily attacked by acidic solutions, is common in sandstones that have lost epidote and hornblende; therefore, stability is far more than a matter of one mineral being more resistant than another. Furthermore, heavy-mineral suites were more varied in calcareous concretions in the Monterey Shale than in the adjacent enclosing beds; hornblende was particularly more abundant in the concretions. Because the concretions formed in situ, a different source for the heavy minerals in the concretions versus the surrounding sandstone was not plausible.

Bramlette later became an outstanding authority on microfossils while he was professor of geology at the University of California at Los Angeles (1940–1951) and Scripps Institution of Oceanography (1951-1962). He was elected to the National Academy of Sciences in

1954. In 1963 he was presented the U.S. Department of Interior Distinguished Service Award, and in 1969 he was awarded the Thompson Medal.

In 1940, Francis John Pettijohn (Paper 18) made a survey of 315 published papers (see Introduction) and drew the following conclusions: (1) The youngest sedimentary rocks contain the greatest variety of heavy minerals; (2) this variety significantly decreases with increasing age of the rocks; and (3) intrastratal solutions in the older rocks is responsible for this difference in mineralogy.

Pettijohn is emeritus professor of geology at The Johns Hopkins University. His textbook, *Sedimentary Rocks,* now in its third edition, has been a standard reference since its first appearance in 1949. He is a member of the National Academy of Sciences and winner of the Penrose, Wollaston, Sorby and Twenhofel Medals.

Pettijohn's view was challenged almost immediately by Paul Dimitri Krynine (Paper 19). Krynine maintained that major differences in mineral composition of rocks of varying ages is due more to changes in provenance, with intrastratal solution playing a locally important role. Unfortunately, he never published a more complete analysis of the problem, which would have been most welcome. We do not know exactly which studies he examined to reach his conclusions. In fact, Paper 19 was never presented; the American Association for the Advancement of Science meeting in which it was scheduled (December 1942) was canceled due to World War II.

Kyrnine (1902–1964) was a native of Russia who emigrated to the United States in 1927. He became the first sedimentary petrographer at the Pennsylvania State University in 1937 and attracted many outstanding students during his long, colorful teaching career. One of the most controversial geologists of his day, Kyrnine was a source of many original ideas. Sadly, much of his work was either unpublished or appeared only in abstracts.

Maria Turnau-Morawska (Paper 20) offers a third perspective on this problem in an excerpt from a longer paper on the uses of heavy minerals; this is the first time any part of this paper has appeared in English translation from the original Polish. She concludes that at the time of Pettijohn's original survey there were not enough data to make a positive conclusion. A more thorough literature search would probably show that the number of identified species of heavy minerals in rocks of all ages would not be significantly different. There may well be more of a given mineral in younger sediments, but the absolute number of species would be largely the same.

Maria Turnau-Morawska was a Polish sedimentologist at the University of Warsaw and coauthor with Andrzei Bolewski of *Petro-*

grafia, a major petrology text in the Polish language. Her last published paper appeared in 1978; she is now deceased.

In the most recently published paper included here, Blatt and Sutherland (Paper 21) point out that fine-grained shales provide a seal around heavy-mineral grains, protecting them from any effects of intrastratal solution. Studies of heavy minerals in fine-grained clastics are not as common as those for sand-size material, but shales may preserve a more accurate picture of the heavy-mineral suites from the source areas. Therefore, they might be not only a significant but also even better source of heavy minerals for provenance studies than the more permeable sandstones, where percolating waters may attack the less resistant minerals.

Harvey Blatt is professor of geology at the University of Oklahoma. He was a Fulbright senior fellow at Würzburg in 1974 and 1975 and associate editor of the *Journal of Sedimentary Petrology* since 1977. He is the author of *Sedimentary Petrology* and coauthor with Gerard Middleton and Raymond Murray of *Origin of Sedimentary Rocks,* now in its second edition.

Berry Sutherland is the associate dean of the College of Social and Behavior Sciences at the University of Texas at San Antonio and continues to teach geology classes.

10

Reprinted from *Tschermaks Mineralog. u. Petrog. Mitt.* **42:**482–490 (1932)

RELIKSTRUKTUREN DETRITISCHER PYROXENE UND AMPHIBOLE.[1]

C. H. Edelman und **D. J. Doeglas** (Amsterdam).

Mit 1 Textfigur und 1 Tafel (XIX).

Der Vorgang diagenetischer Auflösung detritischer Pyroxene und Amphibole wird in tertiären Sanden untersucht, entsprechende Lösungskörper (Reliktstrukturen) werden beschrieben und erklärt. Es wird darauf hingewiesen, daß die beschriebenen Vorgänge zeigen, warum in alten Sedimenten Pyroxen- und Amphibolkörnchen fast vollkommen fehlen.

Zu den akzessorischen Bestandteilen der klastischen Sedimente gehören auch die mafischen Mineralien der Ursprungsgesteine, Pyroxene und Amphibole. Ihre Verbreitung ist aber in rezenten und fossilen Sedimenten keineswegs dieselbe;[2]) während der Gehalt vieler rezenter Bildungen an Pyroxenen und Amphibolen quantitativ nicht unbedeutend ist (in den Bromoformkonzentraten dieser Bildungen herrschen sie öfters über die Menge aller anderen Schwermineralien vor), ist ihr Vorkommen in älteren Sedimenten sehr beschränkt; die quartären und jungtertiären Sedimente nehmen in dieser Hinsicht eine Zwischenstellung ein. Weil es an sich schwer vorstellbar wäre, daß in den älteren Gesteinen die mafischen Bestandteile von Anfang ab gefehlt hätten, wird die Annahme nahegelegt, daß die Pyroxene und Amphibole, welche ursprünglich zum detritischen Material gehört haben, den neuen Bedingungen zum Opfer gefallen sind und jetzt als Verwitterungsprodukte vorliegen.

Aus den Angaben, die über das Vorkommen von Pyroxenen und Amphibolen in der Literatur vorliegen, gewinnt man den Eindruck, daß die Amphibole im allgemeinen die widerstandsfähigeren sind. Augite werden fast nur aus jungtertiären Sedimenten angegeben; detritischer Hypersthen, in schroffem Gegensatz zu der

---

[1]) Vortrag des erstgenannten der Verfasser auf der Dresdener Tagung der Deutschen Mineralogischen Gesellschaft, September 1931.

[2]) Siehe hierüber z. B. *P. G. H. Boswell*, The Rarer Detrital Minerals of British Sedimentary Rocks, Trans. Geol. Soc. Glasgow XVIII, 1927, p. 131.

öfters reichlichen Beteiligung an rezenten Bildungen, ist in allen fossilen Sedimenten als Seltenheit anzusehen.[3])

Der guten Spaltbarkeit zufolge sind Amphibolkörner im allgemeinen flach parallel (110) und nach c gestreckt. Die relativ geringe Härte verursacht öfters gerundete Formen, so daß in vielen Sedimenten die Hornblenden als längliche Scheibchen vorliegen. Die Augite spalten weniger gut und zeigen daher die Entwicklung parallel (110) nicht so deutlich wie die Hornblenden. Wie zu erwarten ist, werden auch Körner, welche flach nach (100) sind, angetroffen. Bei Hypersthen kann die Form der Körner von allen drei Spaltungsrichtungen (110), (010) und (100) beeinflußt werden. Die verschiedenen möglichen Orientierungen kommen alle vor. Sie sind mit Hilfe des Pleochroismus und nach der Lage der optischen Achsen leicht zu unterscheiden.

Sehr merkwürdig ist in vielen Fällen die Entwicklung der kurzen Seite der Körner, welche keineswegs an Abrundung erinnert. Einspringende Winkel sind allgemein; auch ein Auslaufen der Körner in zahllose feine Fasern ist öfters zu beobachten. In der englischen und amerikanischen Literatur werden mit Bezug auf detritische Augite die Bezeichnungen „*cockscomb*“- oder „*hicksaw*“-Struktur verwendet; Bezeichnungen wie „*ragged*“ oder „*frayed*“ sind bei der Beschreibung detritischer Amphibole öfters benutzt worden. Diese vielen Namen beweisen nicht nur, daß die Erscheinung augenfällig und weitverbreitet ist, sondern auch, daß die Deutung der Strukturen Schwierigkeiten verursacht hat. Man hat die gute Spaltung parallel c für die Erscheinung verantwortlich gemacht; eine nähere Betrachtung der Struktur zeigt aber, daß diese Möglichkeit nicht in Betracht kommt. Viele der hierzugehörigen Strukturen sind so zart, daß ein Transport des detritischen Materials sie zweifelsohne zerstört hätte. Aber auch rein morphologisch ist die Spaltung nicht imstande, diese Erscheinungen zu erklären, wie eine Betrachtung der Abbildungen, welche zu dem vorliegenden Aufsatz gehören, ergibt. So ist zum Beispiel gar nicht einzusehen, wie Pyroxenspaltung die zickzackförmige Begrenzung der Augit- und Hypersthenkörner, Tafel XIX, Fig. 3—13, verursacht haben könnte.

Ross, *Miser* und *Stephenson*[4]) haben gezeigt, daß die „*cockscomb*“-Strukturen, welche von ihnen untersuchte Augite aufwiesen,

[3]) *P. G. H. Boswell*, l. c., p. 140.

[4]) U. S. Geol. Survey, Prof. Paper 154 F, pp. 175—202, 1929.

nachträglich durch Resorption, also sekundär, entstanden sind. Aus unseren Untersuchungen geht hervor, daß auch die Erscheinungen an Amphibolen und Hypersthen in dieser Weise zu erklären sind. Die in Umwandlung begriffenen Körner, welche die typischen Strukturen aufweisen, sind daher als Relikte, die Strukturen als Reliktstrukturen zu bezeichnen. Die im Anfang dieses Aufsatzes angedeutete Neigung der Amphibole und Pyroxene zu Umwandlungen bedingt also die Entwicklung dieser Reliktstrukturen. Die Seltenheit der mafischen Bestandteile in älteren Sedimenten und die Strukturen, welche in jüngeren Gesteinen auftreten, entsprechen nur relativ verschiedenen Abbaustadien der massenhaften, in rezenten Bildungen auftretenden Minerale.[5])

Eine genauere Betrachtung dieser Reliktstrukturen zeigt nun, daß die Unregelmäßigkeit vielfach nur scheinbar ist und daß gewisse Gesetzmäßigkeiten auftreten, welche hier näher beschrieben werden.

a) Monokline Pyroxene.

Wie schon oben bemerkt wurde, ist in der normalen Entwicklung der detritischen Augite eine Formentwicklung parallel (110), seltener parallel (100), bevorzugt, wobei die c-Achse öfters deutlich markiert ist. Die Reliktstrukturen zerfallen in drei Typen:

1. Körner nach (110) mit Zickzack-Begrenzung.

Diese am meisten verbreitete Ausbildung konnten wir studieren an einem jungtertiären Kalksandstein von Soekaboemi (W.-Java), dessen Sandgehalt tuffogener Natur ist. Das Gestein enthält auch Hornblende, welche weniger interessant ist.

Die Photographien 3—5, Tafel XIX, geben einen Überblick über die Erscheinungen. Fig. 3 zeigt ein Korn, dessen äußere Begrenzung von zwei Systemen paralleler Linien bedingt ist, die sich unter $\pm 45^{\circ}$ schneiden. Anscheinend sind die beiden Liniensysteme gleichwertig; derartige Körner löschen aber nahezu parallel der einen Richtung aus, was beweist, daß die andere Richtung dem Spaltdurchgang parallel geht. Ein Vergleich mit den von *Greim* und von *Wülfing* beschriebenen Ätzfiguren zeigt eine gewisse Übereinstimmung. Die

[5]) Selbstverständlich ist dies nicht so aufzufassen, daß alle älteren Sedimente ursprünglich reich an mafischen Mineralien gewesen wären; auch in rezenten Bildungen sind große Schwankungen im Mengenverhältnis dieser Mineralien aufzufinden.

Ätzfiguren auf den Spaltprismen sind längliche dreieckige Grübchen, von denen eine der längeren Seiten parallel der Spaltung geht; die aus der Abb. 3 ersichtliche zweite Begrenzungsrichtung könnte dann der kurzen Seite der Ätzdreiecke entsprechen. Die Stellung mit Bezug auf die kristallographischen Achsen ist aus der Textfigur 1 ersichtlich.

Die zweite Richtung verläuft also im spitzen Winkel β. Der Winkel zwischen den beiden die Reliktkörner umgrenzenden Richtungen ist nicht konstant. Eine große Anzahl Messungen ergab Werte schwankend zwischen 45° und 60°, was aber bei der Variation, welche Ätzfiguren gewöhnlich aufweisen, nichts Überraschendes darstellt.

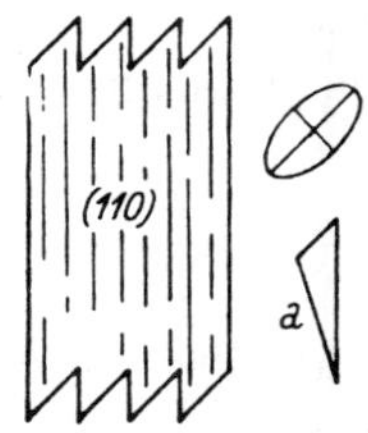

Fig. 1. Schematische Abbildung der Zickzackbegrenzung eines detritischen Augits parallel (110). a = Ätzfigur nach *Greim.*

Fig. 4, Tafel XIX, zeigt eine typische Reliktform, Fig. 5 zahllose sehr feine Kämmchen, welche sich räumlich in viele parallele Ebenen anordnen.

Die Dünnschliffphotographien Tafel XIX, Fig. 1 und 2, sind für die Genese der Reliktstrukturen von Bedeutung. Fig. 1 zeigt noch deutlich den ursprünglichen Umriß des Kristalls. Die weißen Dreiecke, welche die kurzen Seiten des Kristalls aufweisen, sind mit Calcit ausgefüllt. Fig. 2 zeigt unten eine feinere Ausbildung, wobei die Asymmetrie der Erscheinung deutlich ins Licht tritt.

Die von *Ross*, *Miser* und *Stephenson*[6]) angegebenen Umwandlungserscheinungen gehören auch zu dem hier beschriebenen Typus.

2. Körner nach (100) mit Zickzack-Begrenzung.

Die gelegentlich bei detritischen Augiten auftretende Entwicklung nach der Spaltebene der Diallag-Spaltung gibt Anlaß zu einem zweiten Typus, wobei die Kämme einen Winkel von 90° einschließen, der streng symmetrisch ausgebildet ist. Fig. 6, Tafel XIX, zeigt ein derartiges Korn, photographiert aus einem Bromoform-Konzentrat eines jungtertiären Sandsteins vom Mahakam-Delta-Gebiet (O.-Borneo). Diese Ausbildung ist von uns nur sehr selten gefunden worden. Auch hier besteht eine gewisse Übereinstimmung mit den künstlichen Ätzfiguren, welche auf (100) längliche Vierecke bilden.

[6]) *Ross, Miser* und *Stephenson,* l. c., Plate 24.

Die beiden kürzesten Seiten dieser Vierecke schließen ungefähr einen rechten Winkel ein. Wie bei den (110) Körnern scheint auch hier die natürliche Umwandlung die Richtungen zu bevorzugen, welche bei den künstlichen Versuchen untergeordnet auftreten.

3. Körner mit feinfaseriger Endbegrenzung.

Dieser Typus ist gleichfalls selten; von einer Gesetzmäßigkeit ist hier eigentlich nicht mehr die Rede. Die meisten Körner, welche wir aufgefunden haben, zeigten parallele Verwachsung mit rhombischen Pyroxenen, und es scheint uns nicht unwahrscheinlich, daß die durch diese Einlagerungen bedingte mechanische Instabilität der Körner mit der unregelmäßigen feinfaserigen Entwicklung der Reliktformen zusammenhängt.

b) Rhombische Pyroxene.

Bei den rhombischen Pyroxenen kommen, wie schon oben bemerkt, drei Entwicklungsmöglichkeiten der Formorientierung in Betracht: nach (110), (100) und (010), die zu den nachfolgenden Relikttypen Anlaß geben.

1. Körner nach (110) mit Zickzack-Begrenzung.

Einige Sandsteine aus dem Mahakam-Delta-Gebiet lieferten sehr zahlreiche Hypersthene, von denen Tafel XIX, Fig. 12, ein gutes Beispiel liefert. Der Kammwinkel beträgt gewöhnlich 60° bis 70°, kann aber gelegentlich viel kleiner sein. Künstliche Ätzfiguren an Hypersthen konnten nicht zum Vergleich herangezogen werden; einige Körner aber zeigten eine schön entwickelte Anätzung der Oberfläche (siehe Tafel XIX, Fig. 9), welche aus ellipsenförmigen Vertiefungen besteht. Diese Wahrnehmung brachte uns zu der Erklärung der schon oben mitgeteilten großen Variation des Kammwinkels, indem die verschiedenen Berührungsmöglichkeiten dieser Ellipsen zu sehr verschiedenen Winkeln Anlaß geben können. Tafel XIX, Fig. 8, zeigt einen derartigen Hypersthen, bei dem die Spitzen deutlich gebogen begrenzt sind. Es bestehen also Übergänge zwischen diesem Typus und dem unten zu besprechenden faserigen Typus (Typus 6).

2. Körner nach (100) mit Zickzack-Begrenzung nach (110).

Wenn Körner, die Reliktstrukturen des vorigen Typus aufweisen, nach (100) flach genug sind, um nach dieser Fläche orientiert zu sein, entwickeln sich Strukturen, bei denen der Kammwinkel von 60° bis 70° der Orientierung entsprechend verkürzt ist. Tafel XIX, Fig. 7, zeigt eine dachziegelartige Anordnung der Kämmchen; beim Heben und Senken des Mikroskoptubus sieht man, daß die Kämmchen schief zur Bildebene stehen. Fig. 10—11, Tafel XIX, sind Mikrophotographien desselben Korns in zwei Stellungen; Fig. 11 zeigt den (110) Typus auf (110), Fig. 10 auf (100).

3. Körner nach (100) mit Zickzack-Begrenzung nach (100).

Die nach (100) orientierten Hypersthene können auch eine eigene (100) Umwandlung aufweisen; genau so wie bei Augit erkennt man hier eine symmetrisch entwickelte Kammbegrenzung mit einem Winkel von 90°. Tafel XIX, Fig. 6, ist auf dem Bilde nicht von Hypersthen dieses dritten Typus zu unterscheiden. Fig. 13 zeigt noch einen solchen Hypersthen.

4. Körner nach (010) mit Zickzack-Begrenzung nach (110).

Die Erscheinungen sind genau dieselben wie bei Typus 2.

5. Körner nach (010) mit Zickzack-Begrenzung nach (010).

Eine typische (010) Entwicklung wurde nur einmal gefunden; die Kammbegrenzung war scharf, ungefähr dem Korn Tafel XIX, Fig. 8, ähnlich.

6. Körner mit faseriger Endbegrenzung.

Dieser Typus, der durch Übergänge mit den anderen Typen verbunden ist, hat in gewissen Lokalitäten eine große Verbreitung. Die Fasern zeigen aber kaum noch Gesetzmäßigkeiten; bei der Auflösung ist die Richtung der c-Achse derartig resistent, daß daneben keine anderen charakteristischen Richtungen entwickelt wurden.

c) Amphibole.

1. Unregelmäßig begrenzte Körner.

Dieser Typus hat eine sehr große Verbreitung in Sedimenten. Die Umwandlung ist parallel der Spaltfläche vorgedrungen und hat dabei rundliche Formen entwickelt, wie Fig. 20, Tafel XIX, zeigt. Auch Fig. 19 kann als typisch für sehr viele detritische Hornblenden in älteren Sedimenten angesehen werden. Fig. 14 zeigt noch „Ätzhügelchen“ an einer Hornblende; die Übereinstimmung mit den künstlichen Ätzfiguren, welche von *Daly*[7]) beschrieben worden sind, ist nicht sehr groß und auch die äußere Form der Körner dieses Typus wird durch die Richtungen dieser natürlichen Ätzung kaum beeinflußt.

2. K ö r n e r m i t f e i n f a s e r i g e r E n d b e g r e n z u n g.

Die Hornblenden eines sandigen Foraminiferen-Kalksteins von Ketringan (M.-Java), Tafel XIX, Fig. 15—18, zeigen eine wunderbare feinfaserige Endbegrenzung, welche als extremes Endglied einer übrigens kontinuierlichen Übergangsreihe nach dem ersten Typus aufzufassen ist. Die Umwandlung ist parallel c linienartig vorgedrungen und hat als Relikte äußerst feine Nädelchen stehen lassen. Im Dünnschliff sieht man, wie die Körner ursprünglich stark abgerundet waren; der Raum zwischen den Reliktfasern ist mit c h l o r i t i s c h e r Substanz ausgefüllt.

Feinfaserig begrenzte Hornblenden sind sehr verbreitet, wenn auch Strukturen wie die abgebildeten zu den Ausnahmen gehören.

Ergebnisse.

Die Entwicklung der oben beschriebenen Typen der R e l i k t s t r u k t u r e n, die als L ö s u n g s k ö r p e r bezeichnet werden können, scheint zusammenzuhängen mit einer mechanischen Vorbedingung.

Wenn ein Augit nur mechanische Diskontinuitäten nach (110) aufweist, so benützt die von außen nach innen vordringende Umwandlung die Fläche geringsten Widerstands unter Bildung gesetzmäßiger Formen, welche mit Ätzfiguren zusammenhängen. Ist dagegen Neigung zur Bildung von Diallagspaltung vorhanden, so erscheinen Formen des zweiten Augittypus. Wenn mehrere Diskontinuitätstendenzen koexistieren, so wird die gemeinsame c-Achse eine bevorzugte Richtung unter Ausbildung feinfaseriger Reliktformen.

Die Hypersthen- und Hornblendetypen lassen sich in gleichartiger Weise übersehen.

Über die für die Umwandlung verantwortlichen Agenzien gibt das geologische Vorkommen Aufschluß.

---

[7]) *R. A. Daly,* Proc. Am. Ac. Arts and Sc. 34 (1899), p. 311.

Die teilweise ganz jungen, kaum verfertigten Bildungen, welche das Untersuchungsmaterial geliefert haben, schließen die Möglichkeit einer Umwandlung etwa vergleichbar mit Propylitisierung aus. Es kann nur zirkulierendes CO_2reiches Grundwasser die Umwandlung verursacht haben; der ganze Prozeß ist demzufolge nach den gebräuchlichen Definitionen zu der Diagenese zu rechnen.

Über die Umwandlungsprodukte haben wir nur spärliche Wahrnehmungen machen können, weil der Gehalt der Schwermineralien normalerweise so gering ist, daß man sie im Dünnschliff nur ausnahmsweise antrifft, und die Umwandlungsprodukte bei der Laboratoriumbehandlung der Gesteine vernichtet werden müssen, um die Strukturen untersuchen zu können.

In dem Kalksandstein von Soekaboemi (W.-Java) ist Augit durch Calcit verdrängt worden. Der Ketringan-Kalkstein zeigt chloritartige Umwandlungsprodukte. Einer der Sandsteine vom Mahakam-Delta enthält neugebildeten Dolomit, welcher wohl auf Kosten von Hornblende und Hypersthen, welche weitgehende Zersetzung zeigen, entstanden ist.

Ross, *Miser* und *Stephenson* glauben in dem von ihnen untersuchten Beispiel Neubildung von Glaukonit beobachtet zu haben.

Die Verbreitung der Erscheinung ist, wie schon bemerkt wurde, eine besonders große. Die von uns untersuchten Fundorte sind:

1. Deltagebiet Mahakam (O.-Borneo).
   Jungtertiäre Sandsteine mit Augit (Typus 1, 2, 3), Hypersthen (Typus 1, 2, 3, 4, 5) und Hornblende (Typus 1).
2. Soemba.
   Tertiäre Tuffe mit Hypersthen (Typus 6).
3. Ketringan (M.-Java).
   Jungtertiäre sandige Foraminiferenkalksteine mit Hornblende (Typus 2).
4. Soekaboemi (W.-Java).
   Jungtertiäre Kalksandsteine mit Augit (Typus 1) und Hornblende (Typus 1).
5. Djoeloe-Rajeu (N.-Sumatra).
   Jungtertiäre Sandsteine mit Augit (Typus 1), Hypersthen (Typus 6) und Hornblende (Typus 1, 2).

Erklärung zu Tafel XIX.

Fig. 1, 2. Augit mit Reliktstruktur. Dünnschliff eines jungtertiären Kalksandsteins.
Fundort: Soekaboemi (W.-Java).
Vergr. 80 ×.

Fig. 3, 4, 5. Augit aus Streupräparat. Orientiert nach (110) mit Reliktformen nach (110).
Fundort: Soekaboemi (W.-Java).
Vergr. 75 ×.

Fig. 6. Augit aus Streupräparat. Orientiert nach (100) mit Reliktformen nach (100).
Fundort: Mahakam-Gebiet (O.-Borneo).
Vergr. 75 ×.

Fig. 7. Hypersthen aus Streupräparat. Orientiert nach (100) mit Reliktformen nach (110).
Fundort: Mahakam-Gebiet (O.-Borneo).
Vergr. 75 ×.

Fig. 8. Hypersthen aus Streupräparat. Übergang nach dem feinfaserigen Typus.
Fundort: Mahakam-Gebiet (O.-Borneo).
Vergr. 75 ×.

Fig. 9. Hypersthen mit natürlicher Ätzung aus Streupräparat.
Fundort: Mahakam-Gebiet (O.-Borneo).
Vergr. 75 ×.

Fig. 10, 11. Hypersthen aus Streupräparat. Dasselbe Korn in zwei verschiedenen Stellungen. Fig. 10 auf (110), Fig. 11 auf (110). Reliktformen nach (110).
Fundort: Mahakam-Gebiet (O.-Borneo).
Vergr. 75 ×.

Fig. 12. Hypersthen aus Streupräparat. Orientiert nach (110) mit Reliktformen nach (110).
Fundort: Mahakam-Gebiet (O.-Borneo).
Vergr. 75 ×.

Fig. 13. Hypersthen aus Streupräparat. Orientiert nach (100) mit Reliktformen nach (100).
Fundort: Mahakam-Gebiet (O.-Borneo).
Vergr. 75 ×.

Fig. 14. Hornblende aus Streupräparat. Natürliche Ätzung.
Fundort: Mahakam-Gebiet (O.-Borneo).
Vergr. 75 ×.

Fig. 15—18. Hornblende aus Streupräparat. Feinfaserige Reliktformen.
Fundort: Ketringan (M.-Java).
Vergr. 75 ×.

Fig. 19, 20. Hornblende aus Streupräparat. Normal ausgebildete Körner.
Fundort: Mahakam-Gebiet (O.-Borneo).
Vergr. 60 ×.

[*Editor's Note:* In Figure 1 on page 485 of the article the translation for the figure caption is: Schematic illustration of the zigzag borders of a detrital augite on the (110) face. a = etched figure after Greim. In Plate XIX a translation of the key words is as follows: Dünnschliff- thin section; Streupräparat- grain mount; Fundort- locality; Vergr.- enlarged; Atzung- etching; feinfaserig- delicately fibrous.]

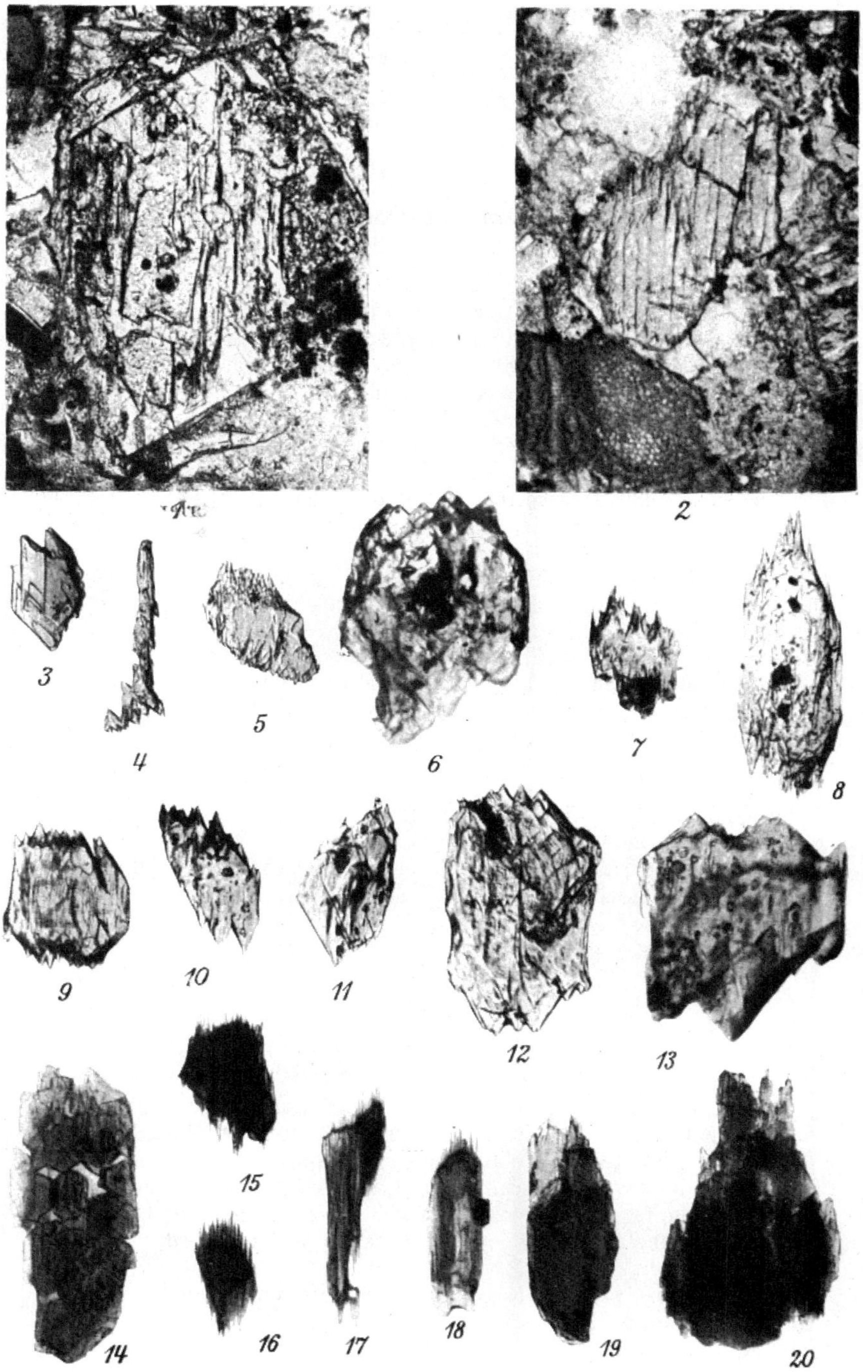

C. H. Edelman
und D. J. Doeglas

Akademische Verlagsgesellschaft m. b. H., Leipzig.

11

Reprinted from *Geol. Mag.* **80:**155–156 (1943)

Solution Etching of Detrital Staurolite

By Geoffrey Bond

ATTENTION has recently been directed, particularly by the writings of Boswell (1942) and Smithson (1941, 1942), to the instability of detrital minerals. In nearly all cases where etching of mineral grains has been described the rocks have been deposited under water, as members of a continuously sedimentary series.

While investigating the heavy mineral suites of sandy formations in Southern Rhodesia, the writer found etched grains of staurolite and kyanite in one sample only. Since this rock is of aeolian origin and is directly overlain by a contemporary flow of basalt it seems worth a short description.

The locality from which the specimen was obtained lies 25 miles north of Bulawayo along the main Victoria Falls Road, being a small roadside quarry 1 mile south of Hilda's Kraal. The rock is a red, flaggy, rather friable, well-jointed sandstone of aeolian origin lying beneath the first lava flow of the Nyamandhlovu series, and is of Karroo age. The hand specimen was taken some 20 feet below the base of the flow.

The scanty heavy suite is made up of zircon, tourmaline, and staurolite (common), and rare apatite, kyanite, sphene, and opaques. Nearly all the staurolite grains show etched terminations. The zircons and tourmalines are beautifully rounded. Except for their etched terminations the staurolites show a sub-rounded form.

These grains extinguish parallel to the slow axis which, therefore, corresponds to the c = Z axis of the crystal. They usually give an excentric " compass needle " isogyre, so that the face on which they lie must contain the point of emergence of the optic axis. Since the optic axial plane of staurolite is (100), the grains must be lying on the (010) face, being terminated by (001). The pleochroism is :—

Z = golden yellow
Y = pale yellow to colourless.

Projections occur only on (001), that is the basal plane of the crystal, and in the first example observed, on one end only. Other grains in the same slide show them at both ends. No grains have been seen with projections on any other face. From their position, and optical orientation of the grain, it can be seen that they are probably bounded by the (101) form, and that they are in perfect optical continuity with the remainder of the grain.

Staurolite possesses two cleavages which are given in Dana as :—

(010) Distinct but interrupted.
(110) In traces.

The first of these determines the orientation of most of the grains of staurolite in these sands, and the traces of it can often be seen on the top surfaces of the grains. The b-axis of uncleaved crystals is usually the shortest, but the (010) cleavage has produced flakes in these sediments which are flattened so that the a-axis is the shortest. Thus the solution planes (Lösungsflachen) of staurolite do not coincide with the cleavages.

Although a large number of slides from the sandy formations has been examined etched grains have only been seen in one sample, from the sandstone immediately underlying the first lava of the Nyamandhlovu group. Except in the lowest member staurolite is rare or absent in samples from this group, but common in Forest Sandstone and Kalahari Sands. It is not inferred that staurolite was originally present, and has been removed by solution. Etched grains represent the first stage in some such process, but the mineral assemblies in this group are quite different from the Forest Sandstone and Kalahari sands, notably in the varieties of tourmaline present. It does suggest that the lava immediately above supplied the agent which attacked the grains of staurolite. In hand specimen the rock is less consolidated than many Forest sandstones in which staurolite grains remain unaffected, as they do in the hard Kalahari quartzite described by Frost (1927), whose slides have been re-examined and found to contain staurolite. The nature of the attacking solution is obscure, but its connection with the overlying basalt seems clear. Other minerals mentioned by Smithson as unstable in sediments and showing etched faces included garnet, kyanite, and apatite. Garnet is absent and the last two extremely rare in this rock, but small kyanites show much raggedness at the ends, though it was not possible to determine the forms produced.

In the Yorkshire Jurassic, solution etching of various minerals is associated with outgrowths on zircon, an effect so far only recorded from Middle Jurassic deposits. In this case an aeolian sand affected by solutions from an overlying basalt, beautiful etching of at least two minerals has been produced, but the most careful search has not shown a single grain of zircon with outgrowths, although it is by far the commonest heavy mineral in the rock. Conditions which will cause etching of staurolite and kyanite by solution are, therefore, not necessarily those which produce outgrowths on zircon.

Acknowledgments

I wish to record my gratitude to Dr. Arnold, Director of the National Museum of Southern Rhodesia, for allowing me to make use of the facilities of the museum during my stay in Bulawayo.

List of Works to which Reference is made

Boswell, P. G. H., 1942. Anniversary Address. *Quart. Journ. Geol. Soc.*, xcvii, lvi.

Dana. *Textbook of Mineralogy*, 4th ed., p. 638.

Frost, A., 1927. Notes on an Occurrence of Quartzite at Gwaai S.R. *S.A. Journal Sci.*, xxiv, 177–181.

Smithson, F., 1941. Alteration of Detrital Minerals in the Mesozoic rocks of Yorkshire. *Geol Mag.*, lxxviii, 97–112.

—— 1942. The Middle Jurassic Rocks of Yorkshire : A Petrological and Palaeographical Study. *Quart. Journ. Geol. Soc.*, xcviii, 1942, 27.

12

Reprinted from *Jour. Sed. Petrology* **23:**106–116 (1953), by permission of the publisher

WEATHERABILITY OF ZIRCON

DOROTHY CARROLL
Bryn Mawr College, Bryn Mawr, Pennsylvania[1]

ABSTRACT

Observations made on numerous grains of zircon from the heavy residues of soils from different areas indicate that zircon becomes corroded under certain conditions, particularly those of strong alkaline leaching associated with the development of a lateritic type of profile. Zircon crystals have a number of weak points, such as inclusions and zoning, which enable corrosion to set in; some of these are illustrated by photomicrographs. It is suggested, following information obtained from spring and ground-water chemical analyses, that Ca-bicarbonate and Na-bicarbonate rich waters may be able to dissolve zircon. Several experiments in the artificial weathering of zircon are described; one of these used a synthetic cation exchange resin. The results, obtained over a short period of time only, were negligible.

INTRODUCTION

The most stable of the heavy minerals found in sedimentary rocks, unconsolidated sediments, and in soils is, by consensus of opinion (Smithson, 1950), zircon, and because of this stability and its almost universal presence in heavy mineral assemblages, it has been made use of by Haseman and Marshall (1945) in studies of soil development. In order to assess the value of zircon for this purpose, it is advisable to inquire more closely into its stability in soils and in other environments, and to distinguish clearly between loss by abrasion and loss by weathering processes.

Many minerals originally considered to be more or less unalterable, are now found, when examined closely, to lose substance and become corroded through the long-continued action of solutions with which they come in contact (Edelman and Doeglas, 1934). In his Anniversary Address to the Geological Society of London in 1941, P. G. H. Boswell stressed the importance of change and loss of mineral grains in sediments by solution as one of the most important to be considered by workers in the field of mineralogical aspects of sedimentary rocks, particularly in regard to the interpretation of provenance and past history of such rocks.

This paper records observations which have been made on the appearance of zircon grains, and their stability, in the heavy residues of soils; it also reports an attempt at the artificial weathering of zircon carried out over a short period of time.

PHYSICAL FEATURES OF ZIRCON

Chemically zircon is an orthosilicate, $ZrSiO_4$, in which part of the Zr is often replaced by Hf, Yt, Ce, etc. The size of the Zr^{4+} ion (0.87 A) leads to a slight instability of the structure, so that when radioactive elements are present a metamict state eventually results because of the weakening of the bond between Zr and the SiO_4 tetrahedra.

Three distinct types of zircon have been recognized by mineralogosts: (1) Normal; (2) Hyacinth or purple zircon; (3) Malacon. A study of the distribution of these types in the Lake Superior area pre-cambrian rocks was made by Tyler *et al* (1940), and Morgan and Auer (1941) have given a useful outline of their optical properties:

(1) Normal type—high refringence and birefringence (<0.05).
(2) Hyacinth type—slightly lower refringence and birefringence (0.017–0.043).
(3) Malacon type—refringence much lower and birefringence ± 0.008.

[1] Now at the United States Geological Survey, Washington, D. C.

The radioactivity increases from (1) to (3). In addition to reducing the refractive index and birefringence, the change to the metamict state increases the volume thereby reducing the specific gravity; furthermore, the material can now take in water which gives it a clouded altered appearance. In reporting zircon from soils and sediments the type should be stated, for the presence of malacon could be mistaken for genuine weathered normal type zircon.

Zircon has a hardness of 7.5, similar to that of staurolite and tourmaline. In abrasion tests reported by Freise (1931) it has a relative "transportwiderstand" of 265, compared with quartz, 245; rutile, 455; garnet, 375; epidote, 320; staurolite, 420; tourmaline, 650–950; and apatite, 275. The general smallness of grain size probably accounts for its persistence along with much more abrasive-resistant minerals.

Zircon has no cleavage, but there is a brittleness which causes the breaking into irregular fragments with conchoidal surfaces. It has been observed that there is often a weakness in the structure at the junction of the prism and pyramid faces.

Zircon crystallizes in the tetragonal system, the principal forms being (100), (110), and (111). In addition it exhibits a fascinating number of varietal forms based on the variation of prism and pyramidal faces, on zoning, and on inclusions and their type and arrangement. Brammall's description (1928) of the varietal features and their use is still the most important paper written on this aspect of zircon.

OBSERVATIONS ON CORROSION OF ZIRCON

Observation has shown that zircon grains can be reduced in size by the ordinary processes of weathering, but the actual processes which produce these results are not yet clear. However, there are a number of ways in which zircon can be reduced in size and eventually disappear.

The type of zircon.—It has already been mentioned that there are three types of zircon and innumerable crystallographic varieties. Corroded grains appear to fall into the following groups, considering only Type (1), Normal zircon:

A. Grains with numerous inclusions are less stable than those with very few. The inclusions may be, as summarized by Brammall (1928):

(a) gas-filled cavities.
(b) stringers, clots, and cloud-like suspensions of opaque dust;
(c) crystalline inclusions of various minerals such as monazite, hematite, quartz, apatite, etc.

B. Grains which are zoned are less stable than unzoned grains. Zoning appears to be more frequent in the hyacinth than in the normal type zircon.

The action of water containing various solvents appears to be:

(i) Penetration into cavities in the mineral; this is done by the removal of surface material of the grains, particularly when the cavities are only covered by a very thin layer of mineral, i.e., when the cavities are very near the surface. This is illustrated in figure 1.
(ii) Penetration of solutions along zoning lines. The zoning appears to be the result of growth similar to that which produces phantom quartz crystals, and the zones are often lined by a deposit of fine dust-like particles which may be hematite, magnetite, etc. The penetration of solutions is probably assisted by these minute particles of foreign minerals. Zoning is due to lamination, and is one of the results of the introduction of material of slightly different chemical composition, including the variation of radioactive elements, during the growth of the crystal. These slight differences in chemical composition may facilitate the removal of mineral substance (fig. 2).
(iii) Zircon is a brittle mineral, which though hard, may crack and fracture.

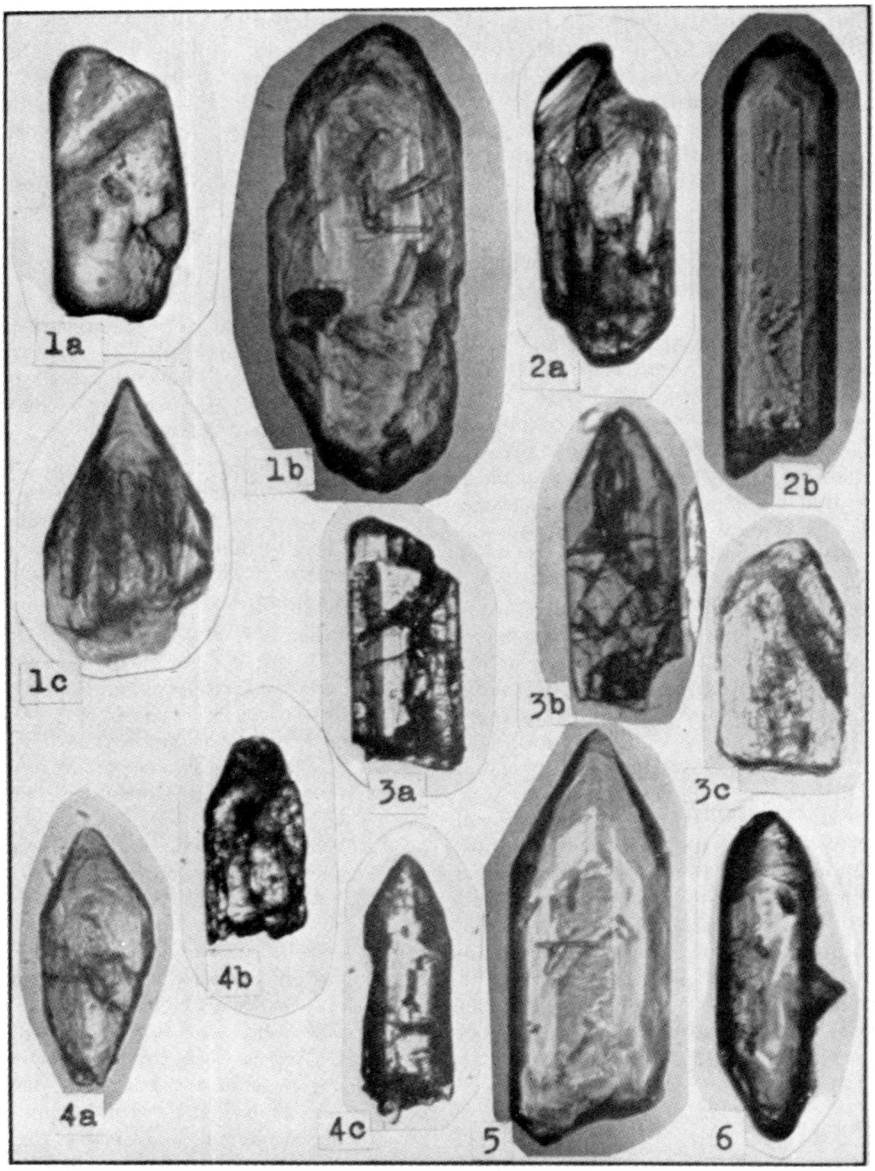

LEGEND FOR FIGURES 1 TO 6

FIG. 1.—Zircon grains showing cavities and the penetration of solutions. 1a. Rounded grain showing open cavities and the start of further corrosion. Residue from soil, Bullsbrook, W. Australia. ×190 (in refractive index of 1.54). 1b. Euhedral grain, alluvial, Colorado. ×185 (in R.I. 1.72). 1c. Pyramidal end of euhedral grain with many cavities. Residue from soil near Bel Air, Maryland. ×140 (in R.I. 1.72).

FIG. 2.—Zircon grains showing corrosion along zones. 2a. Grain showing a late stage in penetration. Residue from soil, Lake Grace, W. Australia. ×150 (in R.I. 1.54). 2b. Early stage of corrosion, zones just visible. Alluvial, Colorado. ×185 (in R.I. 1.72).

Cracks on the surfaces of faces allow the penetration of solutions, and this eventually leads to a removal of material. Hutton (1950, pp. 690 and 696) has described microfissuring in hyacinth grains, and states that this fissuring is caused by increase of volume due to transformation of the mineral to the metamict state by radioactive elements. The grains figured here, however, show cracks in the normal type of zircon (fig. 3).

(iv) The broken ends of prisms appear to be much more susceptible to corrosion than the surfaces of prism and pyramidal faces, and where the latter remain intact, corrosion is at a minimum. The broken ends of prisms when corroded show pointed remnants of zircon (fig. 4) very similar to the remnants of amphibole, pyroxene, epidote, and staurolite which are of quite common occurrence (Edelman and Doeglas, 1934). The etching of garnet is confined to the surfaces and appears to be somewhat different, although the cause is probably similar. There seems to be a fundamental difference in the atomic structure at the ends of the c crystallographic axis when compared with the ends of the a and b axes. In tourmaline it has been observed that it is the antilogous end which is most receptive to the deposition of new material as secondary growth (Alty, 1933), although outgrowths also occur elsewhere (Krynine, 1946).

In many minerals characteristic etch-patterns occur on crystal faces, and appear alike on similar faces, e.g. in quartz, where such patterns are used to determine compound twining. Difference in etch-patterns may cause differences in corrosion of faces in zircon crystals, and an indication of this is shown in figure 5. Long-continued etching by solutions will result in the destruction of the original crystal faces, and their surfaces will be removed.

(v) There is the, as yet, unexplained presence of secondary outgrowths of zircon as described by Trueman (1912), Butterfield (1936), Smithson (1937), Tyler *et al.* (1940), Bond (1948), and Hutton (1950). It would appear that some zirconium was present in the rock and soil solutions, together with silica, and that stagnant conditions occurred that were favorable for the formation of outgrowths on the zircon grains already there. Krynine (1946) has described outgrowths on tourmaline; the titaniferous minerals will crystallize from solutions provided the conditions are suitable; and albite occurs authigenically. An outgrowth on zircon is illustrated in figure 6; it is from a heavy residue from the Jurassic of Yorkshire, England, kindly given me by Dr. F. Smithson.

MECHANISM OF CORROSION OF ZIRCON

In weathering processes with normal and hyacinth types of zircon the grains

---

LEGEND FOR FIGURES 1 TO 6 (continued)

FIG. 3.—Zircon grains with surface cracks. 3a. Broken prismatic grain with cracks filled with clay material (in R.I. 1.72). 3b. Grain with surface cracks and inclusions (in R.I. 1.84). 3c. Badly corroded grain, the corrosion having started from cracks (in R.I. 1.84). All from soil residue, Bel Air, Md. ×140.

FIG. 4.—Zircon grains showing corrosion at broken end of prism. 4a. Euhedral grain with small cavity at position of c axis. 4b. Much corroded grain with irregular pointed remnants. 4c. Similarly corroded grain with possible small outgrowths at end of c axis. All from soil residue, Bel Air, Md. × 140 (in R.I. 1.72).

FIG. 5.—Euhedral grain showing differential removal of surface layers and development of different appearance of faces. Note inclusions; and pock-marks on (110). Alluvial, Colorado. X185 (in R.I. 1.72).

FIG. 6.—Rounded zircon grain with pyramidal outgrowth. Jurassic, Yorkshire, England. Courtesy Dr. F. Smithson. ×200 (in R.I. 1.54).

are reduced by the slow corrosion of the mineral substance which is removed in solution; there is no alteration of the zircon itself whereby an alteration product occurs. The process is apparently similar in action to that which causes the solution of garnet, staurolite, epidote and zoisite, amphiboles, and pyroxenes, but, as is well known, in the latter two minerals there is also the actual change of the mineral, wholly or in part, and in stages, into a chlorite-like mineral. In soils rich in ferromagnesian minerals, both types of alteration can occur together.

It would seem that the zircon reported as altered (e.g. by Goldich, 1938), with cloudy appearance, may be all the malacon type, in which the alteration is caused by radioactivity, and not by weathering. When more observations have been made, it may be found that the alteration by radioactivity has left the remaining mineral substance more in a condition which can be altered by solutions such as occur in the chloritization of amphibole and pyroxene.

Recently Rankama and Sahama (1950) have drawn attention to the fact that although zircon, as the principal repository of zirconium in the lithosphere, appears to be excessively stable, yet there is a zirconium cycle in Nature. The work of Strock (1941) at Saratoga Springs, N.Y. gives evidence of one way in which zirconium may be transported in solution and redeposited.

The mineral waters of Saratoga Springs are remarkably rich in zirconium which Strock (1941) has determined spectrochemically as 35 mg per liter. The source of this Zr is considered to be zircon as there is no other known source for it, nor are any deposits of zirconiferous minerals known in the area from which the spring waters come. The Saratoga waters are strongly carbonated, highly saline, and calcium-rich, and their action has been to dissolve Zr from zircon in the terrain which they drain. Strock further records (1941, p. 868) that "preliminary experiments have confirmed the ability of Ca-bicarbonate solutions to dissolve considerable amounts of Zr from precipitated ZrO_2 and from zircon and eudialyte." Without going into details here, Strock considers that Zr is carried in the Saratoga waters as a stable dicarbonatozirconylate ion, $Ca[ZrO(CO_3)_2]$, in the presence of excess CO_2. This is stable at the acidity of the water (pH 5.5) and Zr is transported in this form; when the water becomes alkaline the Zr is precipitated.

That Strock's ideas may be correct with regard to the solution of zircon by Ca-bicarbonate solutions is suggested by the work of Foster (1950) on the waters of the Atlantic and Gulf Coastal Plains. She finds that, experimentally, the presence of carbonaceous matter such as lignite and soil organic matter, is sufficient to generate CO_2 in sediments, and when this is "dissolved in water enables it to take into solution more calcium carbonate. If base exchange materials are also present to replace Ca with Na, a still greater amount of bicarbonate can be held in solution" (1950, p. 33).

This process, aided by the cation-exchange capacity of the clay-minerals, is probably quite important in soils and may possibly account in part for the corrosion of many minerals, including zircon. It should be remembered that small grains of zircon in a clayey soil may be embedded in aggregates of clay minerals, and that they are passively subjected to any chemical action that may take place in the processes of cation-exchange during and after soil formation. The writer has recently seen zircon grains completely enclosed in clay which penetrated into cracks in the mineral; in one case a broken zircon crystal was held more or less in its natural position by clay. Graham (1941) has explained the mechanism whereby such minerals as feldspars, amphiboles, and pyroxenes, are altered by the presence of acid clay, and such a process, while not achieving the same results with zircon as with more weatherable minerals, must have some effect on the zircon. The results of such corrosion are recorded in the individual grains.

From another angle, Maurice (1949) found that zircon has a wide range of stability in acid solutions, but it is restricted in alkaline solutions (experiments were done at a temperature of 400° C. and under a pressure of 900 atmospheres). Zircon was synthesized at 400° C. in aqueous solutions. After a series of experiments, it was found that if ZrO_2 and SiO_2 are in the presence of each other in dilute acid solutions, the silicate will form and will be stable. In alkaline solutions it was not possible to obtain a pure crop of zircon crystals. It was found that the maximum alkalinity at which quartz and zircon are stable together does not exceed 0.07 N Na_2O. Zircon is readily crystallized in acid solutions, but its field of stability in alkaline solutions is very restricted. These observations were all made at high temperatures and pressures, but it is possible that the same results may be produced under normal temperatures in rocks when the time is unlimited.

In 1948 Fredrickson made some interesting comments on zirconium in laterites, and noted that where zircon grains were protected inside pisolites they remained unchanged, but elsewhere zircons "show corroded irregular outlines indicating that they are gradually going into solution." Bramlette (1941) made the same observations with regard to hornblende, pyroxene, and epidote in concretions in sandstone. Frederickson (1948) considers that Zr is more easily removed in solution than Ti, and that because it does not form secondary Zr minerals, the zirconate ions are adsorbed on the aluminum hydroxide minerals and may substitute for Al in the various aluminous minerals; the ionic potentials are very similar, but the ionic radius of Zr^4 (0.87 A) is much greater than that of Al^3 (0.57 A) or Ti^3 (0.64 A). It could, however, substitute for Fe^2 (0.83 A). The movement of silica in bauxite and laterite formation, resulting, if the conditions remain favorable, in its complete removal from a laterite profile, is caused by alkaline solutions, and it would appear that these same conditions are conducive to the corrosion of zircon. It has been noted by the writer that the most strongly corroded zircons are in soil profiles which have been subjected to laterizing conditions, and that the smallest grains have sometimes completely disappeared. It has been suggested elsewhere (Carroll, 1951) that a shorter period of time is required for the complete laterization of a profile developed on granite, than for one on a basic type of rock, as there is actually less silica to be removed (much SiO_2 is in the quartz in granite, and it is considered to be relatively inert). The zircon grains in old lateritic soils are generally strongly corroded (see fig. 2). In a laterite developed on basalt described recently (Carroll and Woof, 1951), only the smallest traces of zircon remained. In a red soil from the Piedmont region of Maryland (Montalto silt loam), the zircons were strongly corroded (fig. 4, a, b, c) in company with corroded zoisite and amphibole.

The finding of zircon with outgrowths by Butterfield (1936) and Smithson (1937), as well as others, indicates that conditions were favorable for the addition of material from solution, probably without much transport. The sandstones in which these zircons occur were formed under deltaic and estuarine conditions, so that salinity was present when the sandstones were laid down. It is probable, however, that the outgrowths on zircon were formed in the sandstones subsequently to deposition, as suggested by Butterfield in his paper (1936). The outgrowths appear to be similar in origin, but not in appearance, to the zircon outgrowths pictured by Tyler *et al* in the Red Clastic series of Minnesota (1940, pl. 4, fig. 2). Arenaceous beds which are leached by later solutions seem a particularly favorable environment for the production of secondary mineral growths; anatase and brookite are common, tourmaline has been recorded by Krynine (1949), and quartz with outgrowths is characteristic of some sandstones. Sand-

TABLE 1.—*Series A zircon leaching experiment*

	Zircon A Australian concentrate[a]	Zircon B Australian concentrate	Zircon C Alluvial, Colorado[b]
Size, mm.	0.25–0.12	0.12–0.06	0.25–0.12
Weight, gm. original	2.0	2.0	2.0
Weight, gm. final	1.11	2.76	1.80
Difference	−0.89	+0.76	−0.20
Test for Zr:			
Alizarin	nil	nil	nil
Cupferron	trace	trace	trace

[a] Commercial concentrate; monazite removed with Franz Isodynamic Separator. Foote Mineral Co. sample donated by Mrs. L. Dryden.

[b] Alluvial concentrate from Lincoln, Placer Co. Colorado; ferromagnesians removed with Franz Isodynamic Separator. Donated by Dr. H. A. Ireland, University of Kansas, Lawrence, Kansas.

stones commonly contain siliceous solutions which cause the hardening of the rock when it is quarried for building material. Solution of a little zirconium and its association with siliceous solutions in the right proportions under stagnant conditions seem to be all that is necessary to produce zircon outgrowths. Alkalinity is required for the solution of zirconium, but acidity for its crystallization as zircon.

EXPERIMENTS ON THE WEATHERABILITY OF ZIRCON

It was thought possible that some results might be obtained by treating zircon grains in various ways to similate simplified natural conditions in sediments and soils, but the results obtained were practically negligible. However, the procedures are recorded here so that the scope and limitations of the experiments can be made known. There is very little in the literature concerning the solubility of zircon, but experiments done by Bouyoucos (1921) indicate that it is slightly soluble, when freshly crushed, in pure water.

Series A.—Three samples of clean zircon grains were packed at three levels in a column of synthetic cation exchange resin, Amberlite I.R. 120, with reactive capacity of 4 m.e. per gm dry weight. The column was 24 mm in diameter and 140 mm long; it contained 22.5 gm Amberlite (dry weight). The column was connected to a reservoir which delivered 1.7 liters of distilled water per day by dripping slowly through. The leaching was continued with the same water for 135 days, so that a total of 229.5 liters was leached through the column. After 80 days the column was washed through with 25° ml 2N HCl, followed by 500 ml water; this was repeated after another 7 days, and the washings were kept. After 135 days the column was washed through with 2 liters of 0.5 per cent oxalic acid which was also kept. The zircon samples were recovered from the Amberlite of the column by panning and separation in bromoform; the Amberlite is inert and the separation was easily made.

The leaching solutions were reduced in volume and tested for the presence of Zr with the alizarin colorimetric test (Sandell, 1950, p. 641); the results were negative. When a few ml of freshly prepared 6 per cent cupferron solution was added to a 50 ml aliquot of the leaching solutions precipitates were formed, but when ignited, the weights were negligible. The details are given in table 1. The brownish precipitate from the leaching with HCl may have been Fe coming from the Amberlite or from the few iron-bearing minerals left in the zircon concentrates. There were no iron coatings on the Australian zircon, but there may have been some on the Colorado zircon concen-

TABLE 2.—*Series B zircon experiment (Australian zircon concentrate)*

No.	Mixture	Weight after 164 days	Zr test with cupferron
(1)	5 gm zircon 0.5 gm fluorite[a] 25 ml distilled water	very slightly reduced	small white ppt.
(2)	5 gm zircon 25 ml distilled water	very slightly reduced	small white ppt.
(3)	5 gm zircon 25 ml 0.01 N HCl	very slightly reduced	pale yellow ppt. wt. 1 mg.
(4)	5 gm zircon 0.5 gm fluorite 25 ml 0.01 N HCl	very slightly reduced	pale yellow ppt.

[a] Finely crushed pure fluorite crystals.

trate. There was a reduction in grain size of Sample A which accounts for the gain in weight in Sample B; also a small quantity of grains passed the 0.06 mm sieve. The total loss in weight of A plus B is 0.13 gm or approximately a 3 per cent loss in weight; this represents 64 mg zirconium (some of this loss in weight may represent removal of odd grains of rutile left in the sample which contains about one grain per hundred of rutile). The loss on C is 10 per cent; representing 100 mg zirconium; this may be partly made up of odd grains of epidote and amphibole which were not removed electromagnetically. The zircon in Sample C is in fresh, euhedral, unabraded grains. Zircon grains were not lost by adhering to the Amberlite or to the glass column.

If the loss in weight in zircon could be related to the leaching solutions by their content of zirconium, which should be present in a weighable quantity, then this experiment would be regarded as highly satisfactory. It is possible that Zr has entered into an exchange relationship with the Amberlite, and that leaching with HCl and oxalic acid failed to remove it. The use of exchange resins sometimes requires a complexing agent for the removal of certain ions, Zr being one of these. There is an extensive literature on the use of synthetic resins for Zr removal (Kunin and Meyers, 1950) but time was not available to test various methods of removal. It is hoped at a later date to repeat this and other experiments with zircon. This experiment can only be regarded as suggestive at the present time.

Microscopic examination showed that the only effect of the treatment which could be recognized was a reduction of the polish on the surface of the grains in A. and B. There was no recognizable effect on the grains in C.

Series B.—Several 5 gm samples of Australian zircon concentrate were placed in 50 ml erlenmeyer flasks with various mixtures, shaken up, and allowed to stand for 164 days. The mixtures and results are as given in table 2.

Microscopically the grains showed no change except for no. (3) in which they appeared to have an increased number of cracks.

Series C.—Several 2 gram samples of Australian zircon concentrate 0.25–0.12 mm diameter were placed in 50 ml erlenmeyer flasks with various mixtures, shaken up, and allowed to stand for 147 days. The mixtures and results are given in table 3.

Microscopically there is no noticeable change in nos. (1) and (3), but in no. (2) additional cracks appear to have developed on the surfaces and ends of grains.

Without repetition and more rigorous control of experimental details these

TABLE 3.—*Series C zircon experiment* (*Australian zircon concentrate*)

No.	Mixture	Weight after 147 days	Zr loss[a]	Zr test with cupferron
(1)	2 gm zircon 0.5 gm ground lignite 0.5 gm $CaCO_3$ 25 ml distilled water	loss, 0.035 gm	13 mg	small white ppt.
(2)	2 gm zircon 0.5 gm lignite 25 ml distilled water	loss, 0.021 gm	10 mg	very small white ppt.
(3)	2 gm zircon 25 ml 0.3 N $NaHCO_3$	loss, 0.01 gm	5 mg	very small white ppt.

[a] Calculated from weighed loss.

crude experiments cannot be regarded as giving any conclusive results. However minute precipitates were formed with cupferron indicating the presence in solution of Zr, Ti, V, Ta, or Fe. Ti and Fe are the most likely contaminants here. These experiments are merely suggestive of lines which might profitably be followed in examining the processes of corrosion of zircon grains, which undoubtedly does take place particularly in certain soils. Another line of inquiry into the behavior of zirconium under weathering conditions is that of the zirconium in certain amphiboles. On the alteration of these amphiboles the zirconium may go into the circulating solutions; or it may crystallize into some alterations product which has not yet been recognized.

DISCUSSION

Before the zircon or zirconium content of a soil can be safely used as a measure of that soil's development the following observations appear to the pertinent:—

(1) In sedimentary rocks and the soils developed from them, the zircon will be exceedingly stable because all the unstable zircon grains, particularly of type (3) above, will have already been removed, presupposing that the weathering and sedimentation processes have been rigorous, and that zircon was originally present in the distributive province. An example of the probable removal of unstable types of zircon is the Australian commercial zircon produced by concentrating ancient zirconiferous sands. The grains in an average concentrate are all robust, well-rounded, and practically free from inclusions; about one grain in 200 is zoned, and it shows the effect of gradual alteration and removal. Such grains are definitely corroded, and the original sharpness of the crystal faces has been replaced by a more rounded form which is quite different from that produced by abrasive rounding.

(2) It is not yet known with certainty that all igneous and metamorphic rocks have either a constant amount of zircon for any particular rock type, nor constant varietal features. The latter has been proved in several cases (Brammall, 1928; Tyler *et al*, 1940). If there is patchy distribution of zircon either as to quantity or varietal features, then its use as suggested by Haseman and Marshall (1945) is invalidated. The use of varietal features may have greater value than the actual quantity of zircon present.

(3) The weathering and soil-forming processes have an effect on zircon which so far has not been assessed. Frederickson (1948) and Goldich (1938) have noted this, and the writer's collection of heavy residues of soils contains innumerable corroded grains of zircon. The appearance of individual zircon grains under the microscope will give useful information as to the past history of the material in which they occur; for instance, with

loess as a parent material, the zircons will be small, evenly graded, with fresh appearance, and uncorroded, but consideration of the source of the loess is important. If derived from sedimentary rocks the zircon will most likely show some signs of abrasion: a rounded shape, matt or polished surface. If from igneous rocks, the zircon is liable to be in small, euhedral, unabraded grains, and may all be of one or two varieties.

CONCLUSIONS

Observations have shown that zircon grains become corroded, particularly in soils of lateritic type which have been subjected to alkaline leaching for considerable periods of time. All zircon, even that classed as the normal type, is not equally susceptible to corrosion, and reasons are given for this. The literature suggests that zircon may be slightly soluble in Ca-bicarbonate or Na-carbonate. Observations on a number of zircon grains from different localities suggest acid as well as alkaline leaching, as in soil formation, acting through long periods of time, for the production of corrosion and the final removal of zircon. Experiments done with a synthetic cation exchange resin and with various mixtures during a limited time were not conclusive, but should be repeated. The use of zircon as an indicator of changes in a soil profile based solely on the chemical determination of zirconium should be interpreted with caution, but varietal features of the actual zircon grains in a heavy residue from a soil could be used with advantage. Perhaps a combination of chemical and microscopic determinations should be expected to give the most correct and satisfactory results in studies in soil development.

ACKNOWLEDGEMENTS

This investigation was done while holding a Research Fellowship in the Coordination of the Sciences at Bryn Mawr College, Bryn Mawr, Pa., and grateful acknowledgment is made to the Council of the College. Thanks are due to Mrs. Lincoln Dryden for the gift of a sample of Australian zircon concentrate; to Irene Waraksa for the care and skill she showed in making the photomicrographs; and to Dr. George L. Zimmerman for suggesting the use of synthetic cation exchange resin as a weathering agent.

REFERENCES

ALTY, S. W. (1933) Some properties of authigenic tourmaline from Lower Devonian sediments: Am. Mineralogist, vol. 18, pp. 351–355.

BOND, G. (1948) Outgrowths on zircon from Southern Rhodesia: Geol. Mag., vol. 85, pp. 35–40.

BOSWELL, P. G. H. (1942) The stability of minerals in sedimentary rocks: Geol. Soc. London Quart. Jour., vol. 97, pp. lvi–lxxv.

BOUYOUCOS, G. J. (1921) Rate and extent of solubility of minerals and rocks under different treatments and conditions: Michigan Exper. Sta. Tech. Bull. 50.

BRAMMALL, A. (1928) The Dartmoor detritals: a study in provenance: Geologists' Assoc. London Proc., vol. 39, pp. 27–48.

BRAMLETTE, M. N. (1941) The stability of minerals in sandstones: Jour. Sedimentary Petrology, vol. 11, pp. 32–36.

BUTTERFIELD, J. A. (1936) Outgrowths on zircon: Geol. Mag., vol. 73, pp. 511–516.

CARROLL, D. and WOOF, M. (1951) Laterite developed on basalt at Inverell, New South Wales: Soil Sci., vol. 62, pp. 87–99.

CARROLL, D. (1951) Mineralogy of laterites: Australian Jour. Sci., vol. 14, pp. 41–42.

EDELMAN, C. H. and DOEGLAS, D. J. (1934) Über Umwandlungserscheinungen an detritischen Staurolith und anderen Mineralien: Tscherm. Min. Petr. Mitt., vol. 45, pp. 225–234.

FOSTER, M. D. (1950) The origin of high sodium bicarbonate waters in the Atlantic and Gulf Coastal Plains: Geochim. et Cosmochim. Acta, vol. 1, pp. 33–48.

FREDRICKSON, A. F. (1948) Mode of occurrence of titanium and zirconium in laterites: Am. Mineralogist, vol. 33, pp. 374–377.

FREISE, F. W. (1931) Untersuchung von Mineralien auf Abnutzbarkeit bei Verfrachtung im Wasser: Tscherm. Min. Petr. Mitt., vol. 41, pp. 1–7.

GOLDICH, S. S. (1938) A study in rock weathering: Jour. Geology, vol. 46, pp. 17–58.
GRAHAM, E. R. (1941) Acid clay—an agent in chemical weathering: Jour. Geology, vol. 49, pp. 392–401.
HASEMAN, J. F. and MARSHALL, C. E. (1945) The use of heavy minerals in studies of the origin and development of soils: Univ. Missouri Coll. Agric. Res. Bull. 387.
HEVESY, G. VON, and WÜRSTLIN, J. (1934) Die Haufigkeit des Zirconiums: Zeitschr. Anorg. allg. Chemie, vol. 216, pp. 305–311.
HUTTON, C. O. (1950) Studies of heavy detrital minerals: Geol. Soc. Amer. Bull., vol. 61, pp. 635–710.
KRYNINE, P. D. (1946) The tourmaline group in sediments: Jour. Geology, vol. 44, pp. 65–87.
KUNIN, R. and MEYERS, R. J. (1950) Ion exchange resins: John Wiley & Sons, New York.
MAURICE, O. D. (1949) Transport and deposition of the non-sulphide vein minerals, V. Zirconium minerals: Econ. Geology, vol. 44, pp. 721–731.
MITCHELL, R. L. (1944) The distribution of trace elements in soils and grasses: Nutrition Soc. of England & Scotland Proc., vol. 1, pp. 183–189.
MORGAN, J. H. and AUER, M. L. (1941) Optical, spectrographic, and radioactivity studies of zircon: Am. Jour. Sci., vol. 239, pp. 305–310.
RANKAMA, K. and SAHAMA, TH. G. (1950) Geochemistry: Univ. Chicago Press.
SANDELL, E. B. (1950) Colorimetric determination of traces of metals, 2nd ed., Interscience Publishers, New York.
SMITHSON, F. (1937) Outgrowths on zircon in the Middle Jurassic of Yorkshire: Geol. Mag., vol. 74, pp. 281–283.
SMITHSON, F. (1950) The mineralogy of arenaceous deposits: Sci. Prog., vol. 149, pp. 10–21.
STROCK, L. W. (1941) Geochemical data on Saratoga mineral waters—applied in deducing a new theory of their origin: Am. Jour. Sci., vol. 239, pp. 857–898.
TRUEMAN, D. J. (1912) The value of certain criteria for the determination of the origin of foliated crystalline rocks: Jour. Geology, vol. 20, pp. 228–258.
TYLER, S. A., MARSDEN, R. W., GROUT, F. F., and THEIL, G. A. (1940) Studies of the Lake Superior Pre-Cambrian by accessory mineral methods: Geol. Soc. Amer. Bull., vol. 51, pp. 1429–1538.

13

Reprinted from *Great Britain Geol. Survey Bull.* **10:**74–80 (1955), by permission of Her Majesty's Stationery Office

ON THE IMPERSISTENCE OF URANINITE AS A DETRITAL MINERAL

C. F. Davidson and M. E. Cosgrove

Summary.—At Carbis Bay on the north coast of Cornwall an old mine dump with pitchblende (uraninite) in its debris abuts on the shore, and feeds a small pebble bed along high-water mark. A radiometric and mineralogical examination has been made to determine the extent to which pitchblende persists in the pebble bed and also in the sands which form the greater part of the beach. Rounded fragments of weathered pitchblende are readily found among the gravel ; but as the pebbles become comminuted to sand the pitchblende does not survive, disappearing by chemical and physical weathering. It is concluded that alluvial pitchblende or uraninite has only a short life, and that hypotheses on the placer origin of pitchblende-uraninite deposits in the sediments of the Witwatersrand and of Colorado-Utah cannot be maintained.

INTRODUCTION

THE extent to which uraninite or pitchblende survives as a detrital mineral in sediments derived from primary uranium-bearing rock has recently been the topic of discussion arising from divergent views on the origin of the uraninite-gold deposits of South Africa. It has been widely held by mining geologists that the gold and pitchblende which occur in sympathetic association within the conglomerate reefs of the Witwatersrand are both of placer derivation. The evidence contradicting this view has lately been summarized in papers by the senior author (Davidson and Bowie 1951 ; Davidson 1953). An important factor of this evidence is the chemical and physical instability of uraninite or pitchblende during the processes of weathering.

At the sea beach of Carbis Bay on the north coast of Cornwall geological conditions, which are probably unparalleled anywhere in the world, permit a thorough practical investigation into the stability of uraninite as a detrital mineral. At the western end of this small bay the cliffs backing the beach are capped by large waste dumps, in part more than a century old, from the adjacent tin-copper-uranium mine of Wheal Providence. The dumps contain a significant amount of pitchblende and form the parent material of a large lens of gravel close to high-water mark on the sandy beach immediately below. Traced eastward along the direction of the coastal drift and followed seawards towards low-water mark the gravel passes into and is overlain by sand. The present paper reports a study of the behaviour of the pitchblende as it passes from the mine dumps to the gravels, and as the gravels become reduced to beach sand.

The geography of the beach is shown in the accompanying sketch-map (Plate IV).

WHEAL PROVIDENCE

Wheal Providence, the source of the detritus described in this note, was one of the largest mines of the St. Ives district in Cornwall, its workings extending laterally from under the sea in Carbis Bay to a point about half a mile inland, and vertically to a depth of 150 fathoms. The mine was sunk on a group of narrow lodes through metamorphosed killas (slate) into underlying granite ; and as in adjacent Cornish properties of comparable geology copper ore was mined where killas was the wall rock, and tin ore in depth when the lodes entered the granite. Pitchblende mineralization occurs in pockety form in nearby mines (e.g. Wheal Edward and Wheal Trenwith) within the copper zone, most often in shoots where the killas wall-rock is invaded by intrusive greenstone or diabase ; and since a frequent matrix for pitchblende veins found in the spoil dumps at Wheal Providence is a basic igneous rock, similar conditions would seem to have prevailed here also.

The minerals recorded from Wheal Providence are dominantly cassiterite, chalcocite, cuprite, native copper and oxidized copper ores. Pitchblende and torbernite were by no means rare, the former being recorded on the 20- and 40-fathom levels and also in workings extending under the sea on the 30-fathom level. Full particulars of the output of the property are not available, but records show that 10 300 tons of 8 per cent copper ore was produced from 1836 to 1855, and 2340 tons of black tin from 1852 to 1861. About three tons of pitchblende was raised up to 1846, and at a later date odd parcels of a few hundredweights were produced. The dumps were prospected for uranium minerals in 1907, but apparently no further output resulted from these operations.

Carbis Bay is one of the more select holiday resorts of the Cornish coast, and the environs of the mine on the coastal side are now occupied by residential properties which extend over part of the old spoil dumps on the steep slopes overlying the south-west corner of Carbis Bay beach. An investigation of the dumps has been made by our colleague Mr. C. B. Campbell, from whose unpublished report (Campbell 1953) we have derived much of our information. Because the ground is in part built over, and because of the uncertainty regarding the original outline of the ground surface, the amount of the spoil overlying the beach is difficult to calculate ; but it is roughly estimated to contain between 3000 and 8000 tons of rock at an average grade of 0·04 per cent U_3O_8. The constituent rocks are principally tourmalinized granite, schorl rock, metamorphosed killas and greenstone, with occasional quartz veinstuff often rich in chlorite and bearing abundant pyrite. Blocks of killas and greenstone carrying veinlets of pitchblende and sulphide ores of arsenic and copper may be found on searching the dumps with a Geiger counter. The radioactivity over this spoil, measured by a Halross scintillometer, averages about 3000 impulses per second, with higher values in the vicinity of discrete lumps of high-grade pitchblende. This average radioactivity is three or four times greater than that encountered over the pebble beds on the beach below.

THE BEACH DEPOSITS

Immediately below the St. Ives branch railway which has been driven through the spoil dumps at the south-west corner of Carbis Bay, there is a pocket of beach gravel, elongated along high-water mark. The exposed extent of this lens in July 1953 was roughly 500 feet in length and 100 feet in width, but the true size of the gravel bank is greater, being obscured towards its eastern end by a surface layer of sand. The pebbles are derived by the attrition of rock debris which has fallen from the mine dumps above, and they are consequently similar in petrography to the rocks of the dumps, consisting of tourmalinized granite and schorl, greenstone, some baked killas, and vein quartz often of a 'peachy' or chlorite-rich character. The pebbles average about two inches in diameter, ranging upwards to six inches or more. At the west end of the beach large angular boulders torn from the killas cliffs border the shore; but this material does not contribute notably to the pebble bed.

On prospecting the pebble bed with a Geiger counter or scintillometer occasional fragments of pitchblende ore may readily be discovered. These are sometimes as much as six inches in diameter, the larger pieces consisting of killas or greenstone shot through by veins of pitchblende and pyrite, the whole weathering with a rusty limonitic crust. Two varieties of pitchblende have been observed in these veins—one a soft black sooty material and the other a hard and massive mineral with a clean sharp fracture. The smaller pebbles of pitchblende, often with adhering wall-rock, are likewise distinguished by a ferruginous coating, and may show yellow specks of secondary uranium mineralization in cracks and fissures. When broken across, the surface of the uraninite is more often than not coated with a thin yellow paint, probably of zippeite. The thickest pitchblende vein found is about three inches across; in polished section this is seen to consist dominantly of fresh botryoidal pitchblende, with a minor development of sulphides and a little secondary uranium mineralization.

Whilst the pebble bed is less radioactive than the spoil dumps overhead, throughout its whole extent it is of much higher radioactivity than the adjacent beach sands; and indeed wherever a slight increase in radiation is found in the sands, digging has revealed a bank of pebbles at a depth of one or two feet. This relatively high radioactivity of the gravels, which in large part are derived from the wall-rock of the tin-copper-uranium lode, must be assumed to be due to the presence of finely dispersed uraninite or possibly adsorbed uranium. A comparable increase in the radioactivity of the country rock near to uranium lodes has been found in many pitchblende mines overseas. The distribution of radioactivity along the beach can be seen from the scintillometer readings given on Plate IV.

Anomalously high values in the gravels found in the immediate vicinity of pieces of pitchblende have been omitted. The approximate radioactivity of the detritus, and of some rock formations in the neighbourhood, is indicated in the following table :—

Rock-type	*Scintillometer reading (counts per second)*	*Approx. radioactivity (equivalent U_3O_8)*
Mine dumps, Wheal Providence	3000	0·04 %
Pebble beds, Carbis Bay	700–1000	0·01
Heavy mineral sand, Carbis Bay (West)	240–440	0·001 [1]
Beach sand, Carbis Bay (East)	120–160	0·0005 [1]
Beach sand, Hayle	160–220	<0·001
Killas, Carbis Bay	250–400	0·001
Quartz pebble beach, Porthleven	100	<0·001
Tremolite-serpentine, Mullion Cove	70	<0·001
Hornblende-schist, Mullion Cove	50	<0·001
Dunite-serpentine, Mullion Cove	40	<0·001

[1] Laboratory assays

It should be noted that the scintillometer readings are not strictly proportionate to the uranium content, since cosmic radiation provides a larger proportion of the impulses when weakly radioactive sources are being measured than is the case when the radioactivity is more intense. The radioactivity of the killas is due in significant part to its potash content.

Mineralogy of the Beach Sands

A detailed study has been made of the mineralogy of the beach sands, to determine whether pitchblende survives as a detrital mineral. Bulk samples were taken: (i) from sand intimately intermixed with the pitchblende-bearing pebble beds; (ii) from the centre of the beach, near high-water mark; and (iii) from the eastern end of the beach, near low-water mark. Of these samples the first two are rather dark in colour and relatively rich in heavy minerals; the third was found to consist dominantly of dark shell fragments and to possess almost no heavy minerals.

A gravity concentrate was prepared from each of the first two sands using a Wilfley table, and this was further beneficiated by superpanner. Radiometric assays of these products were as follows:—

	Radioactivity as equivalent U_3O_8	
	Carbis Bay West	*Carbis Bay Centre*
Original sample	0·001%	0·0007
Wilfley concentrate	0·004	0·001
Superpanner concentrate	0·01	0·005

After the removal of quite large quantities of magnetite with a hand magnet, the superpanner concentrates were passed through a Frantz isodynamic separator to obtain a magnetic spectrum of the material, and the various fractions collected were then assayed by a beta counter and examined microscopically. Autoradiographs were taken of the more radioactive separates, using nuclear research emulsion plates; and an optical examination was made of the mineral grains found in this way to be of a radioactive character. In each concentrate the radioactivity was found to be due to a little weakly radioactive xenotime with some moderately active monazite. One angular grain of a strongly radioactive mineral believed to be uraninite was isolated in treatment of the Western sample.

In an attempt to isolate further grains of the (?) detrital uraninite, a large bulk of Wilfley table concentrates was treated by the superpanner, only the finest tip of high gravity minerals in which uraninite would be likely to report being collected. This proved to be almost devoid of radioactivity, indicating that uraninite was absent. A further sample of Wilfley concentrate was then sized into three fractions by sieving, and each fraction assayed with the following results:—

Diameter (mm)	*Radioactivity as equivalent U_3O_8*
Less than 0·165	0·008%
0·165–0·40	0·004
Greater than 0·40	0·002

After the removal of magnetite, the most radioactive separate was again fractioned by the Frantz electromagnet, and the most radioactive fractions were autoradiographed. In an intensive search no further grains of uraninite could be observed, the radioactive minerals present being monazite and xenotime only. The possibility that the one uraninite grain recorded was a contaminant from the equipment rather than a constituent of the beach sand cannot be disregarded.

A detailed mineralogical analysis of the two sand samples, as fractionated by the Frantz separator, has been carried out by our colleague Mr. R. K. Harrison, with the following results :—

HEAVY MINERAL CONCENTRATE—CARBIS BAY (WEST)

Key to estimated abundances

tr = trace
o = less than 5%
x = 5–25%
xx = 25–50%
xxx = 50–75%
xxxx = 75–100%

	0·1	0·2	0·25	0·3	0·4	0·5	0·6	0·7	1·35	Non-mags.
Ampères	0·1	0·2	0·25	0·3	0·4	0·5	0·6	0·7	1·35	Non-mags.
Weight %	8·0	5·1	15·4	9·8	13·9	5·8	3·7	3·3	12·4	22·6
1. Magnetite . .	xxxx	x	o							
2. Ilmenite . .		xx	xxx	xx	x					
3. Hematite . .		xx	x	x	x					
4. Goethite . .			x							
5. Garnet . .			x	xxx	x	x	x	x		
6. Siderite . .		xx	x							
7. Epidote . .					x	x				
8. Hornblende . .					x					
9. Monazite . .		tr	o	o	o	o	o	o		
10. Tourmaline . .					xx	xxx	xx	xx	x	
11. Chalcopyrite and other sulphides .							o	o	x	o
12. Xenotime . .							tr			
13. Wolframite . .							o	o	o	
14. Olivine . .							o	o	o	
15. Cassiterite . .							x	xx	xx	xx
16. Zircon . . .									o	x
17. Topaz . . .									o	xx
18. Corundum . .									o	o
19. Apatite . .										o
20. Fluorite . .										o
21. Calcite . .										x
22. Andalusite . .										o

HEAVY MINERAL CONCENTRATE—CARBIS BAY (CENTRE)

Key to estimated abundances

tr = trace
o = less than 5%
x = 5–25%
xx = 25–50%
xxx = 50–75%
xxxx = 75–100%

	0·1	0·2	0·25	0·3	0·4	0·5	0·6	0·7	1·4	Non-mags.
Ampères	0·1	0·2	0·25	0·3	0·4	0·5	0·6	0·7	1·4	Non-mags.
Weight %	5·2	3·1	8·9	8·4	22·0	24·4	4·2	2·6	7·0	14·2
1. Magnetite . .	xxxx	x								
2. Ilmenite . .		x	xx	xx	xx					
3. Hematite . .		xxx	xx	x	x	x				

Ampères	0·1	0·2	0·25	0·3	0·4	0·5	0·6	0·7	1·4	Non-mags.
Weight %	5·2	3·1	8·9	8·4	22·0	24·4	4·2	2·6	7·0	14·2
4. Goethite . .			x							
5. Garnet . .				x	xx	xx	xx	xx		
6. Siderite . .		o	xx	x						
7. Epidote . .			o	x	x	x	o			
8. Hornblende . .				xx	x	x				
9. Monazite . .				o	o	o	o	o	o	
10. Tourmaline . .				x	x	xx	xxx	xx		
11. Chalcopyrite and other sulphides .							o	o	x	o
12. Xenotime . .							tr			
13. Wolframite . .									o	
14. Olivine . .									o	
15. Cassiterite . .								x	x	xx
16. Zircon . . .								o	x	
17. Topaz . . .									o	xx
18. Corundum . .									o	o
19. Apatite . .									x	
20. Fluorite . .										o
21. Calcite . .									o	xx
22. Andalusite . .										o

Mr. Harrison has made the following note on some of the constituent minerals of the concentrates.

RADIOACTIVE MINERALS

Monazite occurs as pale yellow subangular and euhedral grains commonly having a tabular habit. While the average size is 0·15 mm, a few grains range up to 0·24 mm in diameter. The optics are normal; $N_p = 1·790$, $N_g = 1·838$, 2V small.

Xenotime is present as irregular, subangular yellowish grains elongated along the *C* axis, $N_o = 1·72$, $N_e = 1·815$, pleochroism weak with E yellow, O pinkish yellow. The grains contain numerous dark or opaque inclusions.

?*Uraninite.* One small grain of a strongly radioactive mineral occurred in the 0·6 ampère fraction of the West sand. This was mostly black and opaque with some brownish (possibly alteration) material. On the basis of its high radioactivity as displayed on an autoradiograph, the mineral was tentatively described as uraninite.

Zircon is present in only small amounts as colourless tetragonal prisms, generally broken.

NON-RADIOACTIVE MINERALS

Calcite is present as white or red tabular grains which probably represent broken shell fragments.

Cassiterite. Mainly brown angular grains with vitreous lustre.

Corundum occurs as angular grains, mainly as the blue variety—sapphire.

Hematite is fairly common as black angular grains generally with a brilliant lustre—specularite.

Siderite is also common as colourless or stained tabular crystals and rhombohedral cleavage fragments.

Tourmaline occurs in two contrasting forms, as blue, brown or grey crystals and as very rounded elliptical grains.

Wolframite. Black bladed crystals with perfect cleavage and metallic lustre; in thin fragments strongly pleochroic, with Z green-black, X yellow-brown; absorption Z>X; R.I's very high, above 2·09.

The approximate abundance of cassiterite and of sulphides present in the sands of the beach has been estimated from the mineralogical studies to be of the following order :—

	Carbis Bay West		*Carbis Bay Centre*	
	% of sample	*lb. per cu. yd*	*% of sample*	*lb per cu. yd*
Cassiterite	0·0008	0·023	0·0005	0·014
Chalcopyrite and other sulphides	0·00006	0·002	0·00007	0·002

Although these contents are extremely low, the elaborate processes of beneficiation adopted in this study were such that they provided concentrates containing between 25 and 50 per cent cassiterite and upwards of five per cent sulphides. The same processes, supplemented by most delicate tests for radioactivity, failed to produce any concentration whatever of pitchblende or uraninite which, with the doubtful exception of a single sand grain, could not be recognized in any of the products of beneficiation. It would therefore appear to be proven that uraninite has a short life as a detrital mineral in beach deposits.

Conclusions

To many readers this brief study will seem to be labouring the obvious. But the belief in a sedimentary origin for the uraninite present in the gold-uranium reefs of the Witwatersrand, and in the Shinarump and other conglomerates and sandstones of Colorado-Utah, has been maintained by the placer school of geologists in the face of overwhelming geochemical evidence to the contrary, and despite the absence of any analogues to such deposits among present-day sediments. This investigation has shown that when pitchblende-bearing rock debris is fed into a beach at one end, the pitchblende weathers away as the angular detritus becomes rounded into pebbles, and disappears completely as the pebbles become comminuted to sand. If the pitchblende cannot survive transportation of a few hundred yards along a shore-line, it is surely quite outside the bounds of possibility that the same mineral should have survived the prolonged chemical and physical weathering that it would have met with during a long passage from bedrock to the uranium-bearing quartzites and conglomerates of South Africa and of the Western United States.

References

Campbell, C. B. 1953. A reconnaissance of three abandoned mines in Cornwall. *Geol. Surv. Gt. Brit. Rep.* GSM/AED 137 (unpublished).

Davidson, C. F. 1953. The gold uranium ores of the Witwatersrand. *Mining Mag.*, p. 73.

Davidson, C. F., and S. H. U. Bowie. 1951. On thucholite and related hydrocarbon-uraninite complexes, with a note on the origin of the Witwatersrand gold ores. *Bull. geol. Surv. Gt. Brit.*, No. 3, p. 1.

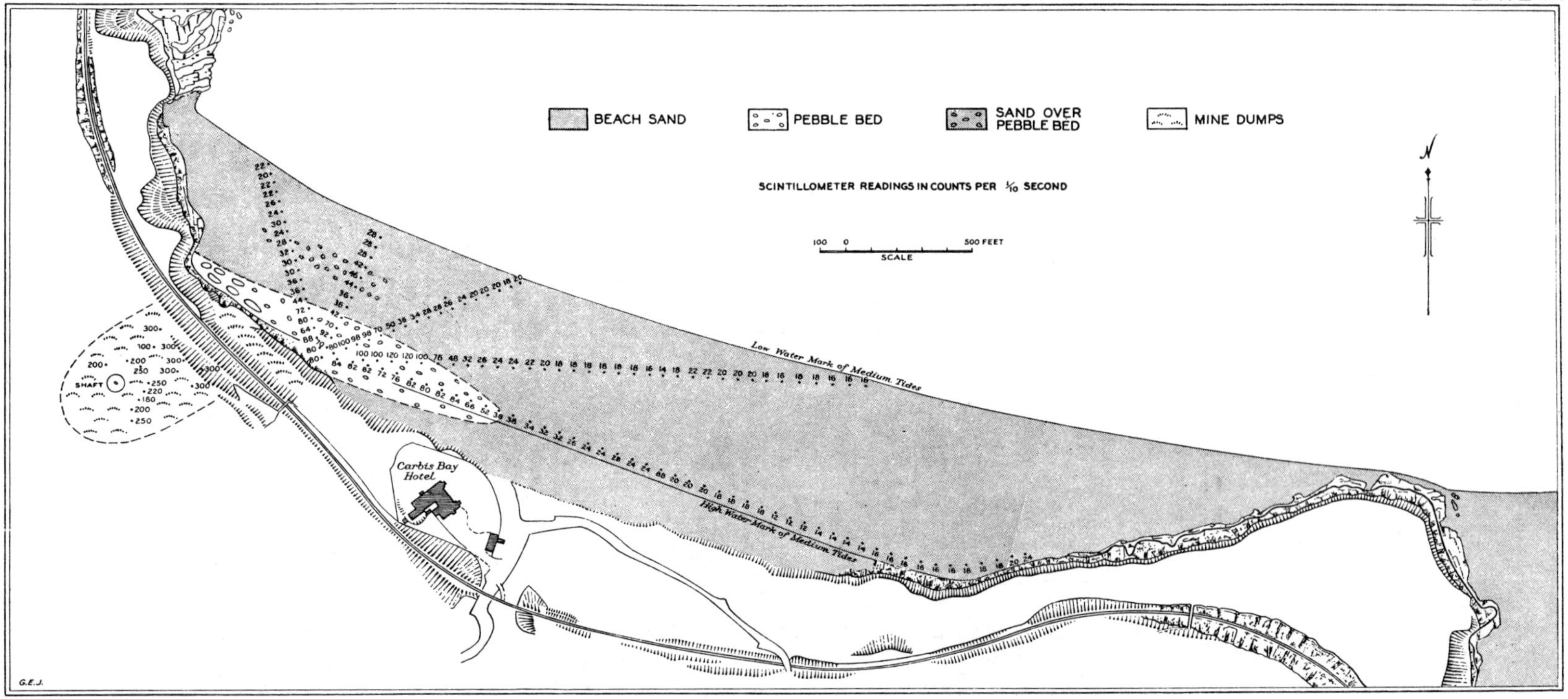

SKETCH MAP OF CARBIS BAY, SHOWING THE DISTRIBUTION OF RADIOACTIVE DETRITUS

14

Reprinted from *Jour. Sed. Petrology* **16:**91–96 (1946), by permission of the publisher

COMPARATIVE RATES OF WEATHERING OF SOME COMMON HEAVY MINERALS

LINCOLN DRYDEN AND CLARISSA DRYDEN
Bryn Mawr College

ABSTRACT

When heavy minerals are liberated by weathering of their parent rock, resistant species show relative concentration and less resistant species decrease in relative abundance or disappear completely. The change in relative abundance of the various minerals from fresh rock to the weathered products is a measure of their comparative resistance to weathering.

Samples of fresh and weathered Wissahickon schist from Pennsylvania and Maryland were analyzed in this way. Zircon proved to be the most resistant mineral considered, garnet the most readily destroyed in weathering. Other common heavy minerals show considerable range in resistance.

The results of this study, particularly the data on weathering of garnet, were used in interpreting the origin of nearby Lower Cretaceous sediments. Applications of the method and results to other fields of geology and to soil science are suggested.

INTRODUCTION

The derivation of detrital heavy minerals is one of the few lines of evidence that bear on the source of the materials in sediments and sedimentary rocks. If certain minerals can have been derived from only one or a few source rocks, their presence in a sediment is a clue to the kinds of rocks being eroded, and to the direction of transport at the time the sediment was formed. Such simple and straightforward interpretation has little value, however, unless based on a considerable body of supporting data. Finding the same heavy minerals in a sediment and a nearby crystalline rock does not prove that the latter was the source rock unless all other reasonably likely sources have been ruled out.

The absence of available heavy minerals may be just as significant as their presence, though often more difficult to explain. A typical case of this kind (to be treated later in more detail) presents the following problem: a crystalline rock is the only known source for a number of distinctive heavy minerals; nearby, a younger sedimentary rock contains most of these minerals, but one or more of them are missing in the sediment. In some instances, the absence of available species cannot be explained; in others, it can be ascribed to destruction during transport, failure to be deposited in a sediment of a particular grain size, or to post-depositional disappearance.

A fourth possibility has been recognized: that the species in question were destroyed in the weathering of the crystalline rock, and were never made available for transport and deposition. However, there are very few published data on this aspect of the problem of non-appearance of available heavy minerals. The writers hope that the present paper will direct attention to this neglected phase of heavy mineral studies.

METHOD AND RESULTS

Most of the crystalline rocks of Pennsylvania and Maryland are covered by their own weathered products, and these weathered products furnish a very large proportion of the material now being carried by streams and deposited as recent sediments. The heavy minerals to be expected in these sediments—or in older sediments of the same area—are those species which have withstood the weathering of the rock which enclosed them. The relative abundance of the different heavy minerals in the fresh crystalline rock, and in its weathered products, is a direct measure of comparative weathering rates of these minerals.

Samples must be carefully chosen if the method is to have any value. The two conditions that must be fulfilled are (a) that the weathered products are *in situ*, and (b), in the case of non-homogeneous crystalline rocks, that the weathered products have been derived from the same rock type as that of the fresh rock sample. For example, it would be difficult to collect significant samples at a locality where a banded crystalline rock outcrops on a steep hillside, unless no more than a "weathered crust" and a fresh sample are taken from one particular band. At the other extreme, especially trustworthy samples can be collected at a hilltop exposure of a homogeneous crystalline rock.

Samples of fresh Wissahickon schist and of its weathered products were collected at 14 localities in southeastern Pennsylvania and northern Maryland.* Consolidated rocks were crushed with mortar and pestle until they yielded a sufficiently large sample passing the ½ mm. sieve and retained on the ⅛ mm. sieve; unconsolidated material was quartered and sieved directly. In both cases, the material on the ⅛ mm. sieve was separated in bromoform (sp. gr. about 2.9), the light fraction was discarded, and the heavy fraction mounted in Canada balsam.

Grain counts of selected heavy minerals present in each sample are given in figure 1. For any one locality the same total grain count is shown for the fresh rock sample and for each of the samples of weathered material. This has been done to emphasize the *ratios* between the numbers of different mineral grains present and to make it possible to see, at a glance, how these ratios change in weathering. For example, at the first locality, in a total of some 140 grains, the ratio of garnet to zircon is about 135:5 in the fresh rock and about 15:125 in the weathered material; at locality 6 the ratios for garnet and staurolite are 75:25 and 50:50, in fresh rock and weathered material, respectively.

The histograms from all 14 localities show that garnet weathers *in situ* more readily and completely than any of the other species counted. In the fresh rock it is commonly clear, with sharp crystal faces or fractures; in the weathered products it shows much pitting and etching, becomes rounded and significantly smaller, and may disappear entirely. (The hypersthene of locality 5 is the only exception; if one sub-soil sample can be trusted, hypersthene disappears "rapidly" after having been initially concentrated with respect to garnet.)

In addition to this conclusion on the weathering of garnet, the histograms show comparative weathering rates of the various mineral species considered; several specific instances are worth mention. (1) Relative to garnet, zircon is concentrated more in weathering than any other species. This result bears out the general impression, imparted by the ubiquity of this mineral in sediments, that it is extremely resistant not only to abrasion but to weathering. (2) Somewhat surprising is the result that (green) hornblende is more resistant than garnet. Edelman and Doeglas (1932) have described features which indicate to them the "rapid" weathering of this mineral; the data presented here do not confirm their results. However, hornblende is rare in nearby, younger sediments, and it may break down rapidly in transport. (3) Staurolite is of intermediate resistance. Some grains in the

* The writers are unable to give an accurate definition of the weathered products, but the following descriptions may help the reader to visualize them: "Weathered rock" still retains its structure intact, but is often weak enough to crumble readily under a hammer blow. "Sub-soil" is rotten rock, that can be readily dug with pick or shovel; it is probably the "C-horizon" and in some cases perhaps the lower "B-horizon" of pedologists. "Soil" is material taken from around the roots of grasses and small plants; it probably corresponds in large part to the "A-horizon" and upper "B-horizon" of pedologists.

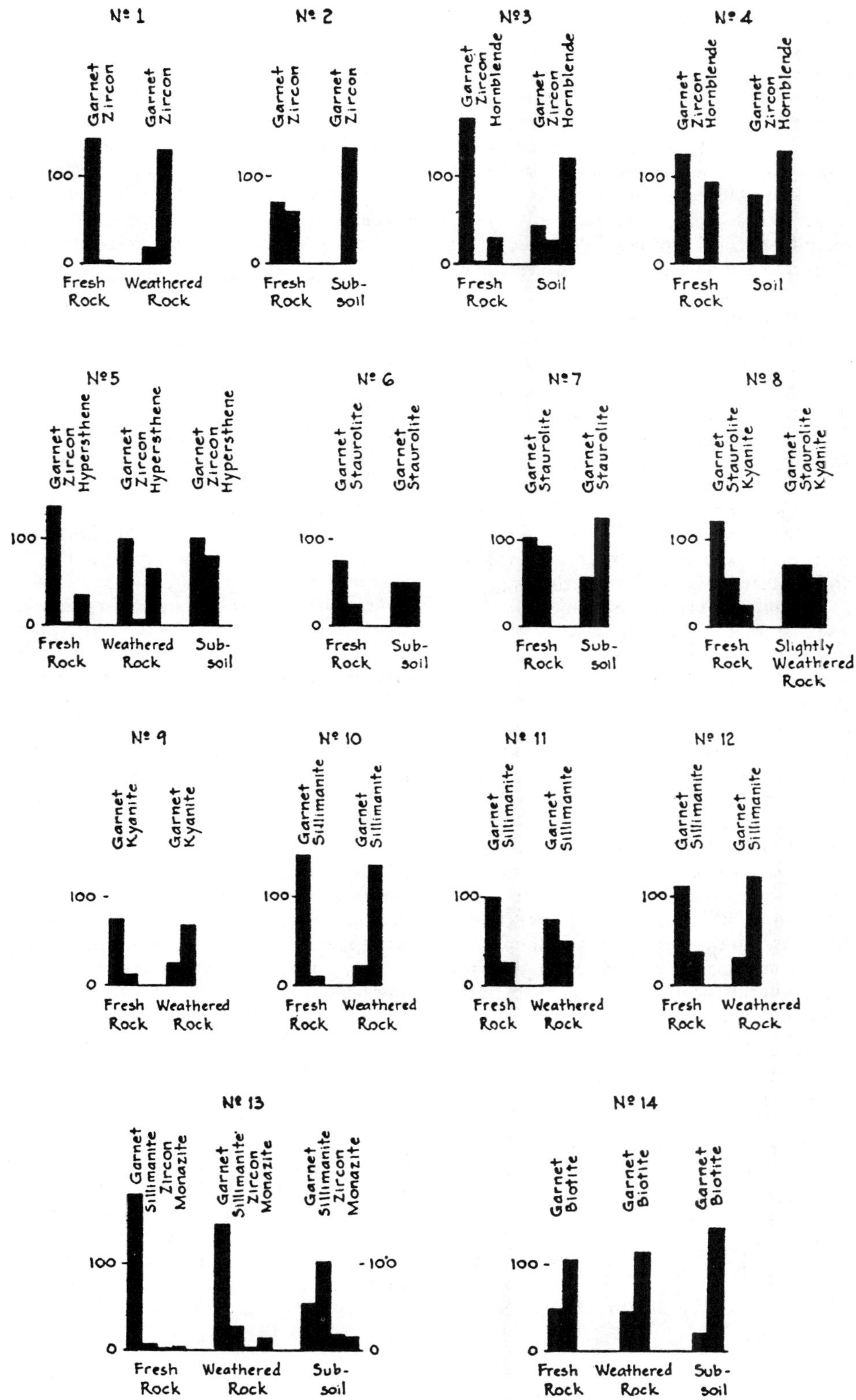

FIG. 1.—Changes in heavy mineral frequencies from fresh rock to weathered products. Numbers above the histograms are localities; the number at the side indicates the height of a histogram block representing 100 grains.

weathered products appear "moth-eaten," but there is an abundance of this species in nearby sediments. Perhaps the results presented underrate its resistance. (4) Kyanite is more resistant than garnet, and sillimanite is much more so. The latter is one of the most resistant species examined, but its probable destruction in transportation may explain its place among the less common heavy minerals in sediments. (5) Biotite is ordinarily thought of as a mineral easily destroyed in weathering, but it shows a concentration in the weathered products from locality 14. There is, however, a marked change in appearance and optical properties during weathering, so that it is perhaps incorrect to call the biotite in both the fresh rock and in its weathered products the same species. (6) On the basis of one series of samples, monazite appears to be quite resistant to weathering.

Though the grain counts of figure 1 are essentially qualitative, in that they do not measure changes in volume or mass, they are probably sufficient for a first approximation to comparative weathering rates of these common heavy minerals of the Wissahickon schist. If, for example, the ratio of garnet to zircon in fresh rock is 5:1 and in weathered rock is 1:5, it is assumed that zircon is 25 times as resistant as garnet. Using the data of figure 1 in this way, and making a rough average of the results from different localities and from different kinds of weathered products, the writers obtained the following values for the resistance of the various minerals compared with the resistance of garnet:

Zircon	100
Tourmaline (no figures available)	80?
Sillimanite	40
Monazite	40
Chloritoid (no figures available)	20?
Kyanite	7
Hornblende	5
Staurolite	3
Garnet (taken as:)	1
Hypersthene	<1?

These figures apply to the Maryland-Pennsylvania region only. Goldich (1938, p. 52) has reported an exactly opposite result (garnet more resistant than hornblende) in the Black Hills. Possibly he dealt with different varieties of the two minerals, or it may be that differences of climate, slope, or vegetation are great enough to explain the disparate results.

BEARING ON ORIGIN OF CERTAIN LOWER CRETACEOUS SEDIMENTS

The Lower Cretaceous clays, sands, and gravels exposed along the inner margin of the Coastal Plain of Maryland and northern Virginia contain a number of distinctive heavy minerals. Among them are large, angular, fresh grains of staurolite and pink tourmaline, and less common kyanite and sillimanite. These minerals are identical in varietal characters with the same species found in the nearby Wissahickon schist and in no other nearby rock older than Lower Cretaceous. The provenance of the heavy minerals in the sediment, then, appears to be well established. But although the schist contains abundant garnet, the Lower Cretaceous sediments contain little or none—a fact which the writers once regarded as an indication that the sediments could not have been derived from the schist (Dryden, Lincoln, and Dryden, Clarissa, 1938).

The absence of garnet from these sediments might be accounted for in a number of ways: (1) By destruction during transportation. Experimentation has shown, however, that garnet possesses more than average durability in transport, and the other species in the sediments show little sign of wear or rounding. (2) Consideration of "grading factors" and "hydraulic values" might explain its absence in sediments of some particular grain size, but samples from silts and from fine and coarse sands show the same lack of garnet. (3) Garnet might have been destroyed after deposition. However, it is found in the same area in similar sedimentary rocks, both older and younger than Lower Cretaceous, and there is no apparent reason why garnet should have disappeared from rocks of this one age alone. Therefore, though the factors enumerated above may be contributory, the writers believe that any one or any combination of them is insufficient to explain the absence of garnet in the Lower Cretaceous sediments.

The only sufficient explanation known to the writers is weathering of a large proportion of the garnet *in situ* in the schist, so that little was made available for transport and deposition during Lower Cretaceous time. The histograms of figure 1 show that garnet is not being completely destroyed in the "present" weathering cycle, and all local streams carry significant quantities of this species. It must be assumed, therefore, that the conditions during Lower Cretaceous time were different from those of the present. Perhaps the climatic and physiographic conditions were somewhat like those of the southern Piedmont of today. In this area, reconnaissance studies indicate that the fresh rocks contain usual quantities of garnet, but the weathered products are deficient in this respect. The streams carry but little garnet, as shown by the small percentages reported by Stow (1939) for the sediments of the James River and by Martens (1935) for the coastal sands derived from the southern Piedmont. In any case, no incredibly long or rigorous weathering is required to explain the lack of garnet in the Lower Cretaceous sediments of Maryland and northern Virginia.

OTHER APPLICATIONS

This method of analysis can be put to a number of uses; a few of the more obvious ones will be mentioned. (1) Certain mineral species are concentrated in the weathered products. For example, fresh Wissahickon schist contains only scattered and rare zircon, but this mineral is common in the overlying soils. Chloritoid behaves similarly, although it is not concentrated to the same degree. In general, rare species resistant to weathering may be found more easily in the weathered products than in the fresh parent rock. (2) In determining the character of the bedrock by study of its overlying soil—as in the work of Dorothy Carroll (1936–17, 1939)—mineral destruction and concentration during weathering must be taken into account. (3) A very similar method is being used by pedologists, with promise of important uses in problems of soil genesis (Cady 1940; Jeffries and White 1940; Marshal 1940; Marshall and Haseman 1942). Pedologists may be expected to furnish valuable data on the effects of soil moisture, slope, drainage, and of vegetation types on weathering of heavy minerals. (4) Finally, the method gives comparative and relative weathering rates for different heavy mineral species. Once these rates have been determined for the important mineral species and varieties, for different climatic and physiographic conditions, reliable interpretations of past conditions will be possible.

REFERENCES

CADY, JOHN G. *1940*. Soil analyses significant in forest soils investigations and methods of determination: 3. Some mineralogical characteristics of Podzol and Brown Podzolic forest soil profiles: *Soil Sci. Soc. Am. Proc.*, vol. 5, pp. 352—354.

CARROLL, DOROTHY. *1936–37*. Some aspects of soil mineralogy: *Roy. Soc. W. Australia*, Jour., vol. 23, pp. 7–12.

———. *1939*. Sand-plain soils from the Yilgarn Goldfield: *Geol. Surv. of W. Australia*, Bull. 97, pp. 161–180.

DRYDEN, LINCOLN and DRYDEN, CLARISSA. *1938*. Stratigraphic importance of heavy mineral studies in the Philadelphia region: *Penna. Acad. Sci., Proc.*, vol. 12, pp. 101, 102.

EDELMAN, C. H. and DOEGLAS, D. J. *1932*. Relikstrukturen detritischer Pyroxene und Amphibole: *Miner. und Petrog. Mitt.*, Bd. 42, pp. 482–490.

GOLDICH, SAMUEL S. *1938*. A study in rock weathering: *Jour. Geology*, vol. 46, pp. 17–58.

JEFFRIES, C. D. and WHITE, J. W. *1940*. Some mineralogical characteristics of limestone soils of different localities: *Soil Sci. Soc. Am., Proc.*, vol. 5, pp. 304–308.

MARSHALL, C. A. *1940*. A petrographic method for the study of soil formation: *Soil Sci. Soc. Am., Proc.*, vol. 5, pp. 100–103.

——— and HASEMAN, J. F. *1942*. The quantitative evaluation of soil formation and development by heavy mineral studies: *Soil Sci. Soc. Am., Proc.*, vol. 7, pp. 448–453.

MARTENS, JAMES H. C. *1935*. Beach sands between Charleston, South Carolina, and Miami, Florida: *Geol. Soc. America*, Bull. 46, pp. 1584, 1593.

STOW, MARCELLUS H. *1939*. Reflection of provenance in heavy minerals of James River, Virginia: *Jour. Sed. Petrology*, vol. 9, pp. 86–91.

15

Reprinted from *Jour. Sed. Petrology* **29:**493-502 (1959), by permission of the publisher

STABILITY OF INDEX MINERALS IN SOILS WITH PARTICULAR REFERENCE TO QUARTZ, ZIRCON, AND GARNET[1]

JAMES D. RAESIDE
Soil Bureau, Department of Scientific and Industrial Research, Dunedin, New Zealand

ABSTRACT

In soil studies a group of index minerals has been generally accepted as sufficiently stable to be used as the basis for evaluating changes in soil profiles. Three of the minerals, quartz, garnet, and zircon, are shown to be susceptible to physical breakdown through strain caused by crystal disorientation or crystalline inclusions. Some varieties of garnet are susceptible to weathering, notably those with high iron content, and both quartz and zircon may be susceptible to solution. These minerals should, therefore, be used only in circumstances where physical or chemical changes can be excluded and where the minerals are free from crystalline inclusions and strain.

INTRODUCTION

Minerals that are resistant to weathering and physical breakdown are of particular interest to the pedologist, for the extent and the manner in which soil parent materials are altered during soil formation, and the movement of soil constituents either in solution or in colloidal form from one horizon to another, can most easily be determined by assuming that certain constituents remain unaltered throughout the history of the soil. In petrographic studies the minerals zircon, tourmaline, quartz, rutile, anatase, garnet, albite and microcline have been generally accepted as resistant to a greater or less extent, and these have, therefore, been proposed and used as reference minerals in studies on soil genesis.

Soil studies based on index minerals relate both to stratification of parent materials and to calculation of gains or losses of soil constituents during soil formation. Both kinds of study rest on the assumption that certain minerals remain unchanged during soil formation. The two studies are also complementary, for in computing gains or losses of soil constituents it is necessary to separate differences due to stratification from those produced by pedological processes.

On many sites it is a comparatively simple matter to detect stratification of parent materials by taking advantage of sharply contrasting mineral assemblages. Where these contrasts are wanting, however, the mineral distribution has to be studied in detail. In these circumstances careful consideration must be given to the stability of the minerals chosen.

Where gains or losses of soil constituents since the start of soil formation are being computed, the composition of the soil at the start of soil formation must be reconstructed. Where the soil rests on unweathered parent material this presents little difficulty, but where the soil is formed on deeply weathered parent material or parent material that was partly weathered before the start of soil formation, the stability of the index minerals is of prime importance.

Three criteria are generally used in soil studies that depend on the stability of the index minerals (Hasemann, 1944; Haseman and Marshall, 1945; Barshad, 1955):

(1) particle size distribution of certain resistant minerals,
(2) ratio of two resistant minerals in one particular size fraction,
(3) particle size distribution in the whole non-clay fraction of the soil

The assumption that any particular index mineral is stable must embrace stability with respect to solution and chemical weathering, and also, because the criteria are based on particle size distribution of the minerals, with respect to physical breakdown of the mineral grains as well. It is not sufficient to assume that the minerals remain unchanged only during the formation of the main profile characteristics of the

[1] Soil Bureau Publication No. 186. Manuscript received July 15, 1959.

soil. The principal profile characters may be acquired early in the history of the soil and the profile may remain almost unchanged for long periods; but the index minerals may be subject to both chemical and physical change during this period of pedogenic inertia.[2]

To a large extent the order of stability of resistant minerals has been determined by statistical data based on their order of persistence in sediments, as, for example, the orders of stability proposed by Goldich (1938) and Pettijohn (1941). But it is by no means certain that a mineral that will remain inert and unaltered in a sediment will be equally so in a soil. Soils are subject, particularly in their upper horizons, to diurnal and seasonal temperature variations, and to prolonged leaching by percolating water, as well as to weathering of greater or less intensity according to the local environment Moreover, changes in particle size distribution without any appreciable weathering (in a chemical sense) are of much greater significance to the pedologist than to the sedimentary petrographer The soil is, in fact, a highly destructive environment, and for a mineral to be accepted as stable for all the soil studies enumerated it must be shown that under the local conditions it is resistant to:

(a) physical breakdown through splitting along cleavage planes or disruption from any other reason,
(b) authigenic outgrowths,
(c) weathering either through solution in the case of oxide minerals, or breakdown to secondary minerals in the case of silicate minerals.

Minerals that are subject to physical breakdown, although their total mass in the soil may remain unchanged, cannot be used in studies that assume constant particle size distribution. Minerals that are subject to weathering have neither a constant mass not constant particle size distribution.

Various minerals have been proposed as index minerals, the most comprehensive list being that of Barshad (1955) who lists zircon, tourmaline, quartz, rutile, anatase, garnet, albite, and microcline. Marshall (1940) had earlier proposed zircon, tourmaline, garnet, anatase, and rutile, and Adams and Matelski (1955) used zircon and tourmaline only. References in petrological literature to the stability of the individual minerals are few in number and for some of the minerals contradictory. The most valuable study is that of Carroll (1953) on the weatherability of zircon.

[2] This term is used in the sense proposed by Bryan and Teakle (1949).

QUARTZ

Quartz is widely distributed in soils of temperate regions as either a dominant or subdominant mineral. In the extensive area of soils derived from loess in the South Island of New Zealand it is co-dominant with feldspar. It is therefore an important and convenient index mineral. Pedological literature contains numerous references to quartz, the consensus of opinion being that it is resistant to weathering. A few writers, however, admit its stability with reservations.

Polynov (1937) states that "Primary quartz is such a stable substance and undergoes so little alteration that it may be considered to be quite passive in the zone of weathering which it leaves in the same form as that in which it entered." Russell (1950) accepts this view and rates quartz as "by far the commonest mineral in most soils and also the most resistant to decomposition." Barshad (1955) claims that the resistant minerals, in which he included quartz and albite, did not undergo significant change during the course of soil formation and that their particle size distribution pattern would remain unchanged.

Joffe (1949) also accepted quartz as one of the final residual products of weathering and held that it could not, under natural conditions of temperature and pressure, undergo further chemical changes, but he noted, nevertheless, that in soils it might be subject to solution and that alkali carbonates and alkali soluble complexes attacked it readily, citing a reference of Glinka's to some studies of Karpinksi on the Hebrewstone soils of Murzinki from which quartz was leached and felspars remained intact. Joffe's attitude to the stability of quartz is thus somewhat contradictory. His further claim that "Conclusive evidence of the solubility of quartz is its presence in river waters" is not tenable. Silica is transported in river waters not as quartz but as monosilicic acid in true solution or as amorphous silica in colloidal form (Krauskopf, 1956), and the consensus of opinion is that most of it is derived from the decomposition of the silicate minerals or from other forms of silica such as opal.

The importance of the time factor in the solution of quartz has been recognised by Merrill (1897) who pointed out that if quartz is soluble, even to the extent of one part in ten thousand, the determining factor in the loss of quartz in solution is the amount of water that passes through the soil. Blanck (1949) also noted that although quartz may be only slightly soluble in water containing dissolved carbon dixodie, i.e., soil water, the process may be of long duration, and given sufficient time it can effect appreciable solution of quartz. Blanck also drew attention to the fact that quartz tended to split into sharp angular fragments, implying that this offset its low solubility.

Physical Stability

The following discussion of the physical stability of quartz refers to "pure" quartz or quartz with a negligible content of impurities such as aluminium and sodium. As the content of these impurities increases the stability of quartz would be expected to decrease.

Quartz possesses no well defined cleavage[3] along which splitting can take place easily, and quartz grains do not break down into cleavage flakes or rods in the same way as the amphiboles, for example. Quartz is, however, susceptible to strain effects which show up distinctly on the microscope stage under crossed nicols as undulose extinction, strain shadows, or irregular and uneven extinction. Strain is expressed in effect as a local dislocation of the crystal lattice and this in turn leads to stresses during thermal expansion. In accordance with its trigonal crystal symmetry, quartz possesses two principal coefficients of expansion, one parallel to the optic axis and the other normal to it. In the case of quartz, both coefficients of thermal expansion (α) are positive, that parallel to the optic axis being 8[4] and that normal to the axis 13 at room temperature. Thermal expansion is therefore likely to cause disruption at the site of local strain. Examination of quartz grains in soils shows that such disruption does, in fact, take place, incipient and well defined fracturing being associated with local strain.

Disruption from thermal expansion may also take place for another reason. Coarse quartz in metamorphic rocks is predominantly vein quartz. Vein quartz is particularly plentiful in the greywackes and schists of the South Island of New Zealand (Turner, 1942). It is made up of aggregates of quartz units with variously oriented optic axes, a thin section showing, under crossed nicols, a mosaic or patchwork of individual units in various stages of extinction. The texture of granulation in vein quartz is generally fine, and may range from fine sand to silt size. Thermal expansion of such an aggregate sets up stresses that may be expected to lead eventually to disruption and separation of the individual components of the mosaic, the surfaces of weakness lying at the contact of the individual grains of the mosaic. These effects may be intensified where the vein quartz also shows strain effects.

Rupture of quartz has been attributed to pressure exerted by inclusions, both gaseous and solid (Iddings, 1906). Recent work (Ingerson, 1955) has shown that appreciable pressure is not exerted by gaseous inclusions at normal pressures, and these cannot therefore be accepted as disruptive agents. Crystalline inclusions, however, would be expected to cause strain if the coefficient of expansion of the crystalline inclusion was less than that of the containing mineral where the containing mineral was formed at temperatures appreciably higher than normal. The coefficient of quartz perpendicular to the z axis is greater than the coefficients of the most common inclusions (see table 1). In quartz of soils in the South Island of New Zealand, crystalline inclusions are mostly confined to primary quartz and are less abundant in vein quartz.

[3] Quartz possesses a difficult cleavage parallel to $10\bar{1}1$ and $01\bar{1}1$ and in traces parallel to $10\bar{1}0$ and 0001, but these may be neglected for the purpose of this discussion.

[4] See table 1 for note on coefficient of expansion units.

TABLE 1.—*Coefficients of expansion of minerals occurring as inclusions* (*Clarke, 1927*)

	Coefficient of expansion αt°[1]		
Almandite ($t=40°C$)	8.37		
Magnetite ($t=40°C$	8.46		
	∥ to z axis		⊥ to z axis
Anatase ($t=40°C$)	8.19		4.68
Hematite ($t=40°C$)	7.61		7.71
Quartz ($t=40°C$)	8.0		13.0
Rutile ($t=40°C$)	9.19		7.14
Tourmaline ($t=40°C$)	9.09		3.79
Zircon ($t=40°C$)	4.43		2.33
	∥ to x axis	∥ to y axis	∥ to z axis
Brookite ($t=17.5°C$)	14.49	19.20	22.04

[1] $\alpha t^{\circ} = \frac{10^6}{l} \frac{\delta l}{\delta t}$ at t°C

where

l = length

and

t = temperature.

Most of the quartz of soils formed on sediments derived from the greywackes and schists of the South Island appears to be vein quartz and most grains exhibit strain. Fissured and incipiently fissured grains are abundant, the fissures being situated at the site of local strain or following the contact between adjacent units in granular quartz. In some grains fissures are filled with birefringent colloid, presumably the illite-montmorillonite colloid of the clay fraction of the soil. The swelling of such colloidal infilling of cracks may accelerate the disruption of an already fissured crystal. The penetration of an iron pan into an apparently homogeneous quartz stone or boulder is effected by the deposition of iron oxides from soil solutions in microfissures (fig. 1). Thin sections show that in some cases the iron hydroxide is deposited in a network of

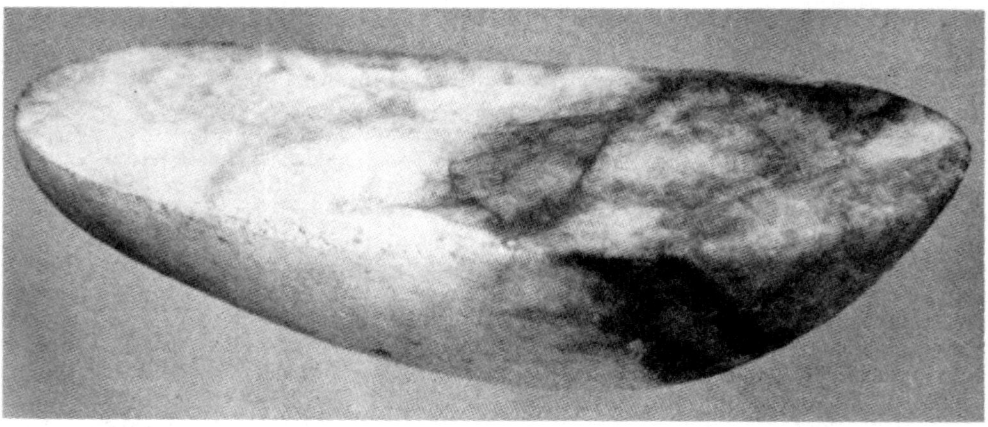

FIG. 1.—Pebble of vein quartz embedded in thin iron pan in podsol soil. The two parallel bands on the lower part (outside) of the pebble show the orientation of the pebble with respect to the iron pan. The polished upper surface of the pebble, a longitudinal section, shows the irregular manner in which the iron bearing soil solutions penetrated the pebble, apparently following the most accessible fissures. Enlarged 3.2×.

microfissures, mostly less than 3 microns wide.

Freezing of water in microfissures in the upper horizons of certain temperate and sub-arctic soils may also accelerate breakdown, but before this process can be effective the crystal must possess microfissures enlarged sufficiently to permit freezing of the contained water.

In thin sections of sandy soils formed on alluvium of mid-Pleistocene age, quartz grains occur in all stages of disintegration. The quartz in sand-sized grains is subrounded and almost exclusively vein quartz. Most of it is finely granular with a texture of granulation ranging from fine-silt sized to medium-silt sized. The coarsely granular quartz mostly shows strain effects as well. Most of the grains are fissured and some of them are completely disintegrated, the shattered fragments of the grain being in most cases separated by colloid matrix.

If the grains had been shattered before transportation they would not have survived the abrasion that rounded them, and if the shattering had taken place during preparation of the thin sections the colloid matrix would not have penetrated the fissures. It can safely be assumed, therefore, that the shattering took place in the soil.

Fragments of quartz rocks in the soils formed on the Moutere Gravels of Nelson Province are sufficiently disintegrated to be readily crushed to a fine sand between the fingers. This may be seen on the Nelson Moutere Road about 1 mile west of Redwoods Valley. In Central Otago examples of the breakdown of quartz may be seen in weathered schist at Chatto Creek, Butcher's Dam, and Galloway.

The disruption of quartz in this way may furnish an explanation of how soils of silt loam and fine sandy loam texture are derived directly from schists and greywackes in which polycrystalline vein quartz aggregates may exceed 1 cm in diameter in the shattered rock.

In rocks that have crystallised at temperatures exceeding the α/β inversion of quartz at normal pressures the transition from the low temperature form to the high temperature form may be a further disruptive force. This change takes place with a volume increase of 3.25 cc per kgm, or 0.86 percent by volume (Sosman, 1928). Fractures or stress acquired during inversion may be expected in the quartz of acid igneous rocks and in certain metamorphic rocks that have exceeded the inversion temperature during metamorphism. Holmquist (1911) claimed that prior strain determined whether quartz shattered on inversion and stated that homogenous quartz apparently survived intact. Wright and Larsen (1908) had already observed that metamorphic quartz was in fact shattered on inversion. It is possible, however, that shattering under these circumstances may be as much due to thermal expansion over the extended temperature range as to stress caused by the inversion. The schists and greywackes of the South Island of New Zealand have not exceeded the inversion temperature (D. S. Coombs, personal communication, 1958), but inversion shattering may be looked for in some of the acid igneous rocks of the North Island. Soil textures, in fact, reveal that the coarse quartz in the older rhyolite ash showers of the Central North Islands, e.g., the older showers from Lake Taupo, breaks down readily to silt sized particles in soils formed thereon, but that the quartz in the latest showers, e.g., the Kaharoa shower, is still intact (N. Z. Soil Bureau, 1954).

Solution

There is considerable literature referring to the solubility of silica but only a limited number of references to the solubility of quartz, many of them conflicting.

Although quartz appears to be the least soluble of the naturally occurring forms of silica as well as the most dense[5] (Moore and Maynard, 1929; Hitchen, 1945; and Krauskopf, 1956), its solubility may nevertheless be sufficient to make it worth taking possible loss from solution into consideration in soil studies. That quartz is appreciably soluble has been accepted for some time. The amount of silica transported annually to the sea which has been cited by Tarr (1917) and Clark (1924) as evidence of the solubility of quartz is, however, not acceptable. Much of this silica is derived from weathering of silicate minerals, and possibly only a small fraction comes from the solution

[5] The new artificial form coesite is more dense.

of quartz. Wahlstrom (1948), however, in a study of the weathering of granodiorite on Flagstaff mountain, recorded that quartz grains in the upper part of the weathered mantle were rounded and embayed and showed the effect of peripheral dissolution.

The solubility of quartz has been studied under a variety of conditions. Matignon and Marchal (1920) showed that appreciable quantities of silica can be dissolved from quartz by cold carbonated water, and Lovering (1923) found that water containing calcium and magnesium carbonates was an effective solvent of quartz, an observation later confirmed by Moore and Maynard (1929). Partial replacement of quartz by carbonates in sediments has been reported by Walker (1957).

Recent views on the solubility of silica have been summarised by Siever (1957). Although most of the recent solubility studies refer to amorphous silica, as for example, those of Alexander, Heston, and Iler (1954), Holt and King (1955), and Krauskopf (1956), values for the solubility of quartz have been reported from 1.6 mgm/100 ml at 20°C (Briscoe, Holt, Mathews, and Sanderson, 1936) to 30 ppm (Lenher, 1921). Siever (1957), using published free energy data, has derived a theoretical range of 7.25 to 14 ppm, which agrees reasonably well with the observed range of values.

It is likely that quartz crystals that have experienced local dislocation may have a greatly increased solubility at the site of the dislocation where the crystal will possess increased free energy and consequently proportionately greater solubility. The relationship between free energy and solubility is expressed by:

$$G = -RT \log k$$

where

G = free energy in kilocalories
T = temperature
R = constant for the particular substance

and

k = partial pressure of silica in solution, or the molecular concentration.

Since this is a logarithmic relationship a small increase in free energy will produce a large increase in solubility. Locally, a hundredfold increase in solubility may not be unlikely. The extension and enlargement of incipient cracks in quartz grains may thus be accelerated appreciably by solution. If the solution in contact with the grain contains any of the molecules known to increase the solubility of silica, e.g., magnesium or calcium carbonates (Lovering, 1923), solution could possibly be a dominant factor in the extension and opening up of incipient fissures.

The solubility of quartz rises with temperature (Gruner, 1930; Krauskopf, 1956) and solution is accordingly more important in the tropics than in temperate regions. Harrison (1933) has shown that in the weathering of granite in the humid tropics crystalline silica is dissolved in the upper horizons of the soil and redeposited lower down in the zone of disintegrating rock. The writer's observations of residual silica in the red and brown loams of North Auckland show that it consists of predominantly rounded unstrained grains.

Solution of quartz in soils changes not only the total amounts of quartz in the sand and silt fraction but also changes particle size distribution. Rate of solution is a linear function of specific surface and the finest particles will be removed completely while little change will have taken place in the largest particles. Jackson and others (1948) have pointed out that quartz solubility is a linear function of specific surface and therefore quartz may be expected and is found to decrease tenfold in quantity for each tenfold decrease in particle size of fraction considered. It was presumably partly for this reason that Jackson reversed Goldich's order of weathering for quartz and muscovite.

Both physical breakdown and solution are therefore likely to produce changes in the particle size distribution of quartz and solution is likely also to reduce its total quantity in a soil. Both processes are quite certainly slow but it must be remembered that except where soils have been constantly renewed because of erosion they approximate in age to that of the landscape on which they are formed, and recent studies in soil genesis have drawn attention to the fact that the age of many widespread soils may be reckoned in many thousands of years. Ruhe and Scholtes (1956) have discussed the age of soils in relation to landscape in Iowa, and demonstrated that many date from Pleistocene times, and Thorp, Johnson and Reed (1951) have shown that some soils of the southern United States date from Pliocene times.

Because of its susceptibility to physical breakdown and solution, quartz can only be accepted as a stable mineral with certain reservations. It may be admissible as an index mineral in young soils, or in semi-arid soils with pH values below 7.0. In older soils or soils receiving higher rainfall careful consideration should be given to the nature of the quartz grains, whether they consist primarily of vein quartz or primary quartz, whether they show strain, or incipient or well defined fractures, and whether they show signs of solution. Quartz that showed any evidence of physical breakdown or solution would not be admissible as an index mineral.

GARNET

Garnets have been rated by some writers as resistant minerals and by others as unstable (Dryden and Dryden, 1946). Dana (1895) noted that garnets containing ferrous iron were particularly prone to disintegration. Milner (1923), Raeburn and Milner (1927), and Goldich (1938), however, ranked garnet as a stable mineral, comparable in stability with zircon and tourmaline. On the other hand, Bayley (1925) interpreted brown spots in the kaolins of North Carolina as chlorite and other micaceous minerals, stained brown with iron, and derived from the weathering of garnets. Allen (1948) believed that these brown spots were due to the weathering of garnets with high iron content and concluded that garnets of this kind were more easily decomposed than those low in iron. Humbert and Marshall (1943) provided supporting evidence for this view from their study of the weathering of acid and basic igneous rocks in Missouri. They showed that lime-iron garnets of the variety melanite and iron and manganese garnets from basic rocks appeared to be more weathered than garnets from granitic rocks.

Bramlette (1929) has pointed out that the etching that is characteristic of much detrital garnet has occurred in the soil since it has been superposed on rounding of the grains. This etching is a result of solution and indicates that local solution may be also a destructive factor.

Anomalous and variously oriented optical zones, which may be due to strain or to inversion to forms that are not strictly cubic, are widespread in detrital garnet, and would lead one to suppose that this mineral also may be subject to disruptive forces expressed along the margins of these zones.

Our present knowledge of the garnet group would therefore support the view that generalisations about the weathering of the group as a whole have no particular validity. Some of the members of the group may be sufficiently stable to serve as index minerals but there seems good reason to exclude garnets high in iron from the list. It is unfortunate that the individual members of the garnet group cannot be satisfactorily separated on the microscope stage from their optical properties alone: a chemical analysis is generally required. Further study of the weathering of the garnet group is therefore desirable before its exact position in the weathering sequence can be settled.

ZIRCON

Zircon has in the past been bracketed with tourmaline as one of the two most resistant minerals in soils. Recent work, however, has shown that zircon is not perfectly stable. Carroll (1953) has pointed out that although zircon has no natural cleavages it is nevertheless a brittle mineral, breaking readily into irregular fragments with conchoidal surfaces. It is apparently structurally weak where the change in direction of growth occurs at the junction of the prism and pyramid faces. Carroll concludes that zircon grains can, in fact, be reduced in size by the ordinary processes of weathering. Grains with numerous inclusions appear to be less stable than grains with few inclusions, and zoned grains less stable than unzoned grains. The hyacinth variety is less stable than normal zircon. It appears also that the disintegration of zircon may be accelerated by the penetration of soil solutions into microfissures and cavities. Microfissures have been attributed by Hutton (1950) to strain induced by the change to the metamict state through the action of radiation, but they may occur apparently just as abundantly in normal zircons.

The mechanism by which rupture takes place near crystalline inclusions in zircon is not obvious. Crystallisation of zircon takes place at temperatures considerably above normal temperatures and strain would thus only be exerted at normal temperatures if the coefficient of expansion of the inclusion was less than that of zircon. Most common inclusions, however, possess coefficients of expansion greater than those of zircon as shown in table 1.

Inclusion cavities that were first filled with fluid under temperatures and pressures that prevailed during crystallisation, will be under reduced pressure at normal temperatures and pressures and are therefore not likely to cause strain (Ingerson, 1955).

An examination of the zircons in the loess soils of the South Island of New Zealand has shown that microfissures are more abundant in grains with inclusions than in grains without inclusions and are particularly abundant and destructive in grains with large inclusions. In many grains containing a large inclusion the orientation of the microfissure is consistent with stress caused by the inclusion. It is also true that at sites of local stress the dislocated lattice will have a greater solubility because of the greater free energy at the site of the stress. This has been discussed more fully with reference to quartz and may explain the solution pits that occur in many zircons in New Zealand soils in close proximity to inclusions. In some grains the solution pit has penetrated to the inclusion and partly destroyed it. The crystalline inclusions that seem to be most productive of solution pits have not been identified with certainty but appear to consist of quartz, apatite and rutile, and iron ores (Brammal, 1928).

Many zircons with abundant inclusions and most crystals with extensive fissures or fractures show abnormal optical properties, presumably as a result of strain, and the crystal shows a biaxial interference figure. In some crystals the figure changes from one part of the crystal to another. Most crystals showing no optical anomalies and containing few or no inclusions show no solution pits and no cracks.

The solubility of normal zircon may possibly be appreciable in certain circumstances. Carroll (1953) concludes that "zircon may be slightly soluble in Ca-bicarbonate or Na-bicarbonate. Observations on a number of zircon grains from different localities suggest acid as well as alkaline leaching, as in soil formation, acting through long periods of time, for the production of corrosion and the final removal of zircon." Carroll also concluded that "The use of zircon as an indicator of changes in a soil profile based solely on the chemical determinations of zirconium should be interpreted with caution, but varietal features of the actual zircon grains in a heavy residue from a soil could be used with advantage. Perhaps a combination of chemical and microscopic determinations should be expected to give the most correct and satisfactory results in studies on soil development." The writer would add the further qualification that zircon should not be used as an index mineral in studies of soil development where the soils have formed on old landscapes where a high proportion of the crystals show abundant crystalline inclusions, local or general optical anomalies, interference figures, or incipient cracking. Possibly the hyacinth variety of zircon should be rejected if many grains show the system of cracks radiating from dark brown or opaque inclusions described by Hutton (1950) and attributed by him to radiation effects.

CONCLUSIONS

An absolute scale of stability for the so-called resistant minerals about which all mineralogists and pedologists would agree is still wanting. It may be that a satisfactory scale can only be prepared for a particular environment. Such a scale, however, would still not be absolute and there could be wide departures from the scale because of physical imperfections in the mineral grains, such as strain or the inclusion of crystalline minerals with coefficients of expansion markedly different from that of the containing mineral, or variations in the composition of the mineral. It must also be remembered that a stable mineral may be weathered to a greater extent in an old soil than a much less stable mineral in a younger soil. In place of a completely satisfactory order of stability for soil index minerals, a somewhat empirical order must be accepted, and the physical and chemical stability of any minerals chosen for a particular study should be carefully assessed in the environment in question before they are regarded as acceptable for reference purposes.

In soil studies that concern particle size distribution of minerals in the silt and sand sized ranges the minerals quartz, garnet, and possibly zircon should not be used as index minerals in the following circumstances:

(1) in soils on old land surfaces. Discrepancies between the texture of the parent material and that of the soil are not uncommon on such landscapes,
(2) in soils that have existed for long periods under high rainfall, especially in the wet tropics,
(3) where the minerals contain abundant inclusions, and especially where the inclusions are large or with markedly different coefficients of expansion from those of the parent mineral grain,
(4) where the mineral grains show extensive strain or incipient fissuring,
(5) where garnet is to be used as an index mineral and the chemical analysis shows it to be a high iron garnet.

REFERENCES

Adams, J. E., and Matelski, R. P., 1955. Distribution of heavy minerals and soil development in Scott silt loam: Soil Sci., v. 79, p. 59–69.

Allen, V. T., 1948, Weathering and heavy minerals: Jour. Sedimentary Petrology, v. 18, p. 38–42.

Alexander, G. B., Heston, W. M., and Iler, R. K., 1954, The solubility of amorphous silica in water: Jour. Phys. Chem. v. 58, p. 453–457.

Barshad, I., 1955, Soil development, *in* Chemistry of the soil. Firman E. Bear, Ed., Reinhold, New York.

Bayley, W. S., 1925, The kaolins of North Carolina: North Carolina Geol. and Econ. Survey Bull. no. 25.

Blanck, E., 1949, Einführung in die Genetische Bodenlehre. Vandenhoeck and Ruprecht, Göttingen, 420 p.

Bramlette, M. N., 1929, Natural etching of detrital garnet: Am. Mineralogist, v. 14, p. 336–337.

Brammall, A., 1928, Dartmoor detritals: a study in provenance: Proc. Geol. Soc. London, v. 39, p. 27–48.

Briscoe, H. V. A., Holt, P. F., Matthews, Janet W., and Sanderson, Phyllis M., 1936, Some new properties of certain industrial dusts: Trans. Inst. Min. and Met., v. 46, p. 291–302.

Bryan, W. H. and Teakle, L. J. H., 1949, Pedogenic inertia, a concept in soil science: Nature, v. 164, p. 969–971.

Carroll, D., 1953, Weatherability of zircon: Jour. Sedimentary Petrology, v. 23, p. 106–116.

Clark, F. W., 1924, The data of geochemistry; U.S. Geol. Survey. Bull. 770, 841 p.

Clarke, J. R., 1927, Density and thermal expansion of chemical compounds in the crystalline state, under atmospheric pressure *in* International critical tables, v. 3, p. 43.

Dana, E. S., 1895, The system of mineralogy of James Dwight Dana. 6th ed., New York.

Dryden, L., and Dryden, C., 1946, Comparative rates of weathering of some common heavy minerals: Jour. Sedimentary Petrology, v. 16, p. 91–96.

Goldich, S., 1938, A study of rock weathering: Jour. Geology, v. 46, p. 17–58.

Gruner, J. W., 1930, Hydrothermal oxidation and leaching experiments; their bearing on the origin of Lake Superior haematite-limonite ores: Econ. Geol., v. 25, p. 697–719.

Harrison, J. B., 1933, The katamorphism of igneous rocks under humid tropical conditions. Imperial Bureau of Soil Science, Harpenden, 79 pp.

Haseman, J. F., 1944, An investigation of the use of heavy minerals in determining the origin and course of profile development in soils. University of Missouri Thesis; Abstract *in* Biol. abstracts, 1945, p. 2171.

Haseman, J. F. and Marshall, C. E., 1945, The use of heavy minerals in studies of the origin and development of soils: Univ. of Missouri Research Bull. no. 387, 75 p.

Hitchen, C. S., 1945, The "solubility " of silica: Econ. Geol., v. 40, p. 361–365.

Holmquist, P. J., 1911, Über die Bildung von Tridymite und Cristobalite in Quartz Ziegeln: Geol. Fören. Stockholm Förh., v. 33, p. 245–260.

Holt, P. F. and King, D. T., 1955, The chemistry of silica surfaces: Jour. Chem. Soc., p. 773–779.

Humbert, R. P. and Marshall, C. E., 1943, Mineralogical and chemical studies of soil formation from acid and basic igneous rocks in Missouri: Univ. of Missouri Res. Bull. no. 359, 60 p.

Hutton, C. O., 1950, Studies of heavy detrital minerals: Geol. Soc. America Bull., v. 61, p. 635–710.

Iddings, J. P., 1906, Rock minerals. John Wiley & Sons, Inc., New York, 548 p.

INGERSON, E., 1955, Methods and problems of geologic thermometry: Econ. Geol., Fiftieth Anniv. v. 341–410.
JACKSON, M. L., TYLER, S. A., WILLIS, A. L., BOURBEAU, G. A., AND PENNINGTON, R. P., 1948, Weathering sequence of clay-sized minerals in soils and sediments: Jour. of Phys. and Colloid Chem., v. 52, p. 1237–1260.
JOFFE, J., 1949, Pedology. 2nd ed. New Brunswick, 662 p.
KRAUSKOPF, K. B., 1956, Dissolution and precipitation of silica at low temperatures: Geochim. et. Cosmochim. Acta, v. 10, p. 1–26.
LENHER, V., 1921, Silicic acid: Jour. Amer. Chem. Soc., v. 43, p. 391–396.
LOVERING, T. S., 1923, The leaching of iron protores: solution and precipitation of silica in cold water: Econ. Geol. v. 18, p. 523–540.
MARSHALL, C. E., 1940, A petrographic method for the study of soil formation: Soil Sci. Soc. Amer. Proc., v. 5, p. 100–103.
MARSHALL, C. E. AND HASEMAN, J. B., 1943, The quantitative evaluation of soil formation and development by heavy mineral studies: Soil Sci. Soc. Amer. Proc., v. 7, p. 448–453.
MATIGNON, C. AND MLLE. MARCHAL, 1920, Action prolongée de acide carbonique sur les silicates et le quartz: Compt. Rend., v. 170, p. 1184–1186.
MERRILL, G. P., 1897, A treatise on rocks, rock-weathering and soils: 1st ed., New York, 411 p.
MILNER, H. B., 1923, The study and correlation of sediments by petrographic methods: Mining Mag., v. 28, p. 80–92.
MOORE, E. S., AND MAYNARD, J. E., 1929, Solution, transportation and precipitation of iron and silica: Econ. Geol., v. 24, p. 272–303.
N. Z. SOIL BUREAU, 1954, General survey of the soils of North Island, New Zealand; N. Z. Soil Bur. Bull. (n.s.) 5, 286 p.
PETTIJOHN, F. J., 1941, Heavy minerals and geologic age: Jour. Geol., v. 49, p. 610–625.
POLYNOV, B. B., 1937, The cycle of weathering. Thomas Murby, & Co., London, 220 p.
RAEBURN, C. AND MILNER, H. B., 1927, Alluvial prospecting. Thomas Murby & Co., London.
RUHE, R. V. AND SCHOLTES, W. H., 1956, Ages and development of soil landscapes in relation to climatic and vegetational changes in Iowa: Proc. Soil. Sci. Soc. Amer., v. 20, p. 264–273.
RUSSELL, E. J., 1950, Soil conditions and plant growth. 8th ed., London, 635 p.
SIEVER, R., 1957, The silica budget in the sedimentary cycle: Am. Mineralogist, v. 42, p. 821–841.
SOSMAN, R. B., 1928, Silica, *in* International critical tables: National Res. Council. McGraw Hill Book Co., Inc., New York.
TARR, W. A., 1917, Origin of chert in the Burlington Limestone: Amer. Jour. Sci., v. 44, p. 409–452.
THORP, J., JOHNSON, W. M. AND REED, E. C., 1951, Some post-Pliocene buried soils of the central United States: Jour. Soil Sci., v. 2, p. 3–19.
TURNER, F. J., 1942, Structural petrology of quartzose veins in the schists of Eastern Otago—Part I: Trans. Roy. Soc. New Zealand, v. 71, p. 307–324.
WAHLSTROM, E., 1948, Pre-formation and recent weathering on Flagstaff Mountain near Boulder, Colorado: Geol. Soc. America Bull., v. 59, p. 1173–1189.
WALKER, T. R., 1957, Frosting of quartz grains by carbonate replacement: Geol. Soc. America Bull., v. 68, p. 267–268.
WRIGHT, F. E. AND LARSEN, E. S., 1908, Quartz as a geological thermometer: Amer. Jour. Sci., v. 27, p. 421–447.

16

Reprinted from *Tschermaks Mineralog. u. Petrog. Mitt.* **41**:1-7 (1931)

Untersuchung von Mineralen auf Abnutzbarkeit bei Verfrachtung im Wasser.

Von Bergingenieur **Dr.-Ing. Fred. W. Freise** (Rio de Janeiro).

Für die wichtigsten Seifenminerale wurden durch Laboratoriumsversuche die relativen Widerstände beim Transport im Wasser ermittelt — auch unter Berücksichtigung des Erhaltungszustandes der betreffenden Minerale — und die Versuchsergebnisse mit Erfahrungen im Gelände verglichen.

Eine langjährige, der Aufsuchung und Untersuchung von Lagerstätten gewidmete Tätigkeit in verschiedenen Gebieten Südamerikas hat dem Verfasser eine große Anzahl von Seifengesteinposten in die Hand gebracht, an welchen die nachstehenden Untersuchungen vorgenommen wurden; die Ergebnisse wurden in einer Laboratoriums-Versuchsanlage mit besonders zu dem Zwecke zusammengestellten Mineralgemischen einer Nachprüfung unterzogen.

Den eigentlichen Anlaß zu diesen Untersuchungen gab dem Verfasser die bei einheimischen Prospektoren aus den alten Gold- und Diamantenbezirken gemachte Beobachtung, daß aus dem Abnutzungszustande der „harten" Minerale einer nur vom Ton befreiten, „halbrein" gewaschenen Seifenprobe auf die Entfernung des anstehenden Gesteins geschlossen wurde, und zwar in vielen Fällen mit verblüffend guter Annäherung, obwohl dem Wäscher durchaus jede eingehendere Kenntnis des geologischen Aufbaus der durchwanderten Gegend fehlte und er sich zum ersten Male am Ort der Waschung befand.

In den Kreis der Untersuchung im Felde sind Gold-, Diamant- und Monazitseifen gezogen worden; die Prüfungen im Laboratorium umfassen auch Minerale aus Platin-, Chromerz- und Zirkonlagerstätten. Über den Mineralbestand der Seifen gibt folgende Zusammenstellung Auskunft:

A) Goldseifen: Magneteisen, Titaneisen, Zirkon, Monazit, Turmalin, Quarz in seinen verschiedenen Abarten, Rutil, Anatas, Granaten, Wolframit;

B) Diamantseifen (außer den bereits genannten Mineralen): Cyanit, Staurolith, Hämatit, Brookit, Topas, Euklas, Chromit, Perowskit, Amphibol und Diopsid;

C) Monazitseifen (außer Mineralen wie bei A und B): Pyrit, Magnetkies, Titanit, Beryll, Chrysoberyll, Flußspat, Apatit, Cordierit, Andalusit, Sillimanit, Spinell, Vesuvian, Epidot, verschiedene Thor- und Cerminerale, verschiedene Glimmer.

Zu all diesen Mineralen kommen noch einige aus Sammlungen entnommene, welche an gelegener Stelle besonders gekennzeichnet sind.

Um für die Feststellung der Abnutzbarkeit der Minerale bei Verfrachtung im Wasser möglichst genau die in der Natur gegebenen Verhältnisse nachzuahmen, wurde nach einer Reihe von Vorversuchen eine Eisenblechtrommel von 50 *cm* Länge und 2 *m* innerem Durchmesser genommen, deren Inneres mit in Zement eingebetteten, ziemlich gleich großen gerundeten Feuersteinbrocken so ausgekleidet war, daß nur diese, nicht aber ihr Zementbindemittel sich an der Abnutzung der Versuchsstücke beteiligen konnten. Die Trommel war mit genau waagrechter Drehachse auf Rollenpaaren gelagert und konnte in langsame Umdrehung versetzt werden, so daß die in die Trommel gegebenen Versuchsstücke nur auf dem tiefsten Teile des Mantels der Trommel herumkollern konnten, aber nicht durch Mitnahme an der Wand und Herabfallen zertrümmert wurden. Beide Enden der Trommel waren durch Kreisringscheiben teilweise verschlossen; an einer Seite konnte durch ein längs einer Mantellinie angeordnetes Rohr Wasser eingeleitet werden, dessen Überlauf an der Gegenseite durch Heber abgenommen wurde. Die Zahl der Drehungen wurde durch ein einfaches Zählwerk bestimmt.

Infolge reichlichen Vorhandenseins von Musterstücken fast jeder gewünschten Zusammensetzung konnte Verfasser mit großen Mineralposten arbeiten und war in der Lage, aus den Mineralen Würfel von bestimmter Kantenlänge herauszuarbeiten. Bezeichnet man diese mit L, nennt s das spezifische Gewicht der Würfelsubstanz in Wasser, H die zu ermittelnde Härte, W den von der Trommel zurückgelegten Weg oder statt dessen Z gleich W : (4 L) die Zahl der Wälzungen, denen das Mineral ausgesetzt war, so ist die später zu wägende Abnutzung auf dem Transport unmittelbar proportional der 3. Potenz von L und den Größen s und W, aber umgekehrt proportional der Härte H; man bekommt also:

$$\text{Abnützung gleich } \frac{L^3 \cdot s \cdot \frac{W}{4 \cdot L}}{H} \text{ oder gleich } (L^2 \cdot s \cdot W) : 4\,H$$

Nimmt man die Abnutzung eines Würfels zur Kugelgestalt an, so erleidet das Probestück einen Stoffverlust von 47·6%, bei Abnutzung zur Zylinderform 21·5%. In den Versuchen des Verfassers wurde auf einen Substanzverlust von etwa 40% hingearbeitet, welcher der Rundung des Versuchswürfels zu einem an den Enden der Mantellinien nach L/4 gegen die Endkreise abgerundeten Zylinder entspricht; in dieser Gestalt treten viele der säulig kristallisierenden Versuchsstücke in den Seifen auf.

Für die in die folgende Zahlentafel 1 aufgenommenen Minerale sind die Ergebnisse der Ermittlung der Größe H — Transportwiderstandsfähigkeit — auf die des dichten Roteisensteins gleich 100 bezogen umgerechnet worden; die Zahlen der Tafel geben zu folgenden Schlüssen Anlaß:

Tafel 1. Relative Transportwiderstandsfähigkeit verschiedener Minerale.

Untersuchtes Mineral	Zusammensetzung; besondere Bestandteile	Spezifisches Gewicht	Härte nach Mohs	Bruchform	Relativer Transport-Widerstand
Quarz		2·65	7	muschl.	245
Chalcedon		2·58	6—7	muschl.	510
Zirkon	71·88 ZrO_2, 25·31 SiO_2	4·65	7·5	uneben	265
Rutil		4·19	6·5	uneben	455
Ilmenorutil	... 13·65 Fe_2O_3	4·18	6·5	uneben	390
Zinnstein		6·82	7	muschl.	360
Korund (Saphir)		3·94	9	uneben	750
Eisenglanz	96·81 Fe_2O_3, 3·11 TiO_2	5·28	5·5	körn.	280
dicht. Roteisen		5·23	6	uneben	100
Titaneisen	58·41 FeO, 35·84 TiO_2	4·88	6	uneben	325
edl. Spinell	27·66 MgO, 70·35 Al_2O_3	3·66	8	muschl.	550
Chromspinell	... 8·18 Cr_2O_3	3·81	8	muschl.	680*
Zinkspinell	41·3 ZnO, 54·7 Al_2O_3	4·19	8	muschl.	620*
Chromit	36·3 Cr_2O_3, 31·73 FeO	4·54	5·5	uneben	475
Magnetit	69·2 Fe_2O_3, 30·8 FeO	4 96	6	muschl.	380
Chrysoberyll		3·72	8·5	muschl.	300
Wolframit	70·8 WO_3, 20·15 FeO	7·45	5·5	uneben	450
Apatit	... 82·93 P_2O_5	3·11	5	muschl.	275
Kalktongranat	... 3·50 FeO	3·96	6—7	muschl.	420
Kalkeisengranat	... 12·6 FeO, 1·21 TiO_2	3·91	6	muschl.	365
id.	... 17·33 FeO	3·86	6·5	muschl.	320
„Kaprubin"	... 0·98 Cr_2O_3	3·78	7	muschl.	410*
Almandin		3·85	6·5	muschl.	375
gem. Olivin		3·39	6·5	muschl.	250
id.	... 22·61 FeO	3·84	7	muschl.	330
Andalusit		3·13	7·5	uneben	220
Disthen		3·62	6·5	muschl.	260
Topas		3·57	8	uneben	390
Perowskit	58·88 TiO_2, 8·15 FeO	3·97	5—6	uneben	620*
Cordierit	mit 0·33 FeO	2·48	7—7·5	uneben	480

1*

Untersuchtes Mineral	Zusammensetzung; besondere Bestandteile	Spezifisches Gewicht	Härte nach Mohs	Bruchform	Relativer Transport-Widerstand
Vesuvian	... 0·68 B_2O_3, 0·12 TiO_2	3·41	6·5	splittr.	330
Orthoklas		2·52	6	muschl.	150
Leucit × ×		2·47	5·5	uneben	280
Diopsid × ×	... 0·68 TiO_2, 1·06 Fe_2O_3	3·19	5–6	uneben	160
gem. Augit		3·31	6·5	muschl.	420
Rhodonit		3·62	6—7	uneben	810*
Epidot	22·7 CaO, 27·2 Al_2O_3 11 Fe_2O_3	3·35	6–7	uneben	320
Staurolith		3·68	7·5	muschl.	420
gem. Beryll		2·69	6·5	uneben	270
Aquamarin		2·72	8	muschl.	465
Turmalin	35·7 SiO_2, 31·3 Al_2O_3, 4 B_2O_3	3·22	7·5	muschl.	850
id.	Abart mit 13·4 B_2O_3	3·17	7·5	muschl.	950
id.	(schwarzer Schörl)	3·20	7·5	muschl.	650
Monazit × ×	36·4 Ce-Oxyde, 19·6 P_2O_5	4·95	5	muschl.	105
id.	(Gangmonazit; 31·1 Ce-Oxyde)	4·91	6·5	muschl.	130
Bauxit	45·11 Al_2O_3, 29·1 Fe_2O_3	2·52	2·5	erdig	110*
id.	62·31 Al_2O_3, 4·77 Fe_2O_3	2·61	3·5	muschl.	160
Zinkblende		3·96	3·5	uneben	350*
Pyrit		5·17	6—6·5	muschl.	500
Psilomelan	93·71 MnO_2	4·24	5–6	muschl.	180*
Bleiglanz	... 1·86 Sb	7·49	2·5	blättr.	75
Hauerit	44·8 Mn, 2·62 Fe	3·47	4	uneben	225*

Die mit * bezeichneten Minerale stammen aus Sammlungen; alle anderen sind von den natürlichen Lagerstätten in frischem Zustande entnommen.

Die Zahlen der letzten Kolonne sind Mittelwerte aus drei Versuchen.

1. Es besteht kein Zusammenhang zwischen der in diesen Versuchen erforschten Eigenschaft und der lang bekannten Härte gemäß der Skala von Mohs; die Versuchseinrichtung wendet sich in erster Linie an die Zähigkeit der Minerale; Beispiel: Chalcedon gegen Quarz.

2. Die Versuche geben nur relative Zahlen; um absolute Größen hinsichtlich des Weges zu bekommen, welchen ein Mineral im Wasser durchwandern muß, um vollständig zerrieben zu werden, müßten die genauen Transporterscheinungen unter Wasser, namentlich auch in bezug auf die Begleitkörper, genauer geprüft werden; für viele Seifenmineralkombinationen begegnet dies unüberwindlichen Schwierigkeiten.

Tafel 2. Zerstörung von natürlichen Mineralgemischen durch Verfrachtung im Wasser.

Zusammensetzung des Mineralgemisches (die Zahlen in Klammern sind Korngrößen in *mm*) Abkürzungen s. Fußnote.	Mineralbefund nach Verfrachtung auf *km*	(die Zahlen sind Substanzverlust und runde Zahlen)
1. Granit mit Monazitschliere: Qu (2—5), Zr (1—3), Mag (2—5), Tt (1—2), Trm (1—3), Mon (2—5), Gr (2—8), Gl, Fds	2·0	Qu, Zr, Trm, Tt fast 0; Gr. 20; Mon 30, Mag 10—15
	5·0	Zr, Trm, Tt rd. 5, Qu 15, Mon 40—50, Mag 30
	8·0	Qu 25—30, Mon fast 100
	12·0	Alle Minerale gerundet, Mon verschwunden
2. An Akzessorien reicher Granit: Trm, Ap, Zr, Py, Tt, Mag, Mon, durchwegs 2—4 *mm* groß; Quarz (1—3), Fds bereits angewittert, Gl großtafelig	1·5	Fds zerfallen, Ap, Mon stark gerundet, andere Akzessorien wenig angegriffen
	5·0	Zr, Trm, Tt, Py etwa 10; Mag ziemlich angerundet, Mon zu $^2/_3$ zerstört
	10·0	Mon verschwunden
3. Typische Diamant-Canga: Mag, Egl, Tt, Qu, Gl, Gold; Bestandteile (3—5) ziemlich gerundet, als Bindemittel Gold	5·0	Gold auf 50% vermindert
4. Sand aus dem oberen Rio Claro, Goyaz: Chromit, Zr, Trm, Top, Chrys, Mag; alle Akzessorien mit scharfen Kanten	10·0	Trm, Zr, Chrys rd. 15% Verlust; Mag 25—30, Chromit und Top fast zerstört
5. Quarzbreccie aus Matto Grosso (?): Qu von 2—4 *mm* Korn von kieseliger Bindemasse umschlossen	25·0	Zu Sand von 1 *mm* Korn zerstört

Abkürzungen: Qu = Quarz, Zr = Zirkon, Tt = Titaneisen, Trm = Turmalin, Mag = Magneteisen, Mon = Monazit, Ap = Apatit, Py = Pyrit, Egl = Eisenglanz, Top = Topas, Chrys = Chrysoberyll, Fds = Feldspat, Gl = Glimmer, Gr = Granat.

3. Für die Bestimmung der relativen Widerstandsfähigkeit gegen Verfrachtung im Wasser eignen sich in erster Linie diejenigen Minerale, welchen dicksäuliger, würfeliger oder derbbrockiger Habitus eigen ist; Minerale von tafel- oder blättrigem Habitus, wie Axinit, Sphen, Magnetkies, Glimmer, entziehen sich der Prüfung in der angegebenen Art.

4. Soweit aus wenigen Fällen — z. B. bei Granat, Rutil — geschlossen werden darf, verändert sich die hier untersuchte Widerstandsfähigkeit mit dem Vorhandensein oder Fehlen gewisser Beimengungen.

Um über die zur Zerstörung einiger Seifenminerale erforderliche Verfrachtungsstrecke einigen Aufschluß zu erhalten, veranstaltete Verfasser folgende Versuche: Von den geeigneten Gesteinen wurden die sich durch größeren Reichtum an Akzessorien auszeichnenden Massen durch wiederholtes langsames Gefrierenlassen und Wiederauftauen zermürbt und dann in die Transporttrommel gegeben; um den Verhältnissen in der Natur möglichst nahe zu kommen, diente als Zugabeflüssigkeit Wasser, dem durch Beimengung von Ton das spez. Gew. 1·08 bis 1·10 gegeben wurde. Nach bestimmten Umdrehungszahlen wurde der Trommelinhalt untersucht; wie sich die verschiedenen Mineralgemische unter diesen Versuchsbedingungen verhalten haben, ist aus der 2. Zahlentafel ersichtlich.

Soweit Monazitlagerstätten in Betracht kommen, ergaben sich die Versuche als sehr gut mit der Beobachtung im Gelände übereinstimmend: über 8 *km* vom Ursprungsorte ist Monazit nur in sehr seltenen Fällen noch in Wasserläufen nachweisbar. Die Beobachtung hinsichtlich des in der Canga befindlichen Goldes — etwa 60 Gramm/Tonne Gestein im untersuchten Muster — dürfte nicht ohne weiteres in die Praxis übertragbar sein, da das Edelmetall nicht nur mechanisch angegriffen, sondern auch durch Humussäuren leicht zerstört wird; in Lagerstätten, welche in dichten Urwäldern durch Wasserläufe angeschnitten werden — in Surinam, Brasil.-Guyana, Goyaz, Espirito Santo kommen solche Fälle oft vor — dürfte die Zerstörung des Lagerstättenbestandes an Gold fast ausschließlich auf den Angriff der Humussäuren zu setzen sein. Die oben gegebenen Zahlen wollen, wie hier nochmals gesagt werden soll, nichts als relative Werte darstellen; zu beachten ist auch, daß selbst die als durchaus „unlöslich" angesehenen Minerale, wenn auch in minimalen Mengen, doch in Wasser löslich sind. Dieserhalb sind zu solchen Versuchen, wie hier berichtet wurden, nur ganz unangegriffene Probestücke, unmittelbar aus dem Innern der Lagerstätte entnommen, zu benutzen; schon durch Jahrzehnte oder noch viel länger der Einwirkung der Atmosphärilien ausgesetzt gewesene Minerale zeigen viel größere Angreifbarkeit auf Transport im Wasser, selbst wenn sie äußerlich vollkommen unverändert scheinen. Hierfür geben die in der 3. Tafel niedergelegten Beobachtungen einigen Beweis.

Tafel 3. Abhängigkeit des Widerstandes gegen Transport im Wasser von dem Erhaltungszustande der Minerale.

Mineral	Erhaltungszustand und Härte nach Mohs		Relativer Transport-widerstand, H gleich 100 f. dicht. Roteisen-stein
Magnetit	vollkommen frisch	6·0	380
	ganz wenig angewittert	6·0	320
	bis zur Rißbildung angewittert	5—6	260
Mangantongranat	vollkommen frisch	5—6	330
	angewittert	5	200—190
Gem. Olivin	frisch	6·5	250
	beginnend zu fasern	4—5	150
Andalusit	vollkommen frisch	7·5	220
	mit Sericitschicht	6—7	130
Topas	ganz frisch	8	390
	mit Steatitrinde	4—5	110—90
Cordierit	vollkommen frisch	7	465
	mit matter Oberfläche	6·5—7	380
	schuppig werdend	5·5—6	280
	„Pinit"	4—5	150
Vesuvian	vollkommen erhalten	6·5	330
	mit Steatitrinde	4	90 u. weniger
Leucit	frisch	5·5	280
	leicht angewittert	3—4	120
Rhodonit	bei beginnender Verwitterung	4	350
Epidot	verwittert eisenreich	5—6	100—160
Gem. Augit	vollkommen erhalten	6·5	400
	mit Chloritschicht bedeckt	4—5	140

[*Editor's Note:* In Table 1 on pages 3-4 of the article the translation for the table head is: Relative Capability of Resistance to Transport of Different Minerals. In Table 2 on page 5 of the article the translation for the table head is: Order of Destruction of Natural Mineral Mixtures During Water Transport. In Table 3 on page 7 of the article the translation for the table head is: Dependence of Resistance Versus Water Transport on the State of Preservation of Minerals (H = 100 for massive Hematite).]

17

Reprinted from *Jour. Sed. Petrology* **11:**32–36 (1941), by permission of the publisher

THE STABILITY OF MINERALS IN SANDSTONE*

M. N. BRAMLETTE
U. S. Geol. Surv., Washington, D.C.

ABSTRACT

Some published data indicate that several minerals, some of which are usually considered relatively stable species, become corroded within fresh and unweathered sandstone strata. This solution or decomposition is evidently due to the action of solutions within the buried strata, and the surficial alteration due to weathering is not included in this discussion. This corrosion of minerals has suggested that some of the less stable ones may have completely disappeared from ancient strata. Examples are presented which show that hornblende and epidote have disappeared from some Miocene sandstones. The mere absence of some relatively unstable mineral, especially some of the ferro-magnesian group, therefore seems doubtful evidence for conclusions regarding the source rocks or the stratigraphic relations of sandstones. These considerations seem especially important for older formations, as the small assemblage of stable minerals, that is usual in early Paleozoic strata, may represent only a survival of the fittest.

The chemical stability of minerals in the source rocks of sediments under the influence of weathering and their physical stability during transportation of the sediments have received considerable attention, and are generally recognized factors influencing the mineral content of sedimentary rocks. Several papers have described a corrosion of detrital mineral grains which was effected within the sandstones, and is not related to surface weathering. Boswell (1933) has reviewed and discussed most of the papers bearing on this stability of detrital mineral grains. As pointed out by him, Thoulet was one of the first writers to consider this subject and to note that the complexity of mineral assemblages in sediments increased upward in the geologic column in consequence of the relatively unstable nature of some minerals such as pyroxene and olivine. Among others, Edelman and Doeglas (1934) have presented evidence of a "diagenetic" corrosion of detrital grains of staurolite, kyanite, garnet, epidote, and titanite, and mentioned their previous paper on similar data for two of the less stable minerals, pyroxene and amphibole. The described examples of solution or corrosion of mineral grains within sandstones do not include much evidence of alteration or conversion to other minerals within the strata, except for some of the titanium minerals. Most of the evidence of changes in the titanium minerals is indirectly inferred only from the common occurrence of authigenic titanium minerals, as is evident from the reviews by Boswell. Evidence that the authigenic anatase was formed from titanite in one occurrence was mentioned by the writer (1934).

The corrosion of certain minerals has suggested that some of these minerals may have disappeared entirely from some ancient strata, but evidence on this inference is ordinarily not apparent. Examples are presented here which show that two of the relatively unstable minerals, hornblende and epidote, have disappeard from some Miocene sandstones of California. These examples indicate a method of securing additional data on this subject.

The examination of calcareous concretions within clastic sedimentary rocks indicates that many of them are formed within the strata, but at an early period before much of the compaction and lithifaction; this may thus be considered a typically diagenetic process. This is especially well shown by the concretions in the Monterey formation of California,

* Published by permission of the Director, Geological Survey, United States Department of the Interior.

examined by the writer. The thin rhythmic bedding extends uninterruptedly through the concretions, though the individual beds are thicker within the concretions, and include preserved diatoms, volcanic glass shards, and other relatively unstable constituents that either are not preserved or are partially altered in the same beds outside the concretions.

Samples from two large calcite-cemented sandy concretions and from the equivalent sand immediately adjacent to the sampled parts of the concretions were collected from the Hambre sandstone (Miocene) in the area south of San Pablo Bay, California. The marked difference in their content of hornblende is indicated in table 1 (first four samples).

The percentages of light minerals (lighter than bromoform) are only approximate determinations, as accurate distinction of the proportions of quartz and alkalic feldspar requires an undue amount of work for the value it may have. The percentages of andesine and volcanic glass shards were more quickly and accurately determined in the refractive index liquids. The only striking difference in the light fractions is that the shards of volcanic glass are quite isotropic and apparently unaltered (index 1.51) in samples from the two concretions, but are much altered in the adjacent sandstone samples and consist of a weakly birefringent aggregate, which shows a mean index near 1.49, and is apparently in major part a zeolite. The hornblende that constitutes nearly half of the heavy mineral fraction in the two samples from the concretions is nearly lacking in the adjacent sandstone samples (the basaltic hornblende entirely lacking), and the total heavy mineral percentages show a corresponding decrease. Other minerals are not significantly different in the four samples, except for the relative increase in their percentages in the samples lacking much hornblende. The absolute quantities of these other minerals is not significantly different, as is evident from a comparison of their percentages with the proportion of total heavy minerals in the four samples.

The evidence of the loss of hornblende from this sandstone suggested some doubt regarding certain interpretations made in a recent heavy mineral study by Cogen (1936). He presented heavy mineral data on some Miocene sandstones on the north side of the Santa Monica Mountains in Los Angeles County. The mineral zones cut across stratigraphic units in such abrupt manner that one of the sandstones characterized by a dominance of hornblende and epidote showed showed these minerals absent in less than one-half mile to the east. Thus the same sandstone unit was included in his mineral zone C at one locality and in his mineral zone B in the locality less than one-half mile distant. Cogen states, "Zone C is distinguished by the presence of hornblende and epidote and the relative proportions of ilmenite, titanite, and garnet. Heavy minerals are comparatively abundant here. Hornblende and epidote are absent from zone B but it possesses the same ilmenite-titanite-garnet proportions characteristic of the underlying zone C. The heavy mineral content of the sediment is small."

A few large concretionary bodies of calcite-cemented sand are present in this sandstone, though they appear to be less dense and impervious than those in the finer-grained Hambre sandstone. Samples were collected from two of these concretionary masses and from the immediately adjacent sandstone from the outcrop of Cogen's samples 13B and 14B. Mineral data on my samples from this locality are shown in the last four samples of table 1. Even the samples from outside the concretions show percentages of certain minerals that differ considerably from Cogen's data, but this is evidently due in large part to the use of different size fractions of the sands in the two studies. The writer used a fraction of smaller maximum size and of a larger range in size than the fraction of $\frac{1}{4}$ to $\frac{1}{8}$ mm. that was used by Cogen. Thus, the much larger percentage of zircon shown here is due to the fact that most of the

TABLE 1. *Percentages of mineral grains in samples of sandstone and in adjacent calcareous concretions*

	Appx. percentages of the light mineral fraction							Percentages of the heavy mineral fraction													
	Quartz	Alkalic feldspar	Andesine	"Aphanitic"	Volc. glass shards	Altered Volc. glass	Percentage of heavy minerals	"Leucoxene"	Black opaque	Chromite	Zircon	Garnet	Tourmaline	Rutile	Titanite	Glaucophane	Andalusite	Epidote and zoisite	Actinolite	Brown hornblende	Green hornblende
San Pablo Bay Area																					
Hambre sandstone, Sample no. 1.......	20	20	35	15		10	1.1	2	19	4	12	3	1	½	10	1	½	37	5		5
Sand from concretion adj. to Sample no. 1.	20	15	30	20	15		2.0	2	13	3	5	3	R	R	6	1	R	17	6	6	38
Hambre sandstone, Sample no. 2.......	15	15	40	15		15	0.7	5	16	5	9	5	2	½	12	2	1	33	5		4
Sand from concretion adj. to Sample no. 2.	20	15	35	20	10		2.2	2	10	3	4	2	2		5	½	1	20	4	5	41
Santa Monica Mountains Area																					
Modelo sandstone, Sample no. 3.......	20	25	40	15			1.0	25	15		20	15	½	½	22			2			
Sand from concretion adj. to Sample no. 3.	20	30	40	10			2.5	5	15		12	5	R		10			53			R
Modelo sandstone, Sample no. 4.......	25	30	35	10			0.8	28	12		18	10	1	½	30						
Sand from concretion adj. to Sample no. 4.	30	25	30	15			1.1	20	17		20	10	½	R	27			6			R

Micas not included, but are abundant in the samples of Modelo sandstone. R = rare = 1 grain several hundred.

zircon crystals are smaller than ⅛ mm., and were, therefore, not included in the more coarse-grained fraction used by Cogen. The "leucoxene" shown in my tabulation includes opaque grains that are coated with leucoxene and also those with a more rusty coating. A more strenuous preliminary treatment with hydrochloric acid may have removed such coating from more of the black opaque grains shown in Cogen's data. Nevertheless the mineral data for the two samples of sandstone are rather similar to those of Cogen from the same locality, which were included in his mineral zone B. However, the mineralogy of the sand within the concretion of sample no. 3 at this same locality contains much epidote and is thus more similar to his mineral zone C.

The disappearance of the epidote from the sandstone and its preservation within the calcified body of concretionary sandstone seems evident. Possibly hornblende was originally more abundant as the rare grains of hornblende present in the concretion are long and delicate "skeletal" forms with cockscomb terminations that suggest corrosion within the strata, as they could have survived very little transportation and are present only in the concretion. Because of the massive character of the sandstone at this locality, there is no evidence that the concretions were formed before much of the compaction of the sediments. Therefore, some mineral alteration may have occurred before these concretions were formed. However, in the concretion sample no. 4, it seems possible that some epidote has been lost from within the concretion as well as from the adjacent sand. It was particularly apparent that this concretion was not as dense and impermeable as other concretions, and the abundant biotite was nearly as much altered in the concretion as in the adjacent sand, though the biotite was black and little altered in the nearby concretion of sample no. 3. Also the table shows that this concretion has a higher percentage of "leucoxene," including the rusty coated opaque grains, than does the other concretion. In fact, even the rusty stain of the weathered surface zone had permeated somewhat into this concretionary sample (no. 4), and this may account for most of the change in the biotite and opaque grains.

The loss of hornblende and epidote that has been described, and many of the occurrences of corroded detrital grains are evidently not the result of weathering processes. Some of these samples came from beneath the surficial zone of a few to 20 or more feet of rusty weathered sandstone, which is obvious in the deep roadcut exposures; the sandstones containing hornblende and epidote (Cogen's zone C) show no appreciable differences in these minerals above and below this surficial zone of greater weathering. The corrosion of more stable mineral grains, such as staurolite and garnet is recognized in core samples from deeply buried strata as well as in the outcrop samples of these strata.

Many additional data are necessary before the relative stability of detrital mineral grains, under various conditions, can be profitably considered. Composition of the solutions within the strata would seem to be one of the important factors, as some sandstones of Eocene and Cretaceous age in California contain minerals such as hornblende and epidote, though these minerals have disappeared in some of the Miocene sandstones described here. It is perhaps significant that apatite, which is so easily attacked by weakly acid solutions, is common in the sandstone samples which have lost hornblende and epidote. This apatite is not shown in the table because the samples were treated with dilute hydrochloric acid, but examination of part of the original samples showed that apatite is one of the most abundant heavy minerals.

This evidence seems to indicate that the mere absence of some relatively unstable minerals cannot be safely used as a basis for interpretations regarding the stratigraphic relations or the source rocks of sandstones. From the evidence of the corrosion of some relatively stable minerals, even garnet under some conditions, it would seem that due consideration should be given the possibility of their complete disappearance in some of the oldest strata. The relatively small assemblages of heavy minerals in many of the early Paleozoic formations, and especially the fact that these small assemblages commonly include only the relatively stable minerals, (zircon, tourmaline, rutile, black opaques, and garnet), suggests that a loss of less stable minerals from within the strata may be the explanation. The presence of some of the less stable minerals in some of these older strata does not necessarily vitiate the general relationship suggested, as factors such as permeability and composition of the solutions would doubtless prevent any great regularity of a relationship showing progressive modification of these minerals in the geologic column. However, the presence of small amounts of one of the less stable minerals, such as pyroxene in early Paleozoic strata, suggests the desirability of careful checking to see that this is not a result of contamination or inclusion of the mineral within larger grains that may have been crushed in the laboratory treatment.

The well-established results of some heavy mineral studies of sediments for interpretations regarding their stratigraphic relations show that the conclusions suggested in this paper do not discredit the value of heavy mineral studies of sediments. However, it seems to emphasize the importance of another one of the many factors that must be considered in any interpretation. The mere absence of certain minerals in ancient sandstones, particularly of the relatively unstable minerals, seems of doubtful significance, and if additional data show that the small suite of relatively stable minerals in the oldest sedimentary strata is largely due to the stability relations of the minerals, it is evident that conclusions regarding the source rocks of these sediments are difficult and of questionable value except as interpreted from the minerals actually present, and the distinctive characteristics of these minerals.

REFERENCES

Boswell, P. G. H., *1933*, Mineralogy of sedimentary rocks: *Thomas Murby & Son*, London, pp. 37–46.

Bramlette, M. N., *1934*, Heavy mineral studies on correlation of sands at Kettleman Hills, California: *Am. Assoc. Petroleum Geologists Bull.*, vol. 18, p. 1575.

Cogen, W. M., *1936*, Heavy mineral zones in the Modelo formation of the Santa Monica Mountains, California: *Jour. Sedimentary Petrology*, vol. 6, no. 1, pp. 3–15.

Edelman, C. H., and Doeglas, D. G., *1934*, Über Umwandlungserscheinungen an detrischens Staurolith und anderen Mineralien: *Tschermaks Min. und Petr. Mitt.*, 45, pp. 225–234.

18

Reprinted from *Jour. Geology* **46:**610–625 (1941), by permission of The University of Chicago Press

PERSISTENCE OF HEAVY MINERALS AND GEOLOGIC AGE

F. J. PETTIJOHN
University of Chicago

ABSTRACT

The apparent increase in complexity of the heavy-mineral suites of arenaceous sediments with decrease in age of the deposit is confirmed by compilation from published records. Such increase in number of mineral species in the younger deposits is due to (*a*) greater complexity of terrane from which the younger sediments were derived or (*b*) disappearance by solution of the less stable species in the older deposits. The latter alternative is shown to be the correct explanation, since the order of persistence of the common rock-making heavy minerals worked out from the tabulation is identical with Goldich's "stability series," which is the order of survival of minerals subject to weathering. The stability series is expanded to twenty-five common constituents of sediments.

INTRODUCTION

Students of heavy minerals have long noted an apparent increase in complexity of heavy-mineral suites with decreasing geologic age. J. Thoulet,[1] for example, was one of the first to make this observation. P. G. H. Boswell,[2] who later reaffirmed Thoulet's generalization, attributed the increased complexity to (1) greater complexity of terrane from which the younger sediments were derived and (2) disappearance by solution of the less stable species. More exact information on persistence of the common heavy minerals in the stratigraphic column is needed before the causes of progressive differences with decreasing geologic age can be ascertained.

The presence of a particular mineral in a sediment requires the presence of that mineral in the parent-rock, its ability to survive weathering of the parent-rock and to resist abrasion and decomposition during transportation of the sediment, its deposition with the sediment (and not its removal by sorting action), and its ability to survive post-depositional solution or decomposition. R. Dana Russell[3] has shown that loss during transportation is not an impor-

[1] "Notes de lithologie sous-marine," *Ann. Inst. océanog.*, Vol. V, Fasc. 9 (1913).

[2] "Some Aspects of the Petrology of Sedimentary Rocks," *Proc. Liverpool Geol. Soc.*, Vol. XIII (1923), pp. 231–303.

[3] "Mineral Composition of Mississippi River Sands," *Bull. Geol. Soc. Amer.*, Vol. XLVIII (1937), pp. 1307–48.

tant factor in the Mississippi River. Most minerals, even the amphiboles and pyroxenes, withstand abrasion well. The general absence of these and other minerals from the older sediments cannot, therefore, be ascribed to their inability to survive transportation. Neither can their absence be attributed to the lack of these minerals in the parent-rock, since they are present in the source rocks of all ages. Since they occur in the younger geologic deposits, it may be assumed that they were also deposited in the ancient sediments. Russell was therefore driven to the conclusion that they disappeared following deposition by intrastratal solution. The purpose of this paper is to investigate intrastratal solution, to estimate the survival or persistence of the more common heavy minerals in sediments, and to measure the increase in complexity of heavy-mineral suites with decrease in geologic age.

TABULATION OF DATA

METHOD OF TABULATION

The data were collected from published papers. Sedimentary petrologists report their findings by listing the minerals observed, by estimating the relative abundance of the several species and giving a verbal description of their abundance, and by counting the grains of a representative sample and calculating the relative frequencies. It is apparent that uniformly reliable information on the abundance of a particular species in the sediments of the various periods is difficult to obtain from the existing literature. The writer, therefore, postulated a relationship between the abundance of a given mineral species and the number of times such species is reported in heavy-mineral studies. A mineral reported in 90 per cent of the formations or areas examined for a given period is probably more abundant than the same mineral reported in but 5 per cent of the published records of another geologic period. Accordingly, the author tabulated the reported occurrences of the more common heavy minerals for each formation or region on which published data were available. From this tabulation it is possible to compute the frequency of occurrence, in percentages, for each species for any desired period. Table 1 summarizes the frequency of occurrence for the more common mineral species.

This method no doubt emphasizes the rarer constituents. A mineral of less abundance, though persistent, receives just as high a frequency-of-occurrence number as a much more abundant but

TABLE 1

FREQUENCY OF OCCURRENCE
(In Percentages)

No.	Mineral	Pre-Keweenawan	Keweenawan	Early Paleozoic	Late Paleozoic	Mesozoic	Tertiary	Pleistocene	Recent
1	actinolite	0	8	3	4	3	7	8	26
2	anatase	50	62	50	65	50	36	28	20
3	andalusite	10	0	0	2	26	49	28	45
4	apatite	100	85	23	45	47	11	59	53
5	augite	10	0	12	14	18	16	62	67
6	biotite	40	38	29	37	66	40	59	57
7	diopside	0	0	3	2	13	3	23	43
8	epidote	70	54	12	24	58	61	95	92
9	garnet	70	62	79	76	95	73	92	95
10	hornblende	70	62	32	12	50	54	90	92
11	hypersthene	0	0	15	12	18	13	36	58
12	ilmenite	100	100	21	57	66	79	77	83
13	kyanite	10	0	6	8	58	63	77	70
14	magnetite	100	15	38	51	71	53	95	85
15	monazite	0	0	3	29	18	24	41	25
16	muscovite	20	0	21	63	55	61	38	38
17	olivine	0	0	0	0	0	0	13	18
18	rutile	50	85	68	96	84	85	77	75
19	sillimanite	0	0	0	0	13	30	23	37
20	sphene	50	15	6	10	26	33	59	58
21	staurolite	10	15	18	43	66	71	69	77
22	topaz	0	0	9	0	16	21	26	28
23	tourmaline	70	77	100	98	89	94	80	93
24	zircon	100	100	100	100	92	93	95	95
25	zoisite and clinozoisite	0	8	0	4	24	20	36	42
26	feldspar	......	......	21	43	32	46	46	53
Total occurrences		10	13	34	51	38	70	39	60

equally persistent mineral. Zircon and hornblende, for example, occur with about equal frequency in the Recent sediments, though hornblende is very much more abundant, as a rule, than zircon in these deposits. The frequency of occurrence, therefore, is not strictly a measure of abundance when two different minerals are compared one with the other. When the frequency of occurrence of a particular mineral in one time period is compared to the frequency of occurrence of the same species in another time period, the two values recorded are believed to bear a direct relation to the relative abundance of that mineral in the sediments of the two periods in question.

RELIABILITY OF DATA

A number of difficulties in tabulation were encountered. Certain species were reported so infrequently, even from Recent deposits, that no adequate or statistically valid conclusions could be drawn. The study was therefore confined to the more common minerals. Data on certain minerals are subject to special error. The micas, for example, are reported by some writers as "light" minerals, by others as "heavy" minerals, and by others omitted altogether. In the latter case it is difficult to know whether the light minerals were simply ignored (as is often the case) and the mica therefore overlooked or whether the mica was truly absent.

Another difficulty resulted from grouping of certain minerals under such designations as "opaques," "black opaques," "iron ores," etc. Since magnetite and ilmenite are probably both present, the frequency of opaques was added to the frequency of each of these two minerals. This correction affected primarily the data from the Tertiary, Pleistocene, Recent, and pre-Cambrian. Excepting for the pre-Cambrian, the trend of the data for these two species is fairly uniform, and the correction seems to be justified. Other species were also occasionally grouped under such headings as "carbonates," "micas," "amphiboles," etc. Data on still other minerals are rendered uncertain by the method of treatment of the sample. An acid digest is commonly used to dissolve carbonate cements or remove iron oxide coatings on the grains. Such treatment makes the frequency of occurrence of the acid-soluble species, notably apatite, open to doubt.

Also apparent are differences in thoroughness of investigation. Some workers report exceptionally complete lists of minerals, while others report only the principal species and pass over the minor constituents. The length of the list reported is partly dependent upon the number of samples examined. Extending the latter increases the chances of finding a grain or two of the rarer species. Fortunately, these difficulties—as well as most of the others—are not progressive in time and would not, therefore, affect the trends observed.

Some trouble arises also from certain species which are in part authigenic. Many writers made no distinction between authigenic and allogenic minerals. In other cases the issue was in doubt. This difficulty applies mainly to the titanium minerals, especially anatase and rutile. No attempt was made in this tabulation to differentiate between the two.

Finally, the question of whether the tabulation is adequately representative of all sediments or not may be raised. Are the reported occurrences well enough distributed geographically and stratigraphically? And do they adequately represent all environments of sedimentation? Since the studies cover mainly European—notably British—and American localities, it is believed that the complex geology of these two regions is fairly representative. The coverage of the stratigraphic column is less satisfactory. Far more work has been done on Recent and Pleistocene deposits than on the older strata. It was necessary, therefore, to group some of the older periods in order to obtain a sufficient record for statistical treatment. The problem of sedimentary environments is less satisfactorily answered. No doubt the Recent and Pleistocene sediments of both Europe and the United States are markedly affected by glacial contributions. These raise the frequency of some mineral species in an unusual degree. While the effect of glacial contributions cannot be disregarded, it seems probable that the records of nonglaciated areas demonstrate, nevertheless, a greater mineral complexity for Recent sediments.

MINERAL PERSISTENCE AND GEOLOGIC AGE

It is apparent from a study of the persistence chart (Fig. 1) that some minerals, notably zircon and tourmaline, show almost no change in frequency of occurrence with decreasing age. Others, also present in sediments of all ages, show an increase in number of times reported with decreasing age. Some of these, such as garnet

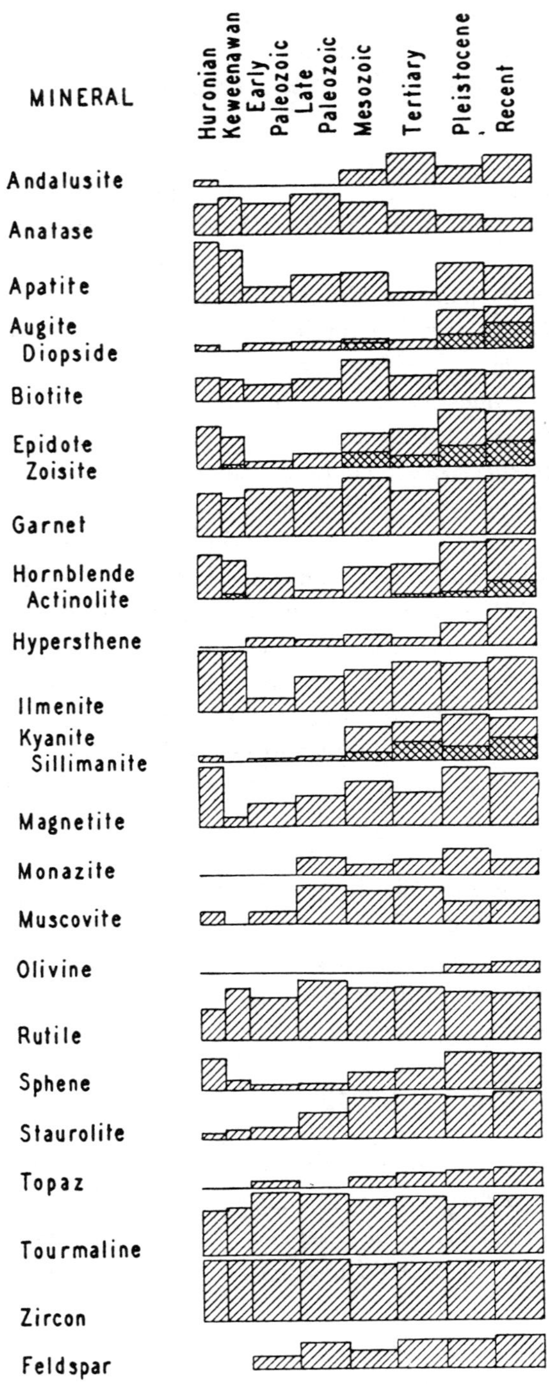

FIG. 1.—Frequency of Occurrence Chart. Based on Table 1

and biotite, show but a feeble increase, while others—notably epidote, hornblende, magnetite, ilmenite, staurolite, sphene, and feldspar—show marked increase. Some minerals are absent in the older deposits and make their appearance late in time and then increase toward the present. In this category fall andalusite, sillimanite, and zoisite. A few, like olivine, occur only in Recent and Pleistocene deposits.

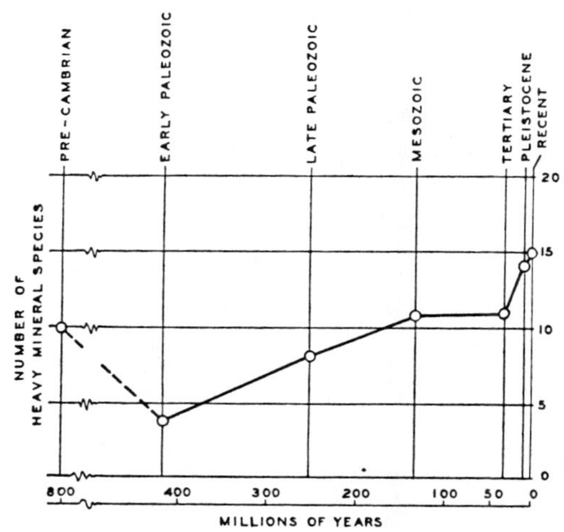

FIG. 2.—Number of heavy-mineral species and age of deposit. The number of species in *more* than half of the reported formations is plotted against the age of the beds investigated. Based on Table 1.

A "negative" persistence is suggested for several minerals. These minerals are more commonly reported from older sediments than from younger deposits. Anatase, in particular, shows a marked decline in frequency of occurrence with decreasing age of the deposit. Rutile shows the same decrease but to a lesser degree. The muscovite record is less clear but possibly also demonstrates a negative persistence.

The pre-Cambrian record, especially the pre-Keweenawan, is anomalous because of the formation *in situ* by metamorphic processes of many mineral species, notably epidote, hornblende, and sphene. The excess of magnetite and ilmenite in the pre-Cambrian is in part due to the assumption that the opaques reported consisted of these two minerals. This is only partly correct.

Inspection of either the table or the persistence diagram shows the increasing complexity in the younger sediments. If the species reported in more than half of the investigated formations of each time period be counted, an increase from 4 in early Paleozoic time to 15

TABLE 2

INDEX NUMBERS OF ABUNDANCE

(Recent = 100)

No.	Mineral	Pre-Keweenawan	Keweenawan	Early Paleozoic	Late Paleozoic	Mesozoic	Tertiary	Pleistocene	Recent
1.......	actinolite	0	31	12	15	12	27	31	100
2.......	anatase	250	310	250	325	250	180	140	100
3.......	andalusite	22	0	0	4	58	109	84	100
4.......	apatite	189	160	43	85	89	21	111	100
5.......	augite	15	0	18	21	27	24	93	100
6.......	biotite	70	67	51	65	116	70	103	100
7.......	diopside	0	0	7	5	30	7	54	100
8.......	epidote	76	59	13	26	63	66	103	100
9.......	garnet	74	65	83	80	100	77	97	100
10.......	hornblende	76	67	35	13	54	59	98	100
11.......	hypersthene	0	0	26	21	31	22	62	100
12.......	ilmenite	120	120	25	69	80	95	93	100
13.......	kyanite	14	0	9	11	83	90	110	100
14.......	magnetite	118	18	45	60	84	62	112	100
15.......	monazite	0	0	12	116	72	96	164	100
16.......	muscovite	53	0	55	166	145	160	100	100
17.......	olivine	0	0	0	0	0	0	72	100
18.......	rutile	67	113	91	128	112	113	103	100
19.......	sillimanite	0	0	0	0	35	81	62	100
20.......	sphene	66	26	11	17	45	57	102	100
21.......	staurolite	13	19	23	56	86	92	90	100
22.......	topaz	0	0	32	0	57	75	93	100
23.......	tourmaline	75	83	107	105	96	101	86	100
24.......	zircon	105	105	105	105	97	98	100	100
25.......	zoisite and clinozoisite	0	19	0	10	57	48	86	100
26.......	feldspar	......	......	40	81	60	87	87	100

in the Recent is at once apparent. The data for all investigated periods are shown graphically in Figure 2. The increase in complexity is greatest in later times, so that the curve rises steeply in Pleistocene and Recent times.

Finally, from the data assembled, an "order of persistence" may be worked out. This is obtained by adding together the frequencies of occurrence in all the time groups (except the pre-Cambrian) for each mineral species. Some minerals, however, are rarer than others, even in Recent deposits. To compensate for this inequality of abundance, index numbers were used. The base period was the Recent, hence the frequency of each species in that period was raised to 100, and that in each of the several older periods was multiplied by a factor which is the ratio of 100 to the observed frequency of occurrence in Recent deposit (Table 2). Andalusite, for example, had a frequency of occurrence in Recent deposits of but 45 per cent. Hence the reported frequencies for the other time periods were all multiplied by the factor 2.22—the ratio of 100 to 45. The order of persistence is as follows:

−3	anatase	10	kyanite
−2	muscovite	11	epidote
−1	rutile	12	hornblende
		13	andalusite
1	zircon	14	topaz
2	tourmaline	15	sphene
3	monazite	16	zoisite
4	garnet	17	augite
5	biotite	18	sillimanite
6	apatite	19	hypersthene
7	ilmenite	20	diopside
8	magnetite	21	actinolite
9	staurolite	22	olivine

The order of persistence is in general accord with opinions held by most investigators. Thoulet's "order of destruction" was olivine, pyroxenes, amphiboles, apatite, feldspar, biotite, muscovite, quartz, rutile, zircon, and corundum. In so far as these minerals occur in the persistence series, it can be seen that they are all in proper order, except for the inverted position of rutile and zircon. The "stability series" of Goldich is muscovite, biotite, hornblende, augite, hypersthene,[4] and olivine. These minerals are in the same order in this series as they are in the persistence series.

[4] Goldich omitted hypersthene, but, since it occurs in the reaction series in the position indicated and since Goldich believed the stability series to be the reverse of the reaction series, it has been added.

Surprising, perhaps, is the position of apatite. The position of this mineral confirms Thoulet's estimate but is opposed to that of H. B. Milner, who considers apatite unstable. Milner[5] also differs

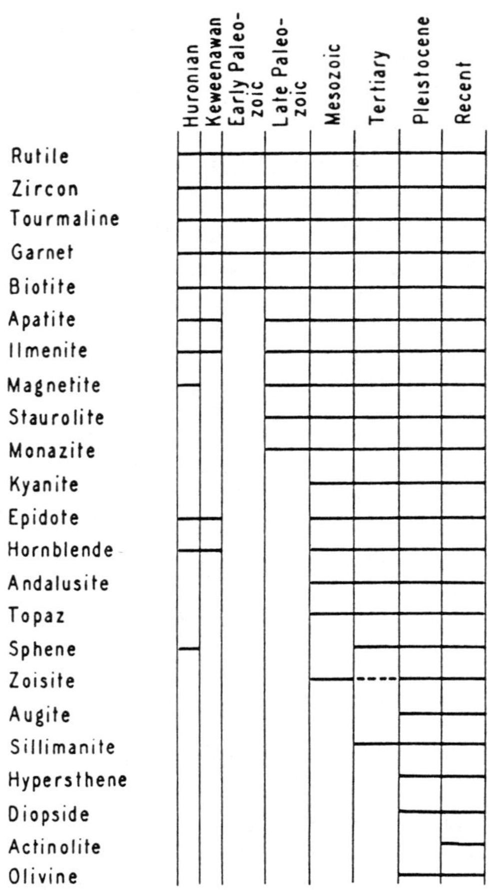

FIG. 3.—Mineral persistence chart. Line indicates presence of mineral in *relative* frequency (Recent = 100) of 50 or greater. Based on Table 2.

from the writer in that he considers magnetite more stable than ilmenite, and in other particulars as well. The writer is confident, however, that the position of the more common minerals is well

[5] *Sedimentary Petrography* (3d ed.; London: Thos. Murby & Co., 1940), p. 499.

established.[6] The positions of the rarer minerals are subject to some revision, since the probable error involved in their frequency of occurrence is large.

DISCUSSION

COMPLEXITY OF SUITE

The two possible reasons for the increasing complexity or "richness" of the mineral suite with decreasing age are (1) progressive increase in complexity of the terrane from which the sediments were derived and (2) loss of less stable minerals by intrastratal solution. The first possible explanation seems to be the less probable one. Several reasons support this view. No doubt each geologic revolution is accompanied by the invasion of the upper crust by magma and by dynamic and thermal metamorphism. Such activity yields new rocks later to be unroofed by erosion and to shed new material to the basins of sedimentation. There is not much evidence to show that these successively younger crystalline source rocks are fundamentally different from those of ancient date. If simplicity of mineral suite is a consequence of re-working older sediments and further elimination of the less stable minerals by abrasion and weathering, why, then, should not sediments become progressively simpler with decreasing age since the newer sediments should contain a larger proportion of re-worked material? Just the reverse is true. It is not reasonable to believe that the older, mineralogically simpler sediments should all be re-worked sediments, while only the Tertiary and

[6] Since this paper was written, Smithson ("The Alteration of Detrital Minerals in the Mesozoic Rocks of Yorkshire," *Geol. Mag.*, Vol. LXXVIII [1941], pp. 97–112) has published on the stability of certain minerals in Triassic and Jurassic rocks. His order is as follows:

Stable............	zircon rutile tourmaline apatite
Somewhat stable...	monazite
Unstable.........	garnet staurolite kyanite
Very unstable.....	ferromagnesian minerals, etc.

It is apparent that this list is in rough agreement with that of the writer. Significant, too, are the high stability ratings given both monazite and apatite.

later sediments are first-cycle deposits. More probably the mineralogical simplicity of the older deposits is due to intrastratal solution.

The correctness of the hypothesis of intrastratal solution is best demonstrated by the persistence order of the important rock-making ferromagnesian minerals. The order is—in reverse order of persistence—(1) olivine, (2) hypersthene, (3) augite, (4) hornblende, (5) biotite, and (6) muscovite. This order is precisely that given by Goldich in his stability series. Since the position in the stability series is a measure of susceptibility to weathering, it seems very improbable that the order of persistence based on the geologic record can mean anything except loss by solution following deposition. Only by the most improbable coincidence can their persistence order be otherwise explained.

Direct evidence of intrastratal solution of even the more stable species has often been described. Etched garnet[7] is often reported as is also etched staurolite. The "hicksaw" and "cockscomb" phenomena displayed by augite, hypersthene, and hornblende have been commented on by several authors[8] and have been shown by C. S. Ross, H. D. Miser, and L. W. Stephenson[9] to be due to post-depositional solution.

F. Smithson,[10] moreover, has shown that the richness of the suite, even in the different parts of the *same* formation, is a function of post-depositional changes rather than an original character. The richness of the suite is inversely related to the authigenic features of the deposit. Where authigenic changes—outgrowths on zircon, etc.—are most prominent, the suite is impoverished; where the authigenic action is feeble, the suite shows a maximum variety of species. These

[7] M. N. Bramlette, "Natural Etching of Detrital Garnet," *Amer. Min.*, Vol. XIV (1929), pp. 336–37.

[8] C. H. Edelman, "Diagenetische Umwandlungerscheinungen an detritischen Pyroxenen und Amphibolen," *Fortschr. Min. Kryst. Pet.*, Vol. XVI (1931), pp. 323–24; C. H. Edelman and D. J. Doeglas, "Reliktstrukturen detritischer Pyroxene und Amphibole," *Min. Pet. Mitt.*, Vol. XLII (1932), pp. 482–90.

[9] "Water-laid Volcanic Rocks of Early Upper Cretaceous Age in Southwestern Arkansas, Southeastern Oklahoma, and Northeastern Texas," *U.S. Geol. Surv. Prof. Paper 154-F* (1929), pp. 175–202.

[10] "Statistical Methods in Sedimentary Petrology," *Geol. Mag.*, Vol. LXXVI (1929), pp. 417–27.

antipathetic relations strongly suggest a causal relation between the chemical activity responsible for authigenic growth and solution of the less stable minerals.

Analogous reciprocal relations appear in the data here tabulated. Anatase is both detrital and authigenic, but since it has a marked negative persistence, it is usually authigenic. The older the deposit, the more probable is its formation. Since it bears a reciprocal relation to sphene, it seems likely that it forms at the expense of this mineral. This relation has been confirmed by P. G. H. Boswell,[11] who noted a similar reciprocal relationship in the frequencies of these two minerals in the various beds of the same age. In deposits of the earlier periods, where anatase is most frequently reported, the associated mineral suite is most restricted in composition.

As a corollary to the conclusions reached above, it should be noted that an impoverished mineral assemblage—such as characterizes the Paleozoic rocks—does not necessarily mean derivation from ancient pre-existing sediments but may mean only that the deposit is of great age and has lost the less stable species by intrastratal solution.

STABILITY OF MINERALS

The stability of minerals depends on a number of factors. Goldich formulated his stability series after a careful study of mineral and chemical changes observed in weathering. He pointed out the relation between the position of the mineral in the stability series and its corresponding position in Bowen's reaction series. Minerals formed at highest temperatures in the most anhydrous magmas are less stable under surface conditions than those formed under lower temperatures in the more hydrous end magmas. The order of persistence determined by statistical summary of the literature enables us to extend the stability series to include many species not studied by Goldich or involved in the reaction principle.

Secondary enlargement is a self-evident criterion of stability just as etched surfaces, etc., are a criterion of instability. Even tourmaline and zircon—the two most stable detrital minerals—exhibit secondary outgrowths. The alkali feldspars, on which persistence

[11] "Petrography of the Sands of the Lias-Inferior Oölite of the West of England," *Geol. Mag.*, Vol. LXI (1924), pp. 246–64.

data are incomplete, presumably would rank high in stability if the secondary-growth criterion is reliable. The reports on feldspar were found sufficiently numerous to justify plotting its frequency of occurrence (Fig. 1). This confirms its high stability. It would be number nine in the stability series. Since quartz is commonly secondarily enlarged, it too should have a high stability. Its ubiquitous nature confirms this conclusion. If, however, the frosted appearance of the quartz in some of the older sandstones is due to solution and is the visual expression of a fine etching (analogous to that produced on glass by hydrofluoric acid), then quartz is not wholly stable. It is never deeply etched, as is garnet, and therefore ranks higher in stability than that very stable mineral.

The negative persistence of a few species, that is, those more abundant in the older deposits than the Recent, is due partly to oversight and partly to authigenic formation. A little zircon and tourmaline, for example, may well be overlooked in the amphibole- and pyroxene-rich glacial suites of the Pleistocene and the Recent sediments derived therefrom. Anatase, on the other hand, may be more frequently reported from the older sediments, because it is largely authigenic. The possibilities for its formation are enhanced with passage of time. In like manner, rutile seems to be authigenic in so many cases that it, too, is slightly more common in the older deposits than in the Recent. The apparent negative persistence of muscovite is less clear, owing perhaps to the uncertain position of muscovite. It was regarded as a light mineral by many investigators.

Boswell has noted a relationship between density and stability of the several forms of composition. Kyanite is both denser and more persistent than either sillimanite or andalusite. Rutile is likewise denser and more persistent than anatase or brookite.

That some minerals have a limited persistence in the geologic record has been noted before. H. H. Thomas,[12] for example, calls attention to the absence of andalusite in older beds and believes that it could survive only for a limited period in sediments. This paper is an attempt to extend this observation of Thomas and others to many more mineral species.

[12] "Detrital Andalusite in Tertiary and Post-Tertiary Sands," *Min. Mag.*, Vol. XV (1909), pp. 241–44.

APPLICATIONS

In a general way it should be possible to determine the geologic age of a formation by a consideration of both number and kind of the heavy minerals independently of fossil content or stratigraphic relations. At present this possibility has only theoretical interest, since many other factors than age govern the presence or absence of a particular mineral species.

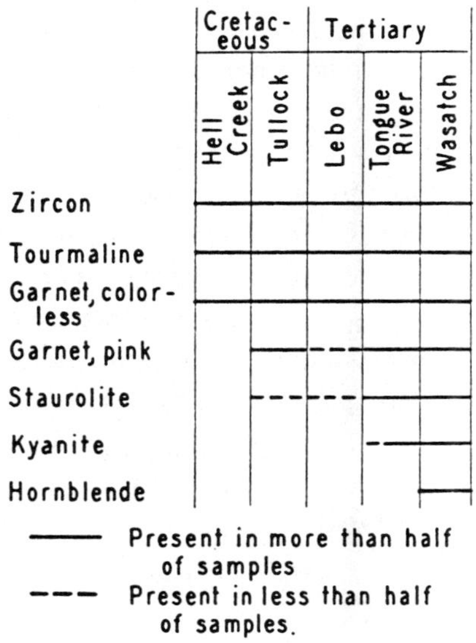

FIG. 4.—Mineral persistence in formations of Big Horn Basin (after M. H. Stow)

Whether a mineral survives intrastratal solution depends on the nature of the mineral, the pH of the interstitial solutions, the permeability of the bed and the velocity of flow of the ground waters, the temperature, and the geologic age of the deposit. In this study only the first and last factors were investigated. The results are, therefore, an average only and not strictly applicable to particular formations or basins of deposition. In a "sealed environment," for example, minerals with low stability may survive longer than more stable species in younger deposits through which a strong flow may have taken place. The order of survival in any given sequence would be the same as the general order here established.

In a study of Cretaceous and Tertiary formations in the Big Horn Basin in Wyoming, M. H. Stow[13] recorded both the relative abundance and the persistence of certain heavy minerals. Figure 4, made from data given by Stow, shows that the order of persistence in this particular section is the same as that worked out by the writer. Stow, however, interprets the appearance of staurolite and kyanite and, particularly, hornblende as a record of new sources of supply. Since the order of appearance is the reverse of that of mineral stability, it seems possible that the order is due to survival ability.

Some writers have noted mineral gradients in ancient deposits. These gradients are in part related to regional dip. The decrease in frequency of some species in a downdip direction (toward the axis of the basin of deposition) and the concomitant increase of other species should be viewed with caution. If one mineral, such as epidote, declines downdip and another, such as garnet, increases, it may be due to loss of the less stable species by solution under the greater cover of sediments (and hence higher temperatures and more active solution) and a *relative* increase of the more stable species.

[13] "Dating Cretaceous-Eocene Tectonic Movements in Big Horn Basin by Heavy Minerals," *Bull. Geol. Soc. Amer.*, Vol. XLIX (1938), pp. 731–62.

19

Reprinted from *Geol. Soc. America Bull.* **53:**1850–1851 (1942), by permission of the Geological Society of America

PROVENANCE VERSUS MINERAL STABILITY AS A CONTROLLING FACTOR IN THE COMPOSITION OF SEDIMENTS

P. D. Krynine

The paucity of unstable minerals in some ancient sediments can be explained by assuming that they are reworked sediments or by the newly proposed theory of intrastratal solution which claims destruction of unstable constituents by postdepositional leaching—oldest sediments attacked longest.

A critical review of the evidence shows that: (1) Intrastratal solution exists only on a limited, local, and erratic scale. It depends upon *absolute leaching intensity* and upon *relative length of time* during which effective circulation is actively maintained. Paleozoic sediments sealed tight by early postdepositional cementation are less leached than highly permeable Pleistocene terraces. (2) Thin-section work shows that most (90%±) sediments poor in unstable minerals are reworked sediments frequently loaded with fragments of shale, slate, phyllite, sandstone, quartzite, low-rank schists, and particularly detrital chert. Sediments rich in unstable minerals generally contain igneous and high-rank metamorphic rock fragments.

These studies place composition of the average clastic sediment at approximately 25% sedimentary derivation, 45% metamorphic (about 30% low-rank), and 30% igneous and ultra-metamorphic.

Schematically lithosphere consists of an upper sedimentary layer, middle low-rank metamorphic, and lower crystalline—high-rank metamorphic and igneous layer. Moderate deformation within source area results only in erosion of upper layer, whereas only profound diastrophism uncovers deepest crystalline zone.

Differences in mineral composition probably reflect intensity of diastrophism; stable minerals (two-cycle sediments) reflect periods of quiescence, and unstable minerals indicate deep erosion following intensive deformation.

Probably provenance changes cause most mineral composition differences locally modified by intrastratal solution.

20

IMPORTANCE OF HEAVY MINERAL ANALYSIS IN SOLVING GEOLOGICAL PROBLEMS*

M. Turnau-Morawska

These excerpts were translated expressly for this Benchmark volume by Waldemere Bejnar, Socorro, New Mexico, from "Znaczenie analizy minerałów cięzkich w rozwiązywaniu zagadnień geologicznych" in Acta Geol. Polonica ***5****:363–364, 371–377, 384–388 (1955), by permission of the publisher, Panstwowe Wydawnictwo Naukowe, Warsaw. Figures 1 and 2 are reprinted from the original publication.*

[*Editor's Note:* In the original, material precedes these excerpts.]

GENERAL OBSERVATIONS

One of the first scientific methods that contributed to the independence of sedimentary petrography is the analysis of the so-called sparse or heavy minerals. These accessory components of sedimentary rocks have a specific gravity greater than 3 and commonly represent a small percent (usually less than 1 percent) of detrital sedimentary components, the rest of which is mostly quartz.

The first description of heavy minerals in sand was done before petrography became an independent branch of geology and before the polarizing microscope was used to study rocks. In 1719 [1718] Réaumur (20) described gold-bearing sands from the Rhone and Rhine Rivers in which he noted the presence of zircon, rutile, epidote, garnet, topaz and corundum. In the next century additional works appeared on the same subject. Only after 1870 did the literature (7, 10, 13, 21, 27) report data on heavy-mineral analyses conducted for the purpose of showing source of detrital material in sands. Fougué (8) analyzed sands of the Santorini Islands, Thoulet (29) studied sands of the Sahara, and Thürach (30) analyzed sands of various ages and emphasized the importance of the shape of the heavy-mineral grains in drawing petrogenetic conclusions. Thürach's work is considered by Boswell (3) as the first serious study of the importance of mineral components in the analysis of the origins of sands.

Today world literature includes several hundred publications concerning heavy minerals in sedimentary rocks. These publications are concerned with refinement of the method of analysis, the heavy fraction

**Editor's Note:* In the Conspectus in *Acta Geologica Polonica* the title of this paper is translated as "Heavy Mineral Analysis and the Solution of Geologic Problems." I have chosen to keep the title as translated by Dr. Bejnar.

of various sediments, solution of geological problems on the basis of qualitative/quantitative relationships, arrangement of these mineral assemblages in sedimentary series, and finally, general considerations relating to pure science as well as to practical applications.

There is no question that the analysis of the heavy fraction of sands and muds has proved important in evaluating their value for industrial purposes. Confirmation of a high concentration of some raw mineral (cassiterite, magnetite, monazite, titanium minerals, zircon) can qualify a sand as an economic metal deposit, but clean sand containing little or no heavy minerals can be used for glass making. However, there is no agreement among petrographers and geologists as to the best method of analyzing the heavy fraction of a deposit for solving geological, i.e., stratigraphic, paleogeographic and sedimentologic problems. Agreement is important not only for scientific purposes but also for practical applications—for example, the exploration for crude oil.

In this article the analysis of heavy minerals will be considered for defining the petrographic character of the source of detrital material and for correlation and discrimination of strata. These conclusions are based on more important publications in the world literature and on observations of my co workers and me.

[*Editor's Note:* Material has been omitted at this point.]

IMPORTANCE OF HEAVY MINERAL ANALYSIS FOR STRATIGRAPHIC PROBLEMS

In the early stages of development of sedimentary petrography, heavy minerals were used to determine the age of sedimentary rocks. In the last few decades the general view has been that heavy minerals cannot be considered as important as organic remains, because the same type of minerals can be found in rocks of different ages. Recently, American petrographers Smithson (26), Pettijohn (17) and Russel (25) emphasized that certain minerals occur rarely or not at all in older sediments, and that the number of mineral species decreases with increasing age of sediments. [*Editor's Note:* Smithson is a British subject.] This change in the quantity of heavy-mineral species that occurs with increasing age of the sediment was illustrated by Pettijohn (16—our Fig. 1), based on data found in the literature. K. Łydka disagrees with Pettijohn, saying that the number of mineral species shows some unimportant variation in sediments of different ages (compare Figs. 1 and 2). However, systematic increases in the number of species occurs only in the Tertiary. K. Łydka has collected a large number of observations from Polish and world literature. The disagreement between the two figures would probably disappear if someone recorded much larger numbers of observations. It is worthwhile to mention that as early as 1913 Thoulet (29) emphasized the increasing number of mineral species in sediments of decreasing age.

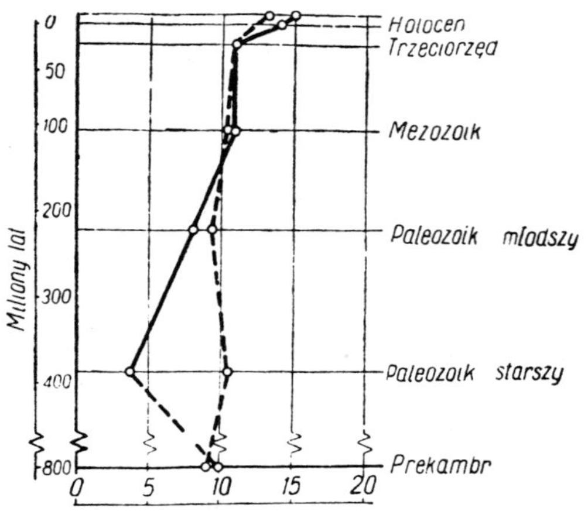

Figure 1. Correlation between the number of heavy-mineral species and age of deposits in millions of years [after Pettijohn (17), 488]. He illustrates this correlation with a diagram represented here by a solid line. On the same figure the dashed line represents the results of K. Łydka's calculations.

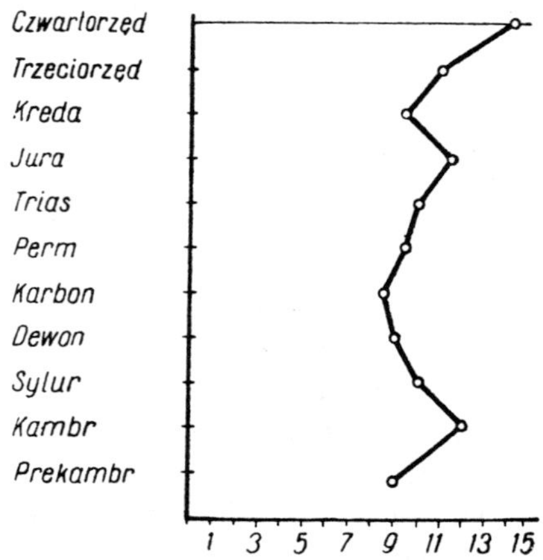

Figure 2. Number of heavy-mineral species in deposits of various ages (as computed by K. Łydka).

A similar view was expressed by Boswell (3), and by Pettijohn, Smithson, and Russel, who attributed this dependence of age to the diagenetic processes of dissolving minerals in sedimentary rocks (17, see Fig. 1). On the other hand, Boswell supposes that the main reason is the increasing petrographic differentiation of the lithosphere with continuing geologic time.

I believe the data gathered so far are not sufficient for making a decisive judgment on this interesting problem. Statistics on the number of mineral species were calculated without taking into consideration the petrographic character of the sediments, depositional conditions and the supposed source of the detrital material. It seems questionable to me that superposition of the various elements shaping the sediments would not eliminate any tendency to regularity in the distribution of heavy minerals. Boswell (3) emphasized that in sediments deposited in geosynclines a tendency exists toward gradual simplification of the composition of the heavy fraction in increasingly younger sediments. According to Pettijohn's scheme, on the other hand, the number of mineral species should decrease in increasingly older sediments. Both tend to eliminate quantitative differences in mineral species in sediments of different ages.

One should not underestimate the fact that authigenic minerals can form during diagenetic processes. Pettijohn (17) indicates that the quantity of anatase and apatite decreases in progressively younger sediments. He also states that both minerals can form by diagenetic processes. Similarly, other heavy minerals in sediments can be authigenic.

At present, we have no reason to believe that analysis of heavy minerals in sedimentary rocks can contribute to solving the problem concerning the age of rocks. However, the majority of current petrographers agree (as I do) that it is possible to separate and correlate the strata of a sedimentary series, designated by age, with the aid of heavy minerals.

Usually, however, applications of this method do not yield good results in the study of terranes composed of multicyclic sediments unless there is new input of material from crystalline rocks. The establishment of succession and correlation of strata can be done by defining quantitative relations in the heavy-mineral fraction or by the degree of rounding of these minerals. However, these studies are elaborate and conclusions must be supported by numerous observations, because there is often doubt and reservation in establishing the quantitative relations.

Differentiation of the heavy fraction took place in the vertical profile of a sedimentary series when the crystalline massif containing abundant nonresistant minerals was eroded and sedimentation occurred in a rapidly sinking geosyncline. Differentiation can be the result of variable materials coming from gradually eroding highlands, in which the petrographic assemblage is seldom uniform. Other influences here may be an aging morphologic terrain and related slow sedimentation as well as more intensive chemical weathering of less resistant minerals. It could also be noted that there is a gradual diminishing of mineral species in the sediments, except for the fact that usually this is compensated for by the diagenetic alteration of minerals in the oldest series. The number of mineral species can actually increase in the vertical profile, from older to younger strata, if rejuvenation of the landscape and uplift of new crystalline terrane occurs as a result of increased tectonic movement. K. Łydka (15) agrees with me that heavy-mineral analysis is important to stratigraphy. He writes that various groups of heavy minerals characterize a rock series formed during various sedimentary cycles.

Tectonic movements, occurring during sedimentation, often cause a change of direction for transported material; during various geologic periods, a sedimentary basin can accumulate detritus from different bordering rock complexes. The work of A. Vatan (33) is a typical example of applying heavy-mineral analyses to describe sedimentary cycles and change of direction of detrital transport during the Cretaceous and Tertiary in Aquitaine. The formations of the upper Cretaceous and Oligocene, deposited at the beginning of the orogeny and during the main phase of diastrophism, were built mostly of detritus weathered from the Pyrenees. This material supplied minerals from granite massifs and their metamorphic aureoles; garnet, tourmaline and zircon are characteristic of this heavy-mineral fraction. However, in the Eocene most of the material came from the erosion of the central massif, which supplied staurolite and kyanite.

In Quaternary stratigraphy, heavy-mineral analysis gives good results. V. Leinz (14) gives the example of the possibility of distinguishing marginal moraines from older and younger glaciations based on olivine content; there is 10 percent in the older moraines and usually less than 1 percent in the younger.

In view of this discussion, it can be said that characteristic differences in qualitative and quantitative heavy-mineral relationships can be found in the vertical profile of some sedimentary series. These relationships permit important conclusions to be reached regarding paleogeographic and sedimentological problems. Moreover, where these relationships are known, there is a basis for distinguishing rocks in a series of various ages.

However, lateral correlation of strata based on heavy-mineral analysis often encounters difficulties. The reason for these difficulties is that the composition of the heavy fraction changes with changes in sedimentary facies. Another reason is segregation of minerals according to their specific gravity, size and shape in sediments of the same age.

Moreover, the major influence on compositional change in the heavy fraction can be attributed to the nature of the depositional process where changes occur in the inflow of the detrital material. This can be most clearly noted in river sediments where the main river channel is supplied by the influx of materials from various directions.

The composition of the heavy fraction, though material may come from the same source, can change laterally depending on grain size. This fact has been confirmed in Polish and world literature (4, 5, 6, 7, 17, 18, 23, 24, 31). Related to this are problems regarding the methods of extracting and analyzing heavy minerals. Baturin (1, 2) points out that the majority of heavy-mineral analyses are based on the fraction 0.05 to 0.25 mm, because this fraction contains the greatest number of minerals and is the most convenient to use. However, according to Preobraženskij (19), data relating the quality and quantity of heavy minerals are a measure of positive value only when such relations of minerals determined in a specific fraction or fractions correspond to the relations in the whole sediment. W. Cogen (6) points out that even with small differences in sediment grain-size, noticeable divergences occur in quantitative relations of heavy minerals.

There are no general rules concerning relationships of the concentration of specific heavy minerals in the various fractions. My own observations on the sediments of the Bug River valley (31) confirm that the quantity of garnet increases with the increase in size of the sand grains between 0.05 and 0.5 mm; however, the amount of amphibole decreases at the same time. Hawkes and Smythe (11) and Pettijohn and Ridge (18) reached the same conclusions. It is easy to explain this fact when one considers the rounded shape of garnet and the flatter shape of amphibole. However, Thiel (28) found that hornblende concentration in the 0.05 to 0.5 mm fractions together increases with increasing grain size.

A difference of opinion exists among petrographers as to whether it is advisable to separate heavy minerals from the separate size fractions or from the entire sediment sample. Many discussions on this subject can be found in the Polish literature (6, 15, 24). [*Editor's Note:* Reference

6 is not a paper from Polish literature. This is a typographical error from the original paper; it is impossible to tell which publication should have been listed here. These discussions conclude that (1) it is not possible to compare the results of the heavy-minerals analyses if they were conducted using different methods and that (2) heavy-mineral separation from the whole sediment sample is best in principle; however, sometimes technical difficulties do occur (fractions above 0.5 mm in grain diameter cannot be analyzed in normal microscopic preparations). In this case it is best to separate minerals from at least two specific size fractions.

Besides problems with selection of methods, there is a basic question of whether the strata correlation test is suitable when one compares rocks of various grain size. Many petrographers and geologists maintain that certain tests are useless and can lead to false conclusions when one considers the differentiation of the heavy-fraction composition that depends on grain size. Griffiths (9), among others, is of the same opinion when considering the importance of petrographic methods in oil exploration.

However, it is worth emphasizing that according to the majority of data, the segregation of heavy minerals, depending on the grain size of a sedimentary facies, leads to changes in the quantitative relations in the composition of the heavy fraction. But it has not been confirmed that any of the principal heavy components completely disappears with this kind of segregation in any given sedimentary facies.

The result of tests on sands and muds of tributaries of the Bug River (31 and unpublished material), based on hundreds of microscopic analyses, leads the author to conclude that garnet and amphibole are the basic components of the heavy fraction of these sediments. The quantitative relations given earlier change depending on the grain size of the sediments, but not only do none of the aforementioned minerals disappear in any sample, but also a preponderance of garnet and amphibole continuously appears as compared to the other components of the heavy fraction. Equally present in almost every sample are heavy minerals accompanying the main components; even the secondary components of the heavy fraction seldom completely disappear.

Based on these data, I conclude that stratigraphic correlation is possible based on heavy minerals, if in a particular sedimentary series there occur friable and less common minerals, so long as these are not inseparably attached to quartz in sedimentary rocks.

Twenhofel and Tyler (32) reached the same conclusions based on studies of beach and shallow water deposits contiguous to Precambrian rocks of a great variety of petrographic character. Results from samples taken over an area of hundreds of kilometers showed that quantitative relations among heavy minerals yielded no correlations. However, these authors are convinced that the presence of certain characteristic heavy minerals of known chemical-physical properties can be important for solving important stratigraphic problems.

In river sediments, where the mineral composition of the heavy fraction can change considerably over small distances as a result of various tributaries from different directions, the petrographer often encounters great difficulty in his attempts at stratigraphic correlation. The problem above was examined earlier by Cayeux (5), later by Krynine (28), Rittenhouse (55) and others. These authors do not reach definite conclusions about stratigraphic correlation of such sediments. A stubborn enthusiast of heavy-mineral analyses can try his luck even in lithologically and genetically complicated terrane, where he can sometimes achieve good results by working systematically.

Based on literature reviewed and my own observations, my conclusions about the usefulness of heavy-mineral analysis for paleogeographic and stratigraphic problems are as follows:

1. Some components of the heavy fraction, particularly components with little resistance to weathering and transportation, can in many cases pinpoint the petrographic character of the source of the detrital material.
2. A systematic analysis of the mineral composition in a vertical profile of a sedimentary series sometimes permits the delineation of strata of different ages. Favorable conclusions can be reached not only when the composition of the heavy fraction changes quantitatively but also when, over time, there occur different characteristic mineral species. Above all, where there are friable minerals, the change in degree of heavy-mineral roundness in layers of various ages can also be of importance in solving stratigraphic problems.
3. Changes in composition of the heavy fraction can sometimes shed light on the course and strength of tectonic movements and allow one to define sedimentation pulses.
4. At present, the least promising application of heavy-mineral analysis appears to be in stratigraphic correlation, because of heavy-mineral compositional changes with sediment facies change and possibilities of mixed assemblages from various sources. However, where characteristic friable minerals appear in a sedimentary series, the systematic analysis of the heavy fraction can sometimes produce favorable results.
5. One should not neglect heavy-mineral analysis of sediment that contains abundant weathered detritus, because one cannot predict if the result of just this particular analysis might produce the key to the solution of important geological problems.

LITERATURE CITED

[*Editor's Note:* Only the references cited in the preceding excerpts are reproduced here. The "in press" reference by Kaszimierz Łydka (15) has not been found in any standard bibliographic reference.]

1. BATURIN V. P. Nekotorye voprosy petrografii v izučenii četvertičnych otłozenij Russkoj platformy. *Trudy Assoc. po izuc. cetvert. perioda,* 3. 1933.
2. ———Paleogeografia po terrigennym komponentam. *ONTI.* 1937.
3. BOSWELL P. G. H. *On the mineralogy of sedimentary rocks. A series of essays and bibliography.* London 1933.

4. BURRI C. Sedimentpetrographische Untersuchungen an alpinen Flusssanden I: Die Sande des Tessin. *Schweiz. Min. Petr. Mitt.*, bd. IX, H. 2. 1929.
5. CAYEUX L. Les roches sédimentaires de France. *Roches siliceuses.* Paris. 1929.
6. COGEN W. M. Some suggestions for heavy mineral investigations of sediments. *J. Sedim. Petrol.*, vol. V. 1935.
7. DAUBRÉE A. Sur la distribution de l'or dans la plaine du Rhin et sur l'extraction de ce métal. *Ann. Mines*, pp. 3-26. Paris, 1846.
8. FOUQUÉ F. Santorin et ses éruptions. 1879.
9. GRIFFITHS J. C. Sedimentary petrography and oil industry. *J. Sedim. Petrol.*, vol. 17, no. 1. 1947.
10. HAÜY R. J. *Traité de minéralogie.* I-ère éd. Paris 1801.
11. HAWKES L. and SMYTHE J. A. Garnet-bearing sands of the Northumberland Coast. *Geol. Mag.*, vol. 68, pp. 345-361, 1931.
12. KRYNINE P. D. Provenance versus mineral stability as a controlling factor in the composition of sediments. *Bull. Geol. Soc. Am.* vol. 53, pp. 1850-1851.
13. LASSAIGNE J. L. Analyse d'un sable titanifère de Madagascar. *Ann. Mines*, VI, pp. 457-458. Paris 1821.
14. LEINZ V. Ein Versuch Geschiebemergel nach dem Schwermineraliengehalt stratigraphisch zu gliedern. Zsch. *Geschiebeforsch.*, bd. 9. 1933.
15. ŁYDKA K. Studia petrogiczne petrograficzne nad permokarbonem krakowskim. *W druku.*
16. PETTIJOHN F. J. Persistence of heavy minerals and geologic age. *J. Geol.*, vol. 49, pp. 610-625. 1941.
17. ———*Sedimentary rocks.* New York 1949.
18. PETTIJOHN F. J. and RIDGE J. D. A mineral variation series of beach sands from Cedar Point, Ohio. *J. Sedim. Petrol.*, vol. 3, no. 2, pp. 92-94. 1933.
19. PREOBRAŽENSKIJ J. O podcetach mineralnogo petrograficeskogo sostava oblemocnych osadocnych porod. *Akad. Beljankinu k semidesiatil. so dnja rozd. Akad. Nauk SSSR*, Inst. Geol. Nauk. 1946.
20. RÉAUMUR R. A. F. Esais de l'histoire des rivieres et des ruisseaux du Royaume qui roulent des Pailletes d'Or. *Mem. Acad. Roy.*, pp. 68-88. 1718.
21. RETGERS J. W. Ueber die mineralogische und chemische Zusammensetzung der Dünensande Hollands. *N. Jb. Miner.* etc. 1895.
22. RITTENHOUSE G. The transportation and deposition of heavy minerals. *Bull. Geol. Soc. Am.*, vol. 54, pp. 1725-1780. 1943.
23. RUBEY W. W. The size distribution of heavy minerals within waterlaid sandstone. *J. Sedim. Petrol.*, vol. 3, pp. 3-29. 1933.
24. RUCHIN L. B. *Osnovy litologii.* Leningrad-Moskva 1953.
25. RUSSEL [Russell] R. D. The size distribution of minerals in Mississippi River sands. *J. Sedim. Petr.* vol. 56, pp. 125-142. 1935.
26. SMITHSON T. [F., not T.] The mineralogy of arenaceous deposits. *Sci. Progress*, January 1950.
27. SORBY H. C. On the microscopical characters of sands and clays. *Monthly Microsc. J.*, XVII, pp. 119-136. 1877.
28. THEIL G. A. Glacio-lacustrine sediments reworked by running water. *J. Sedim. Petrol.*, II, pp. 68-75. 1932.
29. THOULET J. Notes de lithologie sous-marine. *Ann. Inst. Oceanogr.*, V, f. 9, 1913.
30. THÜRACH H. Ueber das Vorkommen mikroskopischer Zircone und Titanmineralien in den Gesteinen. *Verh. Phys. Med. Ges. Würzburg*, XVIII, pp. 203-284. 1884.
31. TURNAU-MORAWSKA M. Utwory rzeczne doliny Bugu między Terespolem a Wyszkowem (Fluvial deposits in the Bug valley between Terespol and Wyszkow). *Biul. I.G.*, 68. 1952.
32. TWENHOFEL W. H. and TYLER S. A. *Methods of study of sediments.* New York and London 1941.
33. VATAN A. Rythmes de sédimentation en Aquitaine au Crétacé et au Tertiaire. *Proc. of Section C. Geol. Congress*, Great Britain 1948, Part IV.

21

INTRASTRATAL SOLUTION AND NON-OPAQUE HEAVY MINERALS IN SHALES[1]

HARVEY BLATT
School of Geology and Geophysics, University of Oklahoma, Norman, Oklahoma
BERRY SUTHERLAND
Department of Geology, University of Houston, Houston, Texas

ABSTRACT

Non-opaque heavy minerals have been examined from contemporaneous sandstone-shale pairs at six horizons in Tertiary sediments of the Texas Gulf Coastal Plain. The total weight of heavy minerals in the shales is much less than in their associated sandstones but the same grain sizes of heavy minerals are present in both rock types. Sediment in the coarse and medium silt sizes contains the same variety of minerals as the very fine sand size.

In these relatively young shales the variety of minor minerals (those forming 0.1–5.0 percent of the non-opaque suite) is only slightly larger than in the contemporaneously deposited sandstones. However, the number of species present only in trace amounts (one or two grains among the 5400 grains counted) is significantly larger in the shales than in contemporaneously deposited sandstones. Also, the sandstones contain much larger percentages of altered grains than do the shales. These results apparently are due to the greater permeability of the sandstones and suggest that shales rather than sandstones should be examined in heavy-mineral studies. Unstable species post-depositionally dissolved in sandy beds may be preserved in contemporaneously deposited shales.

INTRODUCTION

The heavy-mineral fraction of clastic rocks contains nearly all of the mineral species present in these rocks. Consequently, this part of the sediment is of great value for correlation, delineation of petrographic provinces, and identification of the types of crystalline rocks from which the sediment was derived. Unfortunately, many of the more informative but chemically less stable minerals can be dissolved by intrastratal fluids during diagenesis. Although the world-wide importance of this process is still debated (Gazzi, 1965; van Andel, 1959; Krynine, 1942; Pettijohn, 1941), all agree that it does occur at least locally.

At present there usually is no way to be certain that a given heavy mineral suite has not been impoverished or altered by intrastratal solution. It is only possible to be certain that it has, the evidence being the often described saw-toothed aspect of grain outlines. If one is fortunate enough to locate an early-formed concretion within the sandstone being studied, comparison between the heavies in the concretion and those in the surrounding sand may be fruitful (Bramlette, 1941). But such concretions are seldom present where needed.

[1] Manuscript received July 15, 1968.

The chief purpose of the present study was to find a general solution to the problem by means of an "end run" around sandstones and into the nearly impermeable safety of contemporaneously deposited shales. We also hoped to evaluate the potential usefulness of silt-sized heavy minerals in both shales and sandstones for correlation and provenance studies. Silt-sized heavies are nearly always ignored in heavy mineral studies yet it may be that valuable and unique data can be obtained from these grains. Therefore, during the course of this investigation of non-opaque heavy minerals in contemporaneous sandstone-shale pairs three questions were asked: (1) Are a large number of mineral species present in the silt fraction of shales and fine sandstones and are these minerals present in sufficient amounts to be useful? (2) Assuming no intrastratal solution of heavies in either the shales or the sandstones, is the variety of heavies in the sandstones the same as in the associated shales, size for size? If so, correlation by the use of heavies in shales may be possible in down-dip basinal sections where sandstones may be absent and useful fossils frequently are rare. (3) If intrastratal solution has occurred in the sandstones, have contemporaneous shales, because of their relative impermeability, preserved the original suite of minerals deposited with the sand-shale complex?

METHOD OF STUDY

Six sampling sites were chosen from the road log by Eargle and Foust (1962) of Tertiary units on the Gulf Coastal Plain in south Texas. The rocks at five of the six sites are of Eocene age; the sixth is Pliocene. Maximum depth of burial of these rocks probably has not exceeded 2,000 feet. Each outcrop was carefully chosen to obtain a sandstone and a shale or mudstone in direct contact with each other and most of the contacts displayed some gradation between the two lithologies.

Two spot samples were collected at each outcrop, one sample from a sandstone and one from a shale or mudstone. All samples were taken near the contact of the two lithologies and the vertical distance between the two samples never exceeded 12 inches; most were taken within six inches of each other. Care was taken to obtain only fresh samples uncontaminated by soils or weathering products. Four of the six sandstones sampled were partly lithified by opal of Tertiary age; two by microcrystalline calcite of probable recent origin.

In the laboratory, sandstone samples were disaggregated using a rubber pestle. A standard sample splitter was used to obtain approximately 50 grams of sand which was weighed and sieved using screen sizes 0.0ϕ, 1.0ϕ, 2.0ϕ, 3.0ϕ, 4.0ϕ, and 4.75ϕ. The fractions retained on each screen and in the pan were weighed to the nearest 0.01 gram.

Shale and mudstone samples were allowed to dry at room temperature for a period of several weeks. They then were fragmented by pounding with a rubber pestle until green-pea sized aggregates were obtained. Samples were split as above, but a 200-gram split was needed to obtain sufficient amounts of heavy minerals for grain mounts. The ratio weight of heavies per gram of sand/weight of heavies per gram of shale was higher for the sand-shale pairs in which the shales were exceptionally rich in clay and values reached approximately 500/1 for one sand-shale pair. Each split was placed in a peptizing solution of N/100 sodium oxalate for a period of at least two weeks. After soaking, each sample was disintegrated using an ultrasonic vibrator (Gipson, 1963), wet sieved at the same sieve intervals used for the sandstones, dried and weighed. Tap water was used for wet sieving and, therefore, it was necessary to redisperse the pan fractions ultrasonically. Sediment finer than approximately 10 microns was then removed by repeated decantations. Some grains smaller than 10 microns in long dimension were recovered in later heavy mineral separations because the 10-micron size used for decantation was determined from settling velocity tables for spheres of specific gravity 2.65. Also, decantation is an imprecise procedure. However, we found that the lower size limit of positive identification occurs at approximately 10 microns. Refraction effects on smaller particles preclude accurate determination of optical properties. A few mineral species, particularly tourmaline, can be identified down to 5 microns because they possess deep coloration or strong pleochroism. No grains smaller than 10 microns were included in the grain counts.

Heavy mineral separates were prepared from three size fractions of each of the 12 samples. These fractions are 3.0-4.0ϕ, 4.0-4.75ϕ, and 4.75-6.5ϕ (10 microns). These size ranges contain the bulk of the heavy minerals in clastic rocks.

Bromoform and standard laboratory techniques were used for heavy mineral separations of sand-sized grains. Separations of silt-sized grains were made using a centrifuge. No special centrifuge tube was used. The heavy mineral fraction was taken from the centrifuge tube by means of a thin-walled pipette inserted carefully through the crust of light minerals and lowered slowly to the bottom of the tube. During lowering several air bubbles were forced out of the orifice of the pipette by use of a small rubber bulb. This removed the light minerals from the pipette orifice. After slight stirring of the heavies by the pipette they were drawn into its orifice by releasing pressure on the rubber bulb. The pipette was then withdrawn and its contents deposited on glazed filter paper in a glass funnel. Heavies and bromoform adhering to the inside wall of the pipette were removed with acetone and all heavy minerals were soaked in acetone to remove bromoform adhering to grain surfaces. It is important to use glazed filter paper (ordinary ditto paper will do) to line the glass funnel; otherwise the finer-sized heavies become trapped among the fibers of the paper and are very difficult to remove.

Warm dilute hydrochloric acid containing stannous chloride was used to remove iron oxide coatings from the mineral grains and to eliminate carbonate contamination of the heavy-mineral fraction.

Grain mounts were made of each size fraction by imbedding the grains in Lakeside 70 cement. Grain counts were made of each slide along linear traverses to a total of 150 non-opaque heavy-mineral grains. Particular attention was given to grain color, elongation, roundness, alteration, inclusions, overgrowths, and zoning. Only alteration is discussed in the present paper.

Each of the heavy-mineral grains was classified as either altered or unaltered. Altered grains (fig. 1) are those which display either solution embayments, etching and corrosion of surfaces, alteration products, or coatings which remain after acid cleaning. There was no apparent difference between shales and sandstones in proportion of coated grains among altered grains. Therefore, the coatings probably are aluminosilicate precipitates rather than clay coatings, as clay would be dislodged by the ultrasonic treatment to which the shales were subjected.

Heavies which are coated or "weathered" or altered to such an extent that they cannot be identified were tabulated as "alterites." Thus, "alterite" in this work is not synonymous with "altered grain" but is a category equivalent to garnet, topaz, and so forth. Unaltered grains show no evidence of any type of alteration.

DATA AND DISCUSSION

Grain Size Analyses

Figure 2 illustrates the size frequency distributions of the six deltaic sandstones examined. Five are fine-grained; one is medium-grained. Sorting is moderate to poor (σ_I ranges from 0.6 to 1.4) with no apparent correlation between mean grain size and sorting. The light mineral fractions average 72 percent quartz, 16 percent rock fragments, and 12 percent feldspar. The rock fragments are nearly all of certain volcanic origin (glass shards, trachytic felsic fragments) as are many of the feldspar grains (sanidine, zoned plagioclase). However, the proximate source for many of these "unstable" grains may have been older Tertiary sediments.

The amount of sand in the shales ranges from 0.6 to 49.6 percent and averages 17.1. The percentage of sediment coarser than 10 microns, which should include nearly all of the non-clay material in the shales, ranges from 5.9 to 54.9 and averages 28.8 percent. Shaw and Weaver

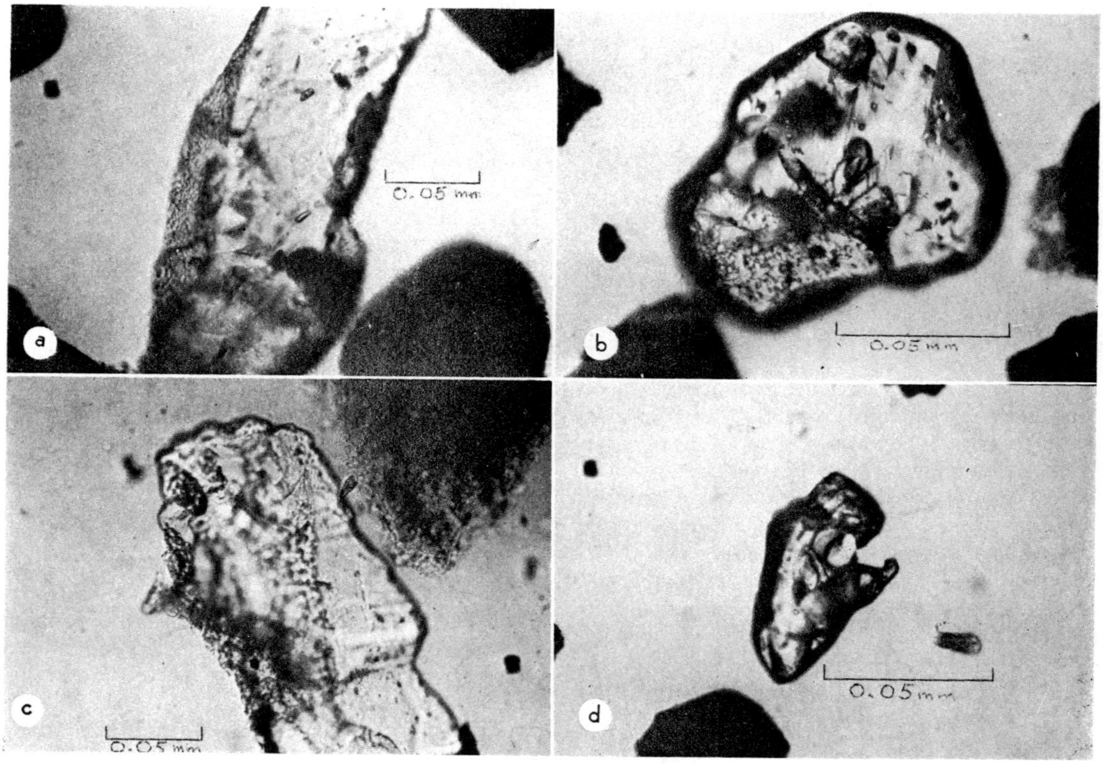

FIG. 1.—Photomicrographs of solution features on heavy-mineral grains. (a) Very fine sand-sized staurolite grain (sample no. 12) showing micro-papillary structure on left side. (b) Very fine sand-sized staurolite grain (sample no. 10) showing intense corrosion on lower left corner. Crystal outlines still clearly visible in upper right. (c) Very fine sand-sized epidote grain (sample no. 6) showing corrosion of borders and hummocky surface (partly out of focus). (d) Coarse silt-sized zircon grain (sample no. 3) showing locally intense corrosion.

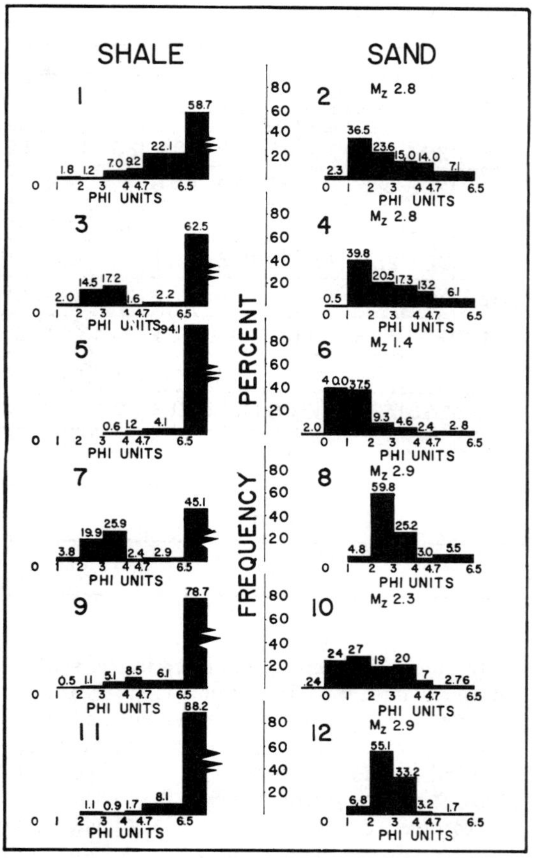

FIG. 2.—Size frequency distribution histograms for the 6 shales and 6 sandstones examined. The contemporaneous shale-sandstone pairs are numbered consecutively; that is, 1 and 2; 3 and 4; etc. M_z is graphic mean size. The value of this parameter is indeterminate for shale samples.

(1965) analyzed 400 shales from the United States and Canada and determined that, on a carbonate-free basis, the non-clay fraction averaged 36.6 percent. Therefore, the mean non-clay content of the shales examined in the present work is comparable to that of the average shale.

Heavy Minerals

Figure 3 illustrates the percentages of non-opaque heavy minerals for each size fraction of each sample. It is apparent that the percentage of each heavy mineral in each size often differs considerably between a sandstone and its associated shale. However, the same scale of variation is present among the sandstones examined despite the fact that five of the six are of the same geologic age (Eocene) and were obtained from a geographically small area. This suggests that the difference in abundance of any one mineral between a sand and a contemporaneous shale or between the nearly contemporaneous sandstones may be due largely to hydraulic causes. Carroll (1957) suggested a combination of hydraulics and chance variation to explain similar variations in heavy-mineral abundances of geographically adjacent samples of recent stream sediment in South River, Virginia.

Topaz is the minor non-opaque heavy-mineral constituent which varies most in abundance and its sporadic occurrence (absent in most slides but as high as 56 percent in one slide) accounts for most of the percentage differences found in the "others" group of trace species in figure 3.

Figure 3 further reveals that the number of heavy-mineral species present decreases only slightly with decreasing grain size down to 10 microns (see also table 7). This result is in marked contrast to the impression one frequently gets from published studies of medium, fine, and very fine sand-sized non-opaque heavy minerals. These studies often leave the impression that with decreasing grain size zircon becomes increasingly abundant and may be the only non-opaque heavy mineral in the silt fraction. The great variety of heavies in the silt fraction in our sediments parallels the results of studies of ancient sediments by Doeglas (1949), Swineford and Frye (1951), and Beveridge (1960); and studies of modern sediments by Haseman and Marshall (1945), Hutton (1950), Game (1964), and Fan (1965). These authors also reported a great variety of heavy minerals in the silt fraction of sediments. Haseman and Marshall reported the modal grain size of heavy minerals to be fine silt-sized in soils they examined. Beveridge reported the modal size of heavies to be coarse silt in several sandstones in California. Fan found the silt fraction of Santa Clara River (California) sands to be more representative of the total sediment composition than either the sand or clay fraction.

The four most abundant non-opaque heavy mineral species in the Tertiary rocks of our study are zircon (shales 47 percent; sandstones, 60 percent), tourmaline (shales 10 percent; sandstones 4 percent), epidote (shales 6 percent; sandstones 6 percent), and garnet (shales 6 percent; sandstones 5 percent). No other mineral except topaz averages more than 3 percent. It averages 8.4 percent because of the flood occurrence (42 percent) in sample 9. Tables 1–4 show, for each of the four abundant mineral species, the relationships among rock type, grain size, and the proportions of unaltered (un) and altered (a) mineral grains. As noted earlier, there is no apparent correlation between lithol-

TABLE 1.—*Percentage of zircon grains in each size fraction of each sample.*

	SHALES										SANDS								
Sample No.	3.0 – 4.0 phi			4.0 – 4.75 phi			4.75 – 6.5 phi			Sample No.	3.0 – 4.0 phi			4.0 – 4.75 phi			4.75 – 6.5 phi		
	un	a	un/a	un	a	un/a	un	a	un/a		un	a	un/a	un	a	un/a	un	a	un/a
1	30.4	17.8	1.7	33.0	20.5	1.6	36.3	13.2	2.8	2	20.4	44.9	0.5	26.4	29.0	0.9	26.8	22.2	1.2
3	18.5	5.3	3.5	49.5	19.8	2.5	46.9	17.8	2.6	4	21.8	51.5	0.4	20.5	24.4	0.8	27.1	24.4	1.1
5	17.4	—	—	40.9	10.1	4.0	52.6	2.4	21.9	6	15.2	33.0	0.5	29.7	32.3	0.9	36.9	33.7	1.1
7	31.0	16.0	1.9	44.7	28.8	1.6	44.8	20.5	2.2	8	32.3	25.8	1.3	30.3	35.0	0.9	24.4	8.6	2.8
9	9.9	3.2	3.1	12.6	5.9	2.1	48.2	9.2	5.2	10	32.3	37.0	0.9	23.8	29.0	0.8	27.0	38.3	0.7
11	30.4	7.6	4.0	33.8	8.1	4.2	44.1	12.5	3.5	12	25.1	38.3	0.7	25.7	40.9	0.6	34.4	42.2	0.8
Σ	137.6	49.9	—	214.5	93.2	—	272.9	75.6	—	Σ	141.7	230.5	—	156.4	190.6	—	176.6	169.4	—
Average	22.9	8.3	2.8	35.8	15.5	2.3	45.5	12.6	3.6	Average	24.5	38.4	0.6	26.1	31.8	0.8	29.4	28.2	1.0
Total Average Unaltered	34.7										26.7								
Total Average Altered	12.1										32.8								
Total Average Un/A	2.9										0.8								

un = unaltered a = altered

ogy and percentage abundance of individual mineral species. Nor is there a consistent relationship between abundance of the four minerals and grain size. Zircon becomes increasingly abundant in the shales with decreasing size, but is essentially constant in amount in the sandstones through the very fine sand, coarse silt, and medium silt sizes. Tourmaline, epidote, and garnet show no significant trend in abundance with changing grain size, either in the shales or in the sandstones. In summary, no readily interpretable relationships are evident among lithology, grain size, and percentage abundance of minerals. The numbers appear to be simply "noise" resulting from random fluctuations.

However, one relationship does stand out. The ratio of unaltered to altered grains (un/a) for all four mineral species is much larger in the

TABLE 2.—*Percentage of tourmaline grains in each size fraction of each sample.*

	SHALES										SANDS								
Sample No.	3.0 – 4.0 phi			4.0 – 4.75 phi			4.75 – 6.5 phi			Sample No.	3.0 – 4.0 phi			4.0 – 4.75 phi			4.75 – 6.5 phi		
	un	a	un/a	un	a	un/a	un	a	un/a		un	a	un/a	un	a	un/a	un	a	un/a
1	10.6	—	—	8.0	0.6	13.3	11.2	—	—	2	2.0	—	—	7.3	—	—	4.4	—	—
3	19.2	1.3	14.8	5.3	—	—	3.4	0.6	5.7	4	—	—	—	4.0	0.6	67	8.6	1.3	6.6
5	11.1	—	—	3.1	—	—	4.8	—	—	6	—	—	—	0.6	—	—	1.3	—	—
7	24.0	1.0	24.0	6.8	—	—	3.3	—	—	8	1.4	0.6	2.3	—	—	—	1.3	—	—
9	1.3	—	—	4.0	—	—	6.0	1.3	4.6	10	0.6	—	—	12.6	0.6	21.0	5.3	2.0	2.7
11	9.1	—	—	20.3	—	—	19.2	—	—	12	4.6	—	—	4.6	—	—	2.0	—	—
Σ	75.3	2.3	—	47.5	0.6	—	47.9	1.9	—	Σ	8.6	0.6	—	29.1	1.2	—	22.9	3.3	—
Average	12.6	0.4	31.5	7.9	0.1	79.0	8.0	0.3	26.7	Average	1.4	0.1	14.0	4.8	0.2	24.0	3.8	0.6	6.3
Total Average Unaltered	9.5										3.3								
Total Average Altered	0.3										0.3								
Total Average Un/A	31.7										11.0								

un = unaltered a = altered

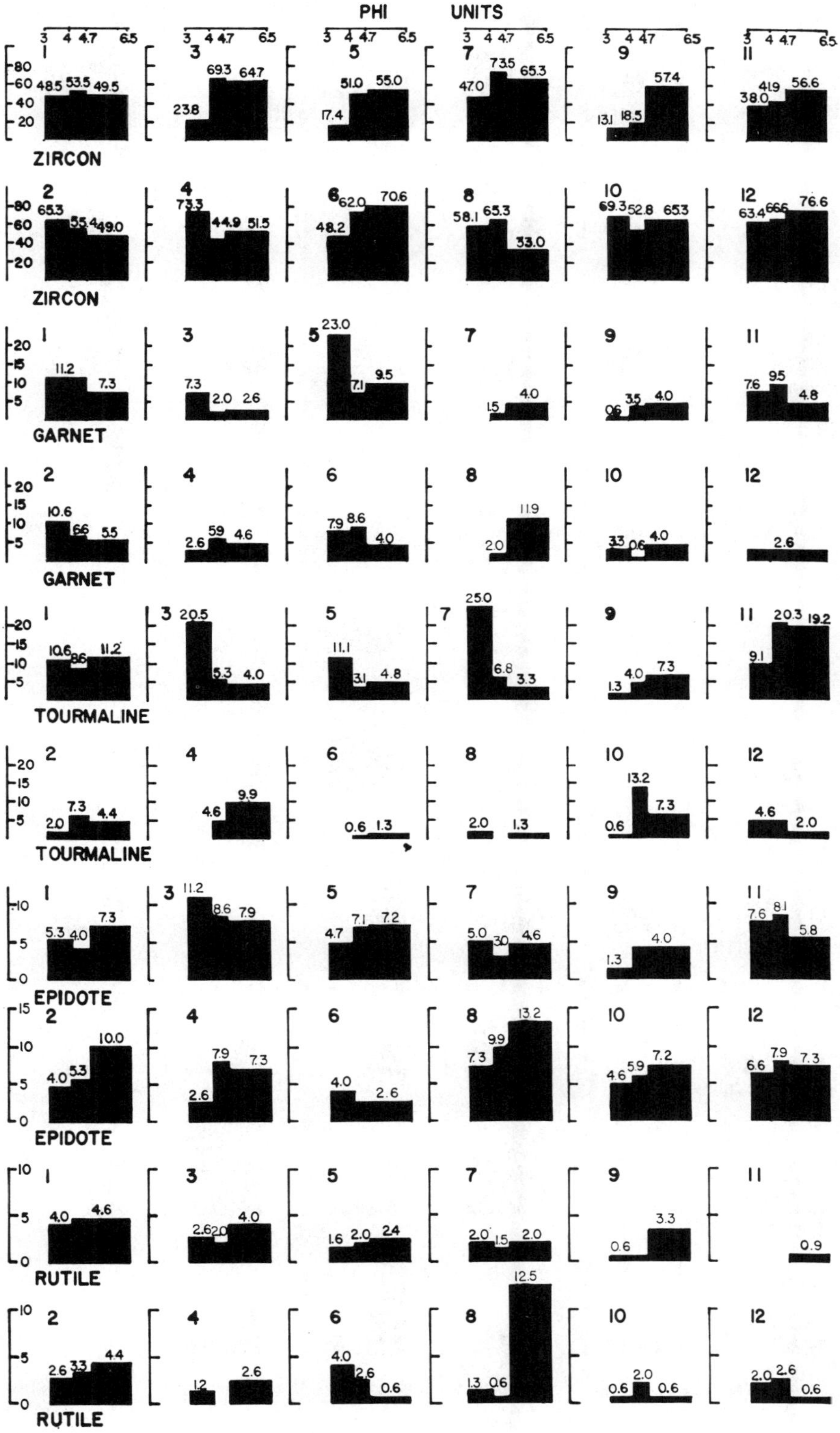

FIG. 3.—Histograms of relative amounts of major and minor non-opaque heavy minerals in the 3 size

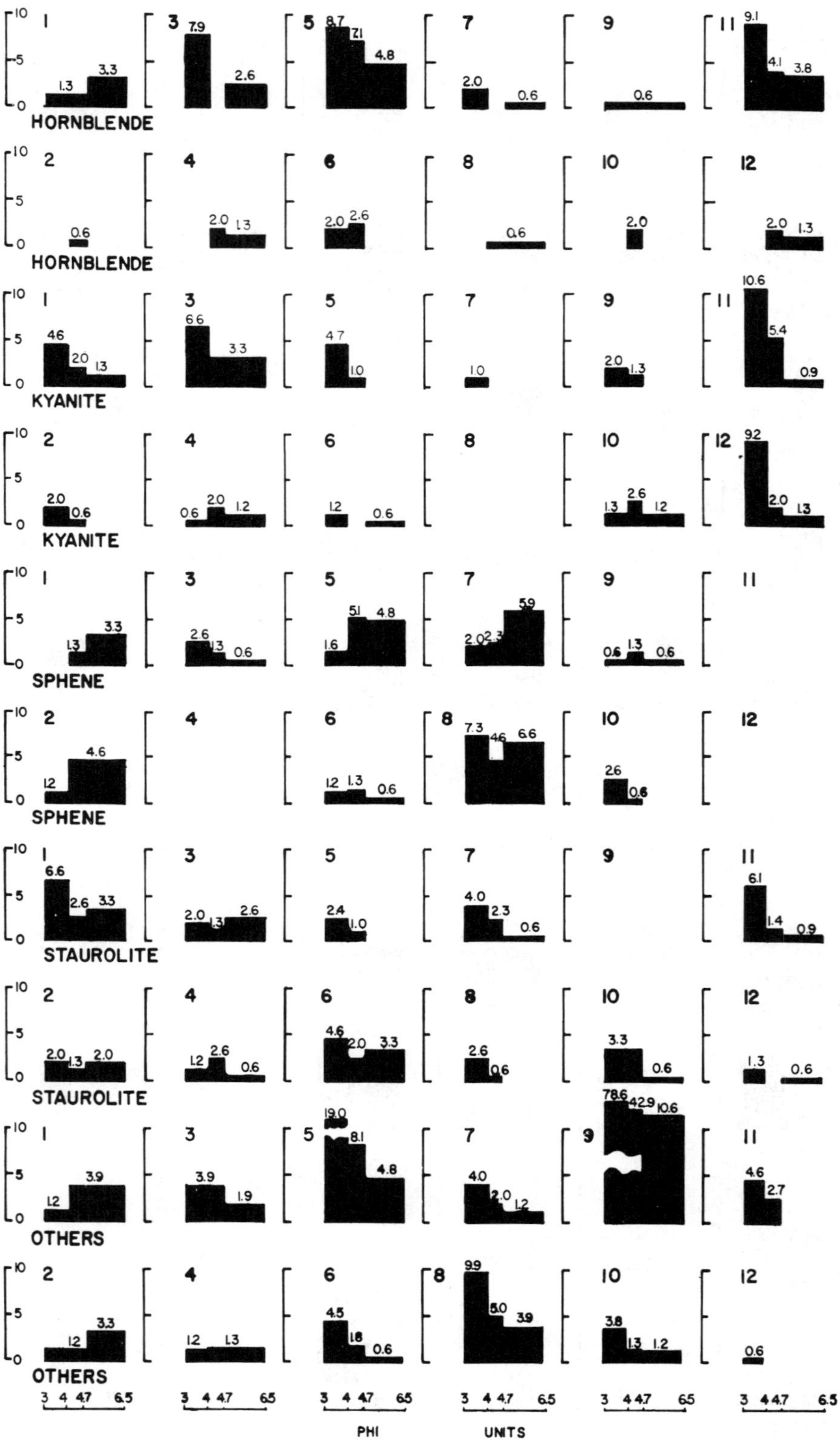

fractions examined. No trends are apparent among lithologies, samples, minerals, or grain sizes.

TABLE 3.—*Percentage of epidote grains in each size fraction of each sample.*

	SHALES										SANDS								
Sample No.	3.0–4.0 phi			4.0–4.75 phi			4.75–6.5 phi			Sample No.	3.0–4.0 phi			4.0–4.75 phi			4.75–6.5 phi		
	un	a	un/a	un	a	un/a	un	a	un/a		un	a	un/a	un	a	un/a	un	a	un/a
1	4.0	1.3	3.1	2.7	1.3	2.1	6.0	1.3	4.6	2	2.7	1.3	2.1	3.3	2.0	1.7	8.9	1.1	8.1
3	9.2	2.0	4.6	6.6	2.0	3.3	5.9	2.0	3.0	4	—	2.6	—	5.9	2.0	3.0	5.3	2.0	2.7
5	4.7	—	—	6.1	1.0	6.1	7.2	—	—	6	3.4	0.6	5.7	2.0	0.6	3.3	1.3	1.3	1.0
7	4.0	1.0	4.0	2.2	0.8	2.8	3.3	1.3	2.5	8	5.3	2.0	2.6	7.9	2.0	4.0	2.6	0.6	21.0
9	1.3	—	—	2.0	2.0	1.0	4.0	—	—	10	1.3	3.3	0.4	4.6	1.3	3.5	4.6	2.6	1.8
11	7.6	—	—	8.1	—	—	5.8	—	—	12	4.6	2.0	2.3	5.9	2.0	3.0	5.3	2.0	2.6
Σ	30.8	4.3	—	27.7	7.1	—	32.2	4.6	—	Σ	17.3	11.8	—	29.6	9.9	—	38.0	9.6	—
Average	5.1	0.7	7.3	4.6	1.2	3.8	5.4	0.8	6.8	Average	2.9	2.0	1.4	4.9	1.6	3.1	6.3	1.6	3.9
Total Average Unaltered	5.0										4.7								
Total Average Altered	0.9										1.7								
Total Average Un/A	5.6										2.8								

un = unaltered a = altered

shales than in the sandstones. The difference between shale and sandstone in ratio for zircon is a factor of 3.6; for tourmaline, 2.9; for epidote, 2.0; for garnet, 2.2. The higher ratio of un/a in shales as contrasted to sandstones is also true of each size fraction considered individually. For example, in the zircons un/a is 2.8 in the 3-4ϕ fraction of shales, but only 0.6 in sandstones; in the 4.0-4.75ϕ fraction, 2.3 in shales but only 0.8 in sandstones; in the 4.75-6.5ϕ fraction, 3.6 in shales and 1.0 in sandstones. In fact, the relationship beween lithology and the un/a ratio is so clear that it is true of nearly every individual sample in each size fraction. Of 72 comparisons (4 minerals × 6 sandstone-shale pairs × 3 sizes) only 9, or 12 percent, fail to have un/a larger in shales than in sandstones. This result seems still more remarkable when one considers that a high proportion of these grains probably are polycyclic in origin and have most

TABLE 4.—*Percentage of garnet grains in each size fraction of each sample.*

	SHALES										SANDS								
Sample No.	3.0–4.0 phi			4.0–4.75 phi			4.75–6.5 phi			Sample No.	3.0–4.0 phi			4.0–4.75 phi			4.75–6.5 phi		
	un	a	un/a	un	a	un/a	un	a	un/a		un	a	un/a	un	a	un/a	un	a	un/a
1	9.2	2.0	4.6	7.9	3.3	2.4	6.0	1.3	4.6	2	7.3	3.3	2.2	4.6	2.0	2.3	3.3	2.2	1.5
3	5.3	2.0	2.7	2.0	—	—	2.6	—	—	4	2.0	0.6	3.3	3.9	2.0	2.0	2.6	2.0	1.3
5	20.6	2.4	8.6	6.1	1.0	6.1	9.5	—	—	6	3.3	4.6	0.7	4.0	4.6	0.9	3.4	0.6	5.7
7	—	—	—	0.7	0.8	0.9	4.0	—	—	8	—	—	—	0.7	1.3	0.5	9.9	2.0	5.0
9	—	0.6	—	—	3.5	—	0.7	3.3	0.2	10	3.3	—	—	0.6	—	—	2.0	2.0	1.0
11	7.6	—	—	8.1	1.4	5.8	4.8	—	—	12	2.6	—	—	2.0	0.6	3.3	2.0	0.6	3.3
Σ	42.7	7.0	—	24.8	10.0	—	27.6	4.6	—	Σ	18.5	8.5	—	15.8	10.5	—	23.2	9.4	—
Average	7.1	1.2	5.9	4.1	1.7	2.4	4.6	0.8	5.8	Average	3.1	1.4	2.2	2.6	1.8	1.4	3.9	1.6	2.4
Total Average Unaltered	5.3										3.2								
Total Average Altered	1.2										1.6								
Total Average Un/A	4.4										2.0								

un = unaltered a = altered

TABLE 5.—*Percentages of unaltered and altered non-opaque heavy minerals in sandstones and shales. For all mineral categories un/a is larger for shales than for sandstones. (Each numerical value is an unweighted average of the percentage in each of the 3 size grades examined.)*

Heavy Mineral Species	SHALES un	SHALES a	SHALES un/a	SANDS un	SANDS a	SANDS un/a
Zircon	34.7	12.1	2.9	26.7	32.8	0.8
Garnet	5.3	1.2	4.4	3.2	1.6	2.0
Tourmaline	9.5	0.3	31.7	3.4	0.3	11.3
Epidote	5.0	0.9	5.6	4.7	1.8	2.6
Staurolite	2:0	—	—	1.7	0.1	17.0
Rutile	2.1	0.1	21.0	2.3	0.2	11.5
Hornblende	3.2	—	—	0.7	—	—
Kyanite	2.4	0.2	12.0	0.8	0.6	1.3
Sphene	1.6	0.2	8.0	1.4	0.6	2.3
Clinozoisite	1.5	—	—	0.9	0.2	4.5
Monazite	0.5	0.1	5.0	0.6	0.3	3.0
Zoisite	0.4	0.1	4.0	0.1	0.1	1.0
Topaz	5.3	3.1	1.7	0.2	0.1	2.0
Alterite	—	6.3	—	—	12.3	—
Others	1.9	—	—	2.3	—	—
Total Percentage	75.4	24.6		49.0	51.0	

un = unaltered a = altered

likely resided in older sandstones before being re-eroded and deposited in the shale or sandstone where we found them. That is, many of the grains have likely been subjected to surface weathering and intrastratal solution in earlier cycles. The fact that we can detect un/a differences of large magnitude in our samples despite "background alteration noise" argues strongly for the importance of intrastratal solution over geologically short periods of time.

TABLE 6.—*Percentage of alterites in each size fraction of each sample. Of 18 comparisons (6 sandstone-shale pairs × 3 size ranges) the percentage of alterites is greater in the shales in only one (4.0-4.75φ fraction of pair 9–10).*

SHALES SAMPLE NO.	3.0-4.0 phi	4.0-4.75 phi	4.75-6.5 phi	SANDS SAMPLE NO.	3.0-4.0 phi	4.0-4.75 phi	4.75-6.5 phi
1	3.3	5.3	2.0	2	7.9	11.2	11.1
3	7.9	3.3	2.6	4	15.8	25.8	16.5
5	4.7	4.1	4.8	6	19.8	11.9	10.6
7	4.0	6.1	8.6	8	9.9	9.2	11.9
9	8.6	25.8	9.9	10	12.5	13.9	10.6
11	3.3	5.4	3.8	12	6.6	9.9	5.3
AVERAGE %	5.3	8.3	5.3	AVERAGE %	12.1	13.7	11.0
TOTAL AVERAGE %	6.3				12.3		

TABLE 7.—*Number of minor and trace non-opaque heavy mineral species in shales and sandstones. (Excluded are zircon, tourmaline, epidote, garnet, and alterite.)*

SHALES Sample Number	3.0-4.0 phi	4.0-4.7 phi	4.7-6.5 phi	SANDS Sample Number	3.0-4.0 phi	4.0-4.7 phi	4.7-6.5 phi
1	8	9	8	2	5	6	6
3	9	6	8	4	4	4	5
5	9	9	4	6	8	6	6
7	10	5	7	8	6	7	5
9	7	6	4	10	6	7	5
11	5	4	4	12	6	3	3
Average	8.0	6.5	5.8	Average	5.8	5.5	5.0
Total Average	6.8				5.4		

It is noteworthy that the most "altered" heavy mineral species is zircon (table 1), which is surprising in view of its accepted ultra-stable character in the sedimentary environment. Probably many of the idiomorphic zircon grains (31 percent of total zircon) are of volcanic origin (see Callender and Folk, 1958) and possibly much of the etching on the surfaces of these grains is of volcanic origin. However, a volcanic origin is not possible for two of the other three abundant minerals (tourmaline, epidote) and is rather unlikely for the third (garnet). The etching on the surfaces of these minerals likely reflects sedimentary processes.

Nine other mineral species were identified in the rocks studied (table 5) and in every instance the un/a ratio in the shales is equal to or larger than the ratio in the sandstones. Also, grains so highly altered or rotted that they could not be identified (alterite; table 6) are twice as abundant in the sandstones as in the shales. There is no correlation between grain size and abundance of alterites.

The "others" group includes grains of beryl, andalusite, apatite, spinel, and corundum, traces of which were found in some samples. It is significant that of 15 occurrences of these rare heavies, 12 were found in shales.

The data presented above have only one reasonable interpretation. Intrastratal solution by underground waters has altered the surface texture and internal structure of mineral grains in the sandstones but has left these features either unchanged or altered only to a minor degree in the shales. Because the rocks studied are geologically young (Tertiary), we would not expect the number of different mineral species to differ greatly, if at all, between shales and sandstones. Table 7 shows the relationship between lithology and number of minor and trace mineral species present (excluding zircon, tourmaline, epidote, garnet, and alterite). There is a suggestion that the average shale contains a larger number of minor species than the sandstones, but the difference is not pronounced enough to have statistical significance. Presumably the magnitude of the difference will increase as intrastratal solution proceeds. A study of Paleozoic or Precambrian sandstone-shale pairs would be useful.

CONCLUSION

Results of this investigation indicate that Tertiary shales on the Texas Gulf Coastal Plain contain appreciable amounts of silt-sized and fine sand-sized non-opaque heavy minerals which can be separated using modifications of standard laboratory procedures. The mineralogy of silt-sized grains can be determined in grains as small as 10 microns and grains of this size contain the same suite of minerals as the very fine sand size. We suggest that non-opaque heavy minerals in shales may be very important for correlation of down-dip basinal sections in cases where useful fossils are absent.

The large amount of intrastratal solution in these young sandstones which have never been deeply buried supports the view of Pettijohn (1941) that intrastratal solution of heavy minerals is of general, rather than local significance. Minerals post-depositionally dissolved in sandstones will probably be preserved in the impermeable shales. The presence and abundance in the fine sand and silt sizes of heavy minerals in shales, coupled with the relatively impervious character of such rocks, suggests that shales rather than sandstones should be examined in heavy-mineral studies of stratigraphic sections.

REFERENCES

BEVERIDGE, A. J., 1960, Heavy minerals in lower Tertiary formations in the Santa Cruz Mountains, California: Jour. Sedimentary Petrology, v. 30, p. 513–537.

BRAMLETTE, M. N., 1941, The stability of minerals in sandstone: Jour. Sedimentary Petrology, v. 11, p. 32–36.

CALLENDER, D. L. AND FOLK, R. L., 1958, Idiomorphic zircon, key to volcanism in the Lower Tertiary sands of central Texas: Am. Jour. Sci., v. 256, p. 257–269.

CARROLL, DOROTHY, 1957, A statistical study of heavy minerals in sands of the South River, Augusta County, Virginia: Jour. Sedimentary Petrology, v. 27, p. 387–404.

DOEGLAS, D. J., 1949, Loess, an eolian product: Jour. Sedimentary Petrology, v. 19, p. 112–117.

EARGLE, D. H., AND FOUST, R. T., JR., 1962, Tertiary stratigraphy and uranium mines of the southeast Texas coastal plain, Houston to San Antonio, via Goliad; p. 225–253 *in* E. H. Rainwater and R. P. Zingula, eds., Geology of the Gulf Coast and central Texas. Houston Geol. Soc., 392 p.

FAN, POW-FOONG, 1965, Recent silts in the Santa Clara River drainage basin, California: A mineralogical investigation of their origin and evolution: Ph.D. thesis, Univ. Calif., Los Angeles, 215 p.

GAME, P. M., 1964, Observations on a dustfall in the eastern Atlantic, February, 1962: Jour. Sedimentary Petrology, v. 34, p. 355–359.

GAZZI, PAOLO, 1965, On the heavy mineral zones in the geosyncline series. Recent studies in the northern Apennines, Italy: Jour. Sedimentary Petrology, v. 35, p. 109–115.

GIPSON, MACK, JR., 1963, Ultrasonic disaggregation of shale: Jour. Sedimentary Petrology, v. 33, p. 955–957.

HASEMAN, J. F., AND MARSHALL, C. E., 1945, The use of heavy minerals in studies of the origin and development of soils: Univ. Missouri Agric. Expt. Sta. Research Bull. 387, 75 p.

HUTTON, C. O., 1950, Studies of heavy detrital minerals: Geol. Soc. America Bull., v. 61, p. 635–710.

KRYNINE, P. D., 1942, Provenance versus mineral stability as a controlling factor in the composition of sediments (abst.): Geol. Soc. America Bull., v. 53, p. 1850–51.

PETTIJOHN, F. J., 1941, Persistence of heavy minerals and geologic age: Jour. Geology, v. 49, p. 610–625.

SHAW, D. B., AND WEAVER, C. E., 1965, The mineralogical composition of shales: Jour. Sedimentary Petrology, v. 35, p. 213–222.

SWINEFORD, ADA, AND FRYE, J. C., 1951, Petrography of the Peoria Loess in Kansas: Jour. Geology, v. 59, p. 306–322.

VANANDEL, TJ. H., 1959, Reflections on the interpretation of heavy mineral analyses: Jour. Sedimentary Petrology, v. 29, p. 153–163.

AUTHOR CITATION INDEX

SUBJECT INDEX

About the Editor

GRETCHEN LUEPKE is a geologist with the U.S. Geological Survey in Menlo Park, California, who specializes in heavy-mineral analysis and sedimentology. She received the B.S. and M.S. degrees from the University of Arizona and joined the Survey in 1967. She is a member of the Society of Economic Paleontologists and Mineralogists, the Geological Society of America, Sigma Xi, and the History of the Earth Sciences Society.